About the Editors

Sunčica Roca

Sunčica Roca is a Research Associate working as a Professional Advisor at the Ruđer Bošković Institute, NMR Centre, Zagreb, Croatia. Her research interests primarily relate to solution-state NMR spectroscopy, particularly the conformational analysis of newly synthesized small organic molecules, metal-coordinated complexes, and natural compounds isolated from plants. She is also involved in the study of the supramolecular interactions of small organic molecules with cucurbituril and various cyclodextrins. To date, she has participated in numerous national and international research collaborations, published numerous scientific papers in peer-reviewed journals, and actively taken part in national and international scientific conferences.

Monika Kovačević

Monika Kovačević is an Associate Professor of Organic Chemistry at the University of Zagreb, Faculty of Food Technology and Biotechnology, Zagreb, Croatia. Her primary research interest is ferrocene peptides derived from various ferrocene scaffolds. Monika Kovačević's scientific work focuses on bioorganometallic chemistry, particularly the study of the conformation of ferrocene peptidomimetics using molecular spectroscopy (NMR, IR, CD) and their resulting biological activities. She also works on isolating natural bioactive compounds and investigating their biological activity. She has received the Award of the Faculty of Food Technology and Biotechnology for a published scientific paper in a leading chemistry journal, support from the Biotechnical Foundation of the Faculty of Food Technology and Biotechnology for young researchers, the National Foundation for Science, Higher Education and Technological Development of the Republic of Croatia, and the KAAD award for her research.

Preface

This Reprint addresses the diversity of metal complexes and its relevance to their synthesis, structural and conformational properties, and biological activities. By highlighting recent research and developments, it aims to provide insight into structure–activity relationships important for catalysis, materials science, and medicinal chemistry. The Reprint is intended for researchers working in coordination chemistry and related interdisciplinary fields.

Sunčica Roca and Monika Kovačević
Guest Editors

inorganics

MDPI

Editorial

Metal Complexes Diversity: Synthesis, Conformation, and Bioactivity

Sunčica Roca [1,*] and Monika Kovačević [2,*]

[1] NMR Centre, Ruđer Bošković Institute, Bijenička 54, 10000 Zagreb, Croatia
[2] Department of Chemistry and Biochemistry, Faculty of Food Technology and Biotechnology, University of Zagreb, Pierottijeva 6, 10000 Zagreb, Croatia
* Correspondence: sroca@irb.hr (S.R.); monika.kovacevic@pbf.unizg.hr (M.K.)

Metal complexes remain central to modern inorganic chemistry due to their structural diversity [1,2] and relevance in various areas, including catalysis [3–5], medicine [6–11], bioinorganic systems [12–16], and supramolecular and materials science [17–20]. The contributions featured in this Special Issue of Inorganics, entitled "Metal Complexes Diversity: Synthesis, Conformation, and Bioactivity", demonstrate how ligand design and synthetic control can be strategically employed to tune molecular geometry, electronic features, and functional properties-factors crucial in determining the stability and activity of metal-based systems.

This Special Issue brings together nine contributions, including one comprehensive review article and eight original scientific papers, providing valuable insights into the synthesis, structure, and reactivity of metal complexes, as well as their conformational and biological properties. The review article by Kostova [21] provides a detailed overview of homo- and hetero-multinuclear iridium (III) complexes with significant cytotoxic activity, discussing their structural characteristics and biological potential as anticancer agents. This comprehensive contribution emphasizes how multinuclearity and the coordination environment affect their biological performance. The eight original scientific papers cover a wide range of topics and metals, showcasing both experimental and theoretical approaches:

- Ralić et al. [22] investigate the biological targets of a $[Pd(dach)Cl_2]$ complex in HeLa cells, identifying proteins involved in ribosomal biogenesis and RNA splicing, thereby linking coordination chemistry with molecular biology.
- Jevtović et al. [23] present a theoretical study on the effects of coordination atoms (O/S/N) on the structure, stability, and protein/DNA binding of nickel (II) complexes with pyridoxal-semi-, thiosemi-, and isothiosemicarbazone ligands.
- Talebi et al. [24] describe the synthesis and structural characterization of a novel azide-bridged polymeric manganese (III) Schiff base complex derived from an allylamine-type ligand, revealing its unique structural and spectroscopic features.
- Nowak et al. [25] report on thionitrosyl complexes of rhenium (II) and technetium (II) with PPh_3 and chelating ligands, providing new insights into their reactivity and allowing new conclusions about similarities and differences in stability, reaction kinetics, and redox behavior between these 4d and 5d transition metals.
- Kovačević et al. [26] investigate ferrocene-based conjugates with hydrophobic amino acids, focusing on their conformational, electrochemical, and antioxidative properties, and highlight their relevance to bioorganometallic chemistry.
- Qiao et al. [27] examine NiFeSe complexes related to the active site of [NiFeSe]-hydrogenases, elucidate electronic and steric effects on oxygen reactivity, and clarify oxidative degradation mechanisms.

- Sánchez López et al. [28] present the synthesis, properties, and electrochemistry of bis(iminophosphorane)pyridine iron (II) pincer complexes to enhance the understanding of pincer-type ligand systems.
- Petkov et al. [29] study di- and trinuclear iron (III) monensinates with different bridging motifs, combining spectral and computational analyses to reveal their structural diversity and bonding characteristics.

This Special Issue provides an integrated perspective on the diversity of metal complexes, emphasizing the importance of molecular design, structure–activity relationships, and interdisciplinary collaboration. The research presented here not only broadens our understanding of complex formation and reactivity, but also opens new pathways for applications in material development and medicinal chemistry.

The Guest Editors would like to express their sincere gratitude to all authors for their valuable contributions, to the reviewers for their constructive and timely evaluations, and to the Editorial Office of Inorganics for their continuous support in the preparation of this Special Issue. We hope that this collection will inspire further studies and collaborations aimed at deepening the understanding of the intricate connections between the synthesis, structure, and bioactivity of metal complexes.

Author Contributions: Conceptualization, S.R. and M.K.; methodology, S.R. and M.K.; validation, S.R. and M.K.; formal analysis, S.R. and M.K.; investigation, S.R. and M.K.; resources, S.R. and M.K.; writing—original draft preparation, S.R.; writing—review and editing, M.K.; supervision, S.R. and M.K. All authors have read and agreed to the published version of the manuscript.

Conflicts of Interest: The authors declare no conflicts of interest.

References

1. Kroll, N.; Theilacker, K.; Schoknecht, M.; Baabe, D.; Wiedemann, D.; Kaupp, M.; Grohmann, A.; Hörner, G. Controlled Ligand Distortion and Its Consequences for Structure, Symmetry, Conformation, and Spin-State Preferences of Iron(II) Complexes. *Dalton Trans.* **2015**, *44*, 19232. [CrossRef]
2. Pell, T.P.; Wilson, D.J.D.; Skelton, B.W.; Dutton, J.L.; Barnard, P.J. Heterobimetallic *N*-Heterocyclic Carbene Complexes: A Synthetic, Spectroscopic, and Theoretical Study. *Inorg. Chem.* **2016**, *55*, 6882. [CrossRef] [PubMed]
3. Gunanathan, C.; Shimon, L.J.W.; Milstein, D. Direct Conversion of Alcohols to Acetals and H_2 Catalyzed by an Acridine-Based Ruthenium Pincer Complex. *J. Am. Chem. Soc.* **2009**, *131*, 3146. [CrossRef]
4. Madec, H.; Figueiredo, F.; Cariou, K.; Roland, S.; Sollogoub, M.; Gasser, G. Metal Complexes for catalytic and photocatalytic reactions in living cells and organisms. *Chem. Sci.* **2022**, *14*, 409. [CrossRef]
5. Lawrence, M.A.W.; Green, K.-A.; Nelson, P.N.; Lorraine, S.C. Review: Pincer ligands—tunable, versatile and applicable. *Polyhedron* **2018**, *143*, 11. [CrossRef]
6. Vigueras, G.; Markova, L.; Novohradsky, V.; Marco, A.; Cutillas, N.; Kostrhunova, H.; Kasparkova, J.; Ruiz, J.; Brabec, V. A photoactivated Ir(iii) complex targets cancer stem cells and induces secretion of damage-associated molecular patterns in melanoma cells characteristic of immunogenic cell death. *Inorg. Chem. Front.* **2021**, *8*, 4696. [CrossRef]
7. Prieto Otoya, T.D.; McQuaid, K.T.; Cardin, C.J. Structural Insights into G-quadruplex Binding by Metal Complexes: Implications for Drug Design. *Med. Chem. Res.* **2024**, *33*, 2001. [CrossRef]
8. Patil, S.A.; Patil, S.A.; Patil, R.; Keri, R.S.; Budagumpi, S.; Balakrishna, G.R.; Tacke, M. *N*-heterocyclic carbene metal complexes as bio-organometallic antimicrobial and anticancer drugs. *Future Med. Chem.* **2015**, *7*, 1305. [CrossRef] [PubMed]
9. Heffern, M.C.; Reichova, V.; Coomes, J.L.; Harney, A.S.; Bajema, E.A.; Meade, T.J. Tuning Cobalt(III) Schiff-Base Complexes as Activated Protein Inhibitors. *Inorg. Chem.* **2015**, *54*, 9066. [CrossRef] [PubMed]
10. Liu, W.; Gust, R. Metal N-heterocyclic Carbene Complexes as Potential Antitumor Metallodrugs. *Chem. Soc. Rev.* **2013**, *42*, 755. [CrossRef]
11. Murillo, M.I.; Gaiddon, C.; Le Lagadec, R. Targeting the Intracellular Redox Balance by Metal Complexes towards Anticancer Therapy. *Front. Chem.* **2022**, *10*, 967337. [CrossRef] [PubMed]
12. Mansour, M.S.A.; Abdelkarim, A.T.; El-Sherif, A.A.; Mahmoud, W.H. Metal complexes featuring a quinazoline schiff base ligand and glycine: Synthesis, characterization, DFT and molecular docking analyses revealing their potent antibacterial, anti-helicobacter pylori, and Anti-COVID-19 activities. *BMC Chem.* **2024**, *18*, 150. [CrossRef]

 https://doi.org/10.3390/inorganics14010017

13. Ceramella, J.; Catalano, A.; Mariconda, A.; D'Amato, A.; Aquila, S.; Saturnino, C.; Rosano, C.; Sinicropi, M.S.; Longo, P. Silver N-Heterocyclic Carbene (NHC) Complexes as Antimicrobial and/or Anticancer Agents. *Pharmaceuticals* **2025**, *18*, 9. [CrossRef]
14. Haas, K.L.; Franz, K.J. Application of metal coordination chemistry to explore and manipulate cell biology. *Chem. Rev.* **2009**, *109*, 4921. [CrossRef] [PubMed]
15. Renfrew, A.K. Transition-Metal Complexes with Bioactive Ligands: Mechanisms for Selective Ligand Release and Applications for Drug Delivery. *Metallomics* **2014**, *6*, 1324. [CrossRef]
16. Catalano, A.; Mariconda, A.; Sinicropi, M.S.; Ceramella, J.; Iacopetta, D.; Saturnino, C.; Longo, P. Biological Activities of Ruthenium NHC Complexes: An update. *Antibiotics* **2023**, *12*, 365. [CrossRef] [PubMed]
17. Eren, N.; Fadaei-Tirani, F.; Severin, K. Stepwise Synthesis of Heterotrimetallic Fe(II)/Pd(II)/Au(I) Coordination Cages. *Inorg. Chem. Front.* **2024**, *11*, 3263. [CrossRef]
18. Zhang, X.; Hou, Y.; Xiao, X.; Chen, X.; Hu, M.; Geng, X.; Wang, Z.; Zhao, J. Recent development of the transition metal complexes showing strong absorption of visible light and long-lived triplet excited state: From molecular structure design to photophysical properties and applications. *Coord. Chem. Rev.* **2020**, *417*, 213371. [CrossRef]
19. Fujisawa, K.; Ishikawa, Y.; Miyashita, Y.; Okamoto, K.-I. Pyrazolate-bridged group 11 metal(I) complexes: Substituent effects on the supramolecular structures and physicochemical properties. *Inorg. Chim. Acta* **2010**, *363*, 2977. [CrossRef]
20. Cecot, G.; Doll, M.T.; Planes, O.M.; Ramorini, A.; Scopelliti, R.; Fadaei-Tirani, F.; Severin, K. Cages vs. Prisms: Controlling the Formation of Metallosupramolecular Architectures with Ligand Side-Chains. *Eur. J. Inorg. Chem.* **2019**, *2019*, 2972. [CrossRef]
21. Kostova, I. Homo- and Hetero-Multinuclear Iridium (III) Complexes with Cytotoxic Activity. *Inorganics* **2025**, *13*, 156. [CrossRef]
22. Ralić, V.; Davalieva, K.; Gemović, B.; Senćanski, M.; Nešić, M.D.; Žakula, J.; Stepić, M.; Petković, M. [Pd(dach)Cl_2] Complex Targets Proteins Involved in Ribosomal Biogenesis, and RNA Splicing in HeLa Cells. *Inorganics* **2025**, *13*, 215. [CrossRef]
23. Jevtovic, V.; Rakić, A.; Alshammari, O.A.O.; Alhar, M.S.; Alenezi, T.; Rakic, V.; Dimić, D. Theoretical Study of the Effects of Different Coordination Atoms (O/S/N) on Crystal Structure, Stability, and Protein/DNA Binding of Ni(II) Complexes with Pyridoxal-Semi, Thiosemi, and Isothiosemicarbazone Ligand Systems. *Inorganics* **2024**, *12*, 251. [CrossRef]
24. Talebi, A.; Salehi, M.; Jesus, A.J.L.; Kubicki, M.; Fausto, R.; Golbedaghi, R. A New Azide Bridged Polymeric Manganese (III) Schiff Base Complex with an Allylamine-Derived Ligand: Structural Characterization and Activity Spectra. *Inorganics* **2024**, *12*, 234. [CrossRef]
25. Nowak, D.; Hagenbach, A.; Sawallisch, T.E.; Abram, U. Thionitrosyl Complexes of Rhenium and Technetium with PPh_3 and Chelating Ligands—Synthesis and Reactivity. *Inorganics* **2024**, *12*, 210. [CrossRef]
26. Kovačević, M.; Roca, S.; Jadreško, D.; Mrvčić, J.; Hanousek Čiča, K.; Čakić Semenčić, M.; Barišić, L. Conformational, Electrochemical, and Antioxidative Properties of Conjugates of Different Ferrocene Turn-Inducing Scaffolds with Hydrophobic Amino Acids. *Inorganics* **2024**, *12*, 195. [CrossRef]
27. Qiao, Y.; Xu, E.; Hao, Y.; Yang, X.; Ni, M. Electronic and Steric Effects on Oxyge Reactivities of NiFeSe Complexes Related to O_2-Damaged [NiFeSe]-Hydrogenases'Active Site. *Inorganics* **2024**, *12*, 163. [CrossRef]
28. Sánchez López, N.; Nunez Bahena, E.; Ryabov, A.D.; Sutra, P.; Igau, A.; Le Lagadec, R. Synthesis, Properties, and Electrochemistry of bis(iminophosphorane)pyridine Iron(II) Pincer Complexes. *Inorganics* **2024**, *12*, 115. [CrossRef]
29. Petkov, N.; Tadjer, A.; Simova, S.; Cherkezova-Zheleva, Z.; Paneva, D.; Stoyanova, R.; Kukeva, R.; Dorkov, P.; Pantcheva, I. Synthesis, Spectral Characterization, and Structural Modelling of Di- and Trinuclear Iron(III) Monensinates with Different Bridging Patterns. *Inorganics* **2024**, *12*, 114. [CrossRef]

https://doi.org/10.3390/inorganics14010017

inorganics

MDPI

Article

[Pd(*dach*)Cl$_2$] Complex Targets Proteins Involved in Ribosomal Biogenesis, and RNA Splicing in HeLa Cells

Vanja Ralić [1,†], Katarina Davalieva [2,*,†], Branislava Gemović [3], Milan Senćanski [3], Maja D. Nešić [1], Jelena Žakula [4], Milutin Stepić [1] and Marijana Petković [1,*]

1 COHERENCE Centre, Department of Atomic Physics, VINČA Institute of Nuclear Sciences, National Institute of the Republic of Serbia, University of Belgrade, 11000 Belgrade, Serbia; vanja.ralic@vin.bg.ac.rs (V.R.); maki@vin.bg.ac.rs (M.D.N.); mstepic@vin.bg.ac.rs (M.S.)

2 Research Centre for Genetic Engineering and Biotechnology "Georgi D Efremov", Macedonian Academy of Sciences and Arts, 1000 Skopje, North Macedonia

3 Department for Bioinformatics and Computational Chemistry, VINČA Institute of Nuclear Sciences, National Institute of the Republic of Serbia, University of Belgrade, 11000 Belgrade, Serbia; gemovic@vin.bg.ac.rs (B.G.); sencanski@vin.bg.ac.rs (M.S.)

4 Department of Molecular Biology and Endocrinology, VINČA Institute of Nuclear Sciences, National Institute of the Republic of Serbia, University of Belgrade, 11000 Belgrade, Serbia; pozegaj@vin.bg.ac.rs

* Correspondence: katarina@manu.edu.mk (K.D.); marijanapetkovic@vin.bg.ac.rs (M.P.)

† These authors contributed equally to this work.

Abstract: This study aims to investigate the effect of the Pd(II) complex on HeLa cells using computational biology and proteomic analysis. [Pd(*dach*)Cl$_2$]-treated HeLa cells were subjected to comparative proteomics analysis using label-free data-independent liquid chromatography-tandem mass spectrometry (LC-MS/MS). In parallel, the informational spectrum method (ISM) was used to predict potential protein interactors of the [Pd(*dach*)Cl$_2$] complex in HeLa cells. Proteomics analysis revealed 121 differentially abundant proteins (DAPs). Enrichment analysis of Gene Ontology (GO) annotations revealed ATP hydrolysis and RNA/protein binding as the top molecular functions and RNA splicing and protein–RNA complex organization as the top biological processes. Enrichment analysis of altered canonical pathways pointed out spliceosome and ribosome pathways. The top hub proteins with potential regulatory importance encompassed ribosomal proteins, translational and transcriptional factors, and components of the ribosome assembly machinery. ISM and cross-spectral analysis identified the nucleoplasm and sensor of the single-stranded DNA (SOSS DNA) complex. Proteome analysis showed that [Pd(*dach*)Cl$_2$] targets proteins involved in ribosomal biogenesis and RNA splicing, whereas theoretical prediction implies also potential effect on p53 signaling pathway, and thus, alterations of the expression of regulatory proteins involved in cell survival and proliferation. These findings underscore the potential of Pd(II) complexes as anti-cancer agents, warranting further exploration and detailed functional validation.

Keywords: Pd(II) complex; HeLa cervical cancer cells; proteomics; bioinformatics; spliceosome; ribosome

1. Introduction

Cervical cancer is still one of the most common malignancies among women worldwide, despite advances in screening and treatment strategies [1,2]. With 600,000 new cases and 340,000 deaths annually, cervical cancer is the fourth most common cancer in women, after breast, colorectal, and lung cancer [1–3]. Chemotherapeutic agents based on platinum complexes, such as cisplatin, have been widely used; however, their effectiveness is often limited by toxicity and resistance mechanisms. In the search for alternative metal-based therapies, palladium(II) complexes have attracted attention due to their structural similarity to platinum compounds and potentially improved pharmacological profiles.

Among Pd-complexes, dichloro(1,2-diaminocyclohexane)palladium(II) (further in text: [Pd(*dach*)Cl_2]) has shown promising anti-cancer activity. In a previous study, we demonstrated that this [Pd(*dach*)Cl_2] complex induces cytotoxic effects and macromolecular alterations in proteins, suggesting its potential as an anti-cancer agent [4]. Namely, the secondary structure of proteins in HeLa cells was significantly altered upon the action of the [Pd(*dach*)Cl_2] complex, particularly the α-helix-containing proteins, and this change was dependent on the concentration of [Pd(*dach*)Cl_2] complex.

The present data indicate that Pd(II) interacts primarily with nucleic acids [5,6], but the precise mechanism of action remains unclear. Furthermore, the binding kinetics and stability of the Pd complexes formed with S, O, or N-donor molecules [7–9] differ in vivo vs. in vitro.

Proteomics approaches offer a powerful means to characterize drug-induced protein expression changes and global pathway regulation. Still, they have not been fully exploited in the investigation of the mechanism of the action of the transition metal complexes on cancer cells [10].

The cellular effects of the [Pd(*dach*)Cl_2] complex were investigated by determining cell viability after treatment with increasing concentrations of the complex. Our previously published results [4] demonstrate that the highest tested concentration of the [Pd(*dach*)Cl_2] complex decreased HeLa cell viability by 22%, indicating moderate cytotoxicity to this cancer cell line compared to other anti-cancer drugs. However, this property of the [Pd(*dach*)Cl_2] complex enabled us to study protein changes and identify potential target proteins that can be used to enhance the efficacy of anti-cancer therapy.

In this study, we applied label-free, data-independent LC-MS/MS acquisition coupled with ion mobility to investigate the proteomic alterations in HeLa cells treated with [Pd(*dach*)Cl_2]. In addition, we applied computational methods to explain the signaling pathways affected mainly by drug treatments [11], as in our previous work [12]. An additional possibility is to calculate the informational spectrum of a compound of interest, which enables the identification of specific target molecules [13]. Our analysis revealed significant modulation of pathways associated with the spliceosome, ribosome, and the pathways involved in cancer development and progression. The identified proteins and processes may represent potential targets for further, more in-depth investigation.

2. Results

2.1. Predicting the Protein Targets for [Pd(dach)Cl_2] by Computational Biology

The sequencing of the HeLa cell's 9462 baseline proteins was screened for potential interactions with the Pd(II) complex using cross-spectral analysis [14]. An IS (information spectrum) of the Pd(II) complex generated by the ISM showed a dominant peak at a frequency of 0.402 (Figure 1A), which was used for further cross-spectral analysis [13,15].

Cross-spectral analysis identified 106 proteins that were subsequently analyzed using the bioinformatics resource DAVID for gene enrichment analysis to determine enriched pathways and GO terms [16]. The Functional Annotation Chart revealed that the GO terms with the highest statistical significance were nucleoplasm and SOSS complex (Figure 1B, p-value after Benjamini–Hochberg correction < 0.05). The nucleoplasm contains a substantial portion of proteins organized as nuclear bodies, which are involved in the regulation of chromosome structure, DNA damage repairs, RNA processing, and epigenetic gene regulation. SOSS protein complex is involved in maintaining genomic stability [17]. Three identified proteins from the SOSS complex are SOSS C (gene name: INIP), SOSS B1 (gene name: NAPB2), and SOSS B2 (gene name: NAPB1). Figure 1C shows the interaction network of these proteins. Protein SOSS B1, encoded by the NAPB2 gene, consists predominantly of β-secondary structures organized into a curved beta-barrel (Figure 1D) [18]. Detailed inspection of the secondary structure of SOSS B1 showed the presence of two helices and seven beta strands (Figure 1E).

A

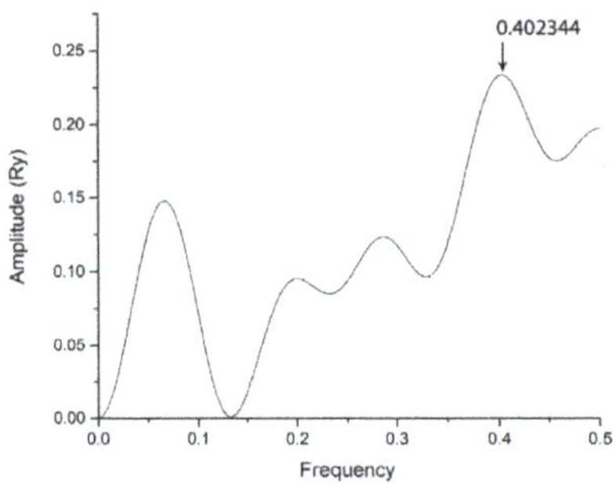

B

Term	p-value	Benjamini
nucleoplasm	1.1×10^{-5}	2.3×10^{-3}
protein binding	2.1×10^{-5}	4.5×10^{-3}
SOSS complex	1.5×10^{-4}	1.7×10^{-2}

C

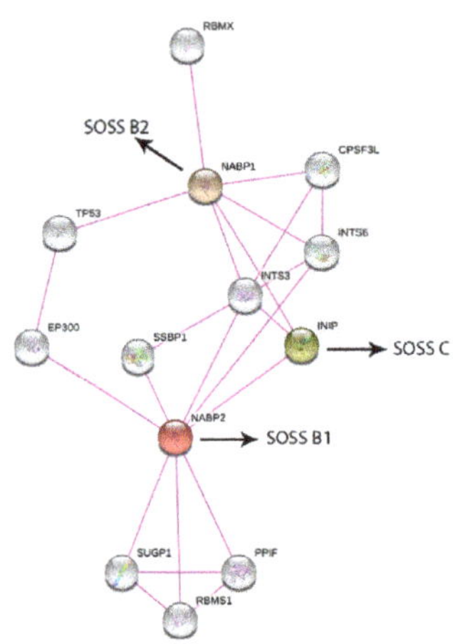

D

E

Secondary structure type	Position
Helix	7-9
Beta strand	20-26
Beta strand	33-35
Beta strand	40-45
Beta strand	48-53
Helix	56-61
Beta strand	66-69
Beta strand	93-98
Beta strand	100-104

Figure 1. Characterization of the [Pd(*dach*)Cl_2] and bioinformatics analysis. (**A**) The IS of [Pd(*dach*)Cl_2] complex where the arrow indicates the peak position for potential interactions with HeLa cell proteins. Functional Annotation Chart (**B**) showing the Gene Ontology (GO) terms with the highest statistical significance (p-value after Benjamini correction < 0.05). Interaction network (**C**) of NAPB1, INIP, and NAPB2, retrieved from the STRING database, including only experimentally verified interactions. NABP1 (orange), NABP2 (red), and INIP (green) correspond to SOSS complex subunits B2, B1, and C, respectively. Crystal structure and secondary structure distribution of SOSS complex subunit B1 are shown in (**D**) and (**E**), respectively [18,19].

2.2. Differentially Abundant Proteins as a Result of [Pd(dach)Cl_2] Treatment

A proteomic analysis was performed to identify target-specific proteins unique to the effects of the [Pd(*dach*)Cl_2] complex in the HeLa cervical carcinoma cell line.

The 1384 identified proteins by comparative proteomics analysis were filtered to remove reverse sequences and proteins identified on one peptide, as well as yeast alcohol dehydrogenase (ADH). The final dataset contained 1255 proteins identified on two or more peptides with 121 proteins that had statistically significant differences in abundance (ANOVA ≤ 0.05) between [Pd(*dach*)Cl_2]- treated HeLa cells and controls (Supplementary Table S1). From the 121 differentially abundant proteins (DAPs), 57 showed increased abundance and 64 showed decreased abundance in [Pd(*dach*)Cl_2]- treated HeLa cells compared to controls. The differential expression is illustrated by volcano plot and hierarchical clustering (Figure 2A,B).

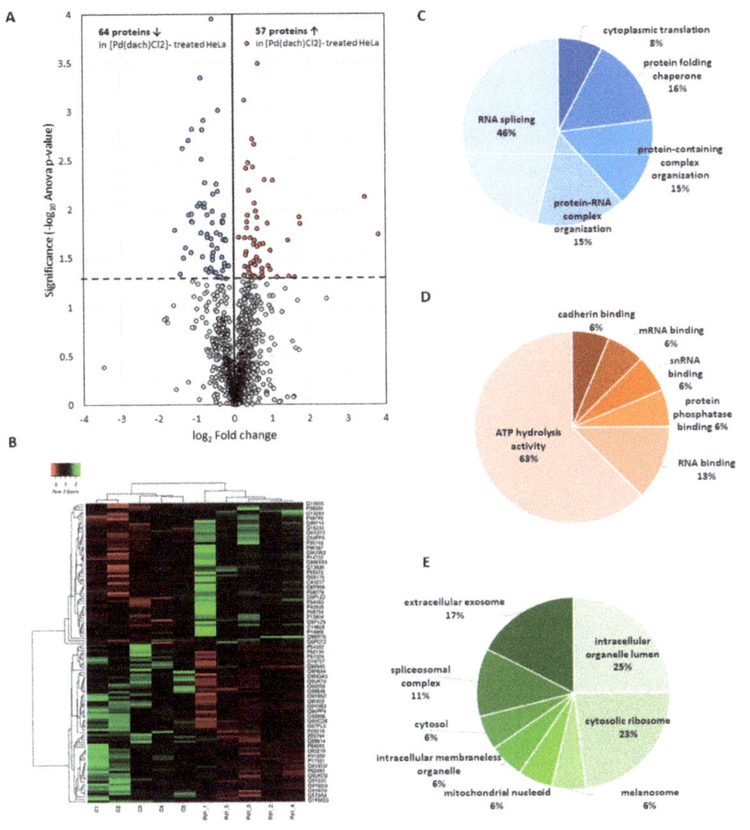

Figure 2. Proteins with altered abundance as a result of [Pd(*dach*)Cl_2] treatment of HeLa cells. (**A**) Differential expression analysis between [Pd(*dach*)Cl_2]-treated HeLa cells and controls is represented by the volcano plot. Statistically significant altered proteins in [Pd(*dach*)Cl_2]-treated HeLa cells compared to the control ($p < 0.05$) are above the dashed horizontal line. Proteins with increased abundance are indicated in red and those with decreased abundance in blue. (**B**) Heatmap of protein expression in [Pd(*dach*)Cl_2]-treated HeLa cells where higher protein abundances are presented in green, and lower ones in red. The samples are shown in columns, and the rows indicate proteins. The clustering method applied was average linkage, and the distance measurement method used was Spearman rank correlation. (**C**–**E**) Enrichment analysis of GO annotations associated significantly with the differentially abundant proteins: (**C**) biological processes, (**D**) molecular functions, and (**E**) cellular compartment.

The most prominent biological process associated with the DAPs was RNA splicing (46%), followed by protein-containing complex organization, protein–RNA complex organization, protein folding chaperone, and cytoplasmic translation (Figure 2C). In terms of the DAP's molecular functions, the majority, or 63%, had ATP hydrolysis activity, while the remaining functions involved different types of RNA and protein binding (Figure 2D). DAPs were mainly intracellular, with 25% located within the intracellular organelle lumen, 23% located within the cytosolic ribosome, 11% in the spliceosomal complex, 12% in different organelles, and 6% as free in the cytosol (Figure 2E). Only 17% of the DAPs were part of the extracellular exosome.

2.3. Altered Pathways in HeLa Cells Due to [Pd(dach)Cl$_2$] Treatment

Enrichment analysis of canonical pathways using the KEGG database showed two associated pathways: Spliceosome (hsa03040) and Ribosome (hsa03010) (Figure 3A). Analysis with the Reactome Pathways database showed more than 20 significantly associated pathways, of which the highest significance showed metabolism of RNA, SRP-dependent co-translational protein targeting to membrane, translation, eukaryotic translation termination, Nonsense Mediated Decay (NMD) independent of the Exon Junction Complex (EJC), formation of a pool of free 40 S subunits, mRNA Splicing—Major Pathway, and rRNA processing (Figure 3B).

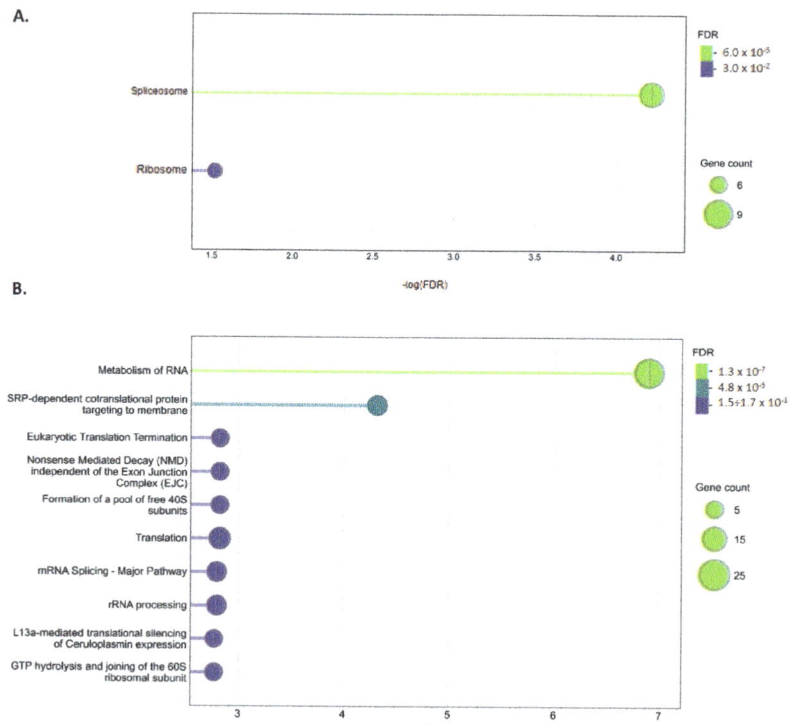

Figure 3. Enrichment analysis of the cellular pathways related significantly with the DAPs according to (**A**) KEGG and (**B**) Reactome Pathways databases. The charts include only terms with statistical significance corrected for multiple testing using the Benjamini–Hochberg procedure (FDR $\leq$ 0.05).

To further investigate the functional significance of DAPs, we constructed a protein–protein interaction network (Figure 4). This network showed that DAPs have significantly increased interactions among themselves compared to the same size and degree of distribution random set of proteins (PPI enrichment p-value = 1.56×10^{-9}). Using this network, we have identified key hub genes using the Cytoscape plug-in cytoHubba v0.1 (Figure 5). The top 25 hub genes were ranked based on the Maximal Clique Centrality (MCC) approach, which prioritizes proteins with high connectivity within the network. As shown in Figure 5, the hub genes are color-coded from highly essential (red) to crucial (yellow), reflecting their potential regulatory importance. The group of highly essential proteins constitutes ribosomal proteins (RPL4, RPS9, RPL32, RPL36AL, RPS15A, RPS25), translational and transcriptional factors (ETF1, EDF1, DRG1), components of the ribosome assembly machinery (MRTO4), components of the signal recognition complexes that target secretory proteins to the rough endoplasmic reticulum membrane (SPCS2, SRP68) and several heat shock proteins involved in correct folding of proteins and degradation pathways (HSPA8), regulation of proteins involved in cell cycle control and signal transduction (HSP90AB1), and ATPases (ATP5F1B, VCP). These hub genes may serve as critical mediators of the observed cellular response and represent potential targets for further investigation.

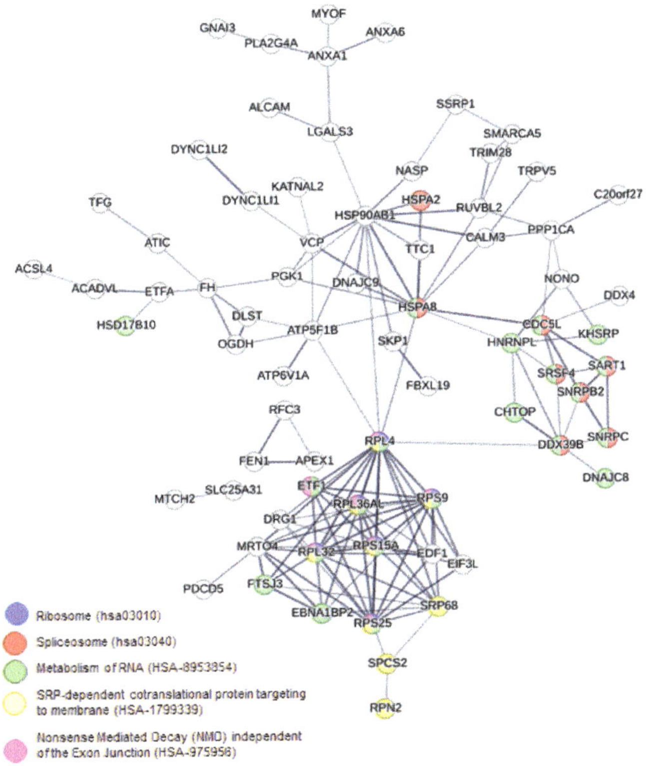

Figure 4. Protein–protein interaction network of the DAPs. The network was obtained using the STRING database. Proteins involved in the top significantly enriched pathways, according to the KEGG and Reactome pathway databases, are colored according to the given legend. The line thickness indicates the strength of data support.

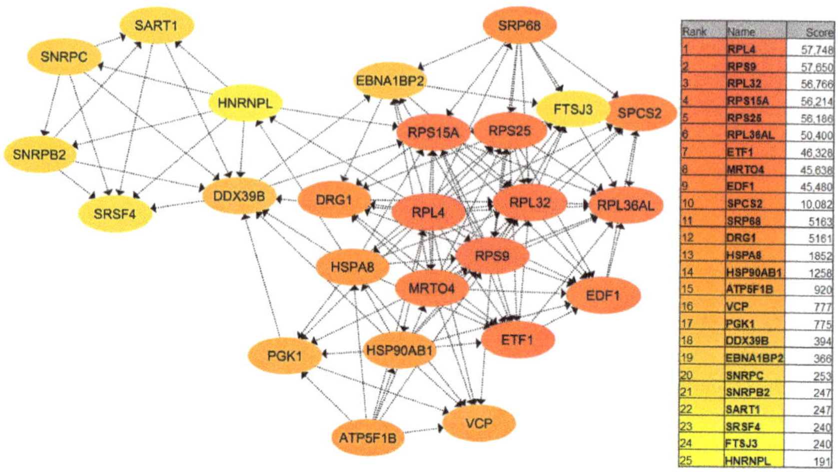

Figure 5. Identification of hub genes in the protein–protein interaction network of DAPs using Cytoscape plug-in cytoHubba. The graph represents the top 25 hub genes ranked by the MaxMCC approach, with scores marked from highly essential (red) to crucial (yellow).

3. Discussion

The potential protein targets of [Pd(*dach*)Cl_2] in HeLa cells, as predicted using the information spectrum of the Pd(II) complex and cross-spectral analysis, were identified as the nucleoplasm and SOSS complex. The nucleoplasm is a complex fluid that contains RNA and proteins, and substantial fraction of nuclear bodies, namely nucleoli, speckles, and Cajal bodies. Many processes and mechanisms that undergo in the nucleoplasm remain to be revealed; however, nucleoli and speckles have been verified to represent active sites for ribosomal synthesis and RNA splicing, respectively [20]. Therefore, we might speculate that [Pd(*dach*)Cl_2] targets proteins involved in ribosomal biogenesis and RNA splicing in HeLa cells. The computational method also suggested that the SOSS protein complex (Sensor of Single-stranded DNA) could be a potential protein target of the studied Pd(II) complex. The interaction network of SOSS complex predicted as Pd(II) complex interactors includes TP53 (tumor suppressor protein, p53) and EP300 (histone acetyltransferase, p300), which are involved in the cellular response to DNA damage and apoptosis [21,22]. In particular, p53 has been considered to be one of the classical types of tumor suppressors, also known as the "guardian of the genome", and its inactivation leads to tumor growth and metastasis [23]. Changes in the p53 signaling pathway are one of the possible mechanisms by which Pd(II) complexes induce the inhibition of cell migration [24], significantly affecting HeLa cell viability. In addition to having a role in maintaining genomic stability and responds to DNA double-strand breaks [25], SOSS complex also plays a role in RNA –DNA interactions by promoting RNA degradation and interacting with the RNA polymerase II enzyme, which is crucial in the transcription process [26].

The comprehensive analysis of the proteomic alterations in HeLa cells resulting from the [Pd(*dach*)Cl_2] treatment showed a strong potential for elucidating its cytotoxic effect on cancer cells. The analysis of the significantly enriched GO annotations of 121 proteins with altered abundance revealed a strong association with cytoplasmic translation, with a particular point toward RNA splicing and protein–RNA complex organization. Furthermore, KEGG pathway enrichment analysis revealed that DAPs play a significant role in the "Spliceosome" and "Ribosome" canonical pathways.

The spliceosome is a ribonucleoprotein complex in charge for RNA splicing, creating a continuous mRNA sequence that can be translated into protein. This complex ribonucleoprotein structure is essential for the accurate and efficient production of mature mRNA in eukaryotic cells. Disruptions in this pathway can lead to aberrant splicing, potentially resulting in the production of dysfunctional or non-functional proteins. Functional link between splicing anomalies and the evolution of cancer is well observed by a number of studies [27]. In vivo inhibition of the spliceosome is shown to impair survival and lead to metastatic affinity of MYC-dependent breast cancers [28], indicating that components of the spliceosome may be therapeutic targets. A recent study that analyzed the altered pathways in HeLa cells based on gene, protein, and phosphoprotein expression [29] revealed many spliceosome genes as overexpressed in HeLa cells, further confirming that overactivation of the spliceosome could play a significant role in the development, progression, and maintenance of cancer.

Our comparative proteomics approach identified nine DAPs that are part of the spliceosome, of which five up-regulated (HSPA8 (HSP73), SNRPB2 (U2B″), DDX39B (UAP56), CDC52 (CDC5), SRSF4 (SR)) and four down-regulated (SNRPC (U1C), SART1 (Snu66), RBMXL3, HSPA2) in [Pd(*dach*)Cl_2]-treated HeLa cells compared to controls (Figure 6). For some of the up-regulated proteins, there is extensive data about their involvement in tumorigenesis. To name a few, the up-regulation of Heat shock 70 kDa protein 8 (HSPA8) is closely related to tumorigenesis and tumor progression in various cancer types [30,31]. SNRPB2, a component of the U2 snRNP, is implicated in various cancers, as it is significantly up-regulated and associated with a poor prognosis [32]. DExD-box helicase 39B (DDX39B), which participates in many steps of RNA metabolism beyond mRNA splicing [33], is highly correlated with tumor development and progression in breast [34], colorectal [35], prostate [36], and renal cancers [37] and melanoma [38].

Of particular interest for elucidating the effects of [Pd(*dach*)Cl_2] on HeLa cells are spliceosome proteins that were down-regulated or repressed as a result of the treatment. One of these is RBMXL3, belonging to the U12-type spliceosomal complex that processes U12-type introns. The U12-type introns are rare (comprising <0.5% of introns) and processed by a specialized minor spliceosome distinct from the major (U2-type) spliceosome [39]. These introns are often spliced less efficiently, and disruptions in their processing can limit gene expression and impact cellular processes unrelated to the direct function of the host gene [40]. The reduced expression of RBMXL3 may contribute to widespread post-transcriptional changes following [Pd(*dach*)Cl_2] exposure and could exert a potential therapeutic effect against cancer cells, which are worth further investigation.

We have also observed altered levels of HSPA2, the spliceosome-interacting chaperone protein, which has been previously implicated as a cancer biomarker across diverse tumor types [16]. It was among a small set of genes (alongside SNRPB and LSM7) used to build machine-learning models for pan-cancer classification, reflecting its broad relevance in spliceosomal integrity and possibly tumor cell vulnerability [41]. The lowered expression of HSPA2 in our data suggests that [Pd(*dach*)Cl_2] may interfere not only with the core spliceosome but also with auxiliary proteins critical for its dynamic assembly and function. Such disruption can shift the precise stoichiometry required for standard spliceosome assembly and function in HeLa cells, modifying the proportion of isoforms [29]. Dysregulated splicing can produce protein isoforms with altered localization, function, or stability—some of which may promote oncogenesis, resistance to apoptosis, or cell cycle progression [42]. Some of these changes might be responsible for the observed cytotoxic effect of [Pd(*dach*)Cl_2] on cancer cells [43].

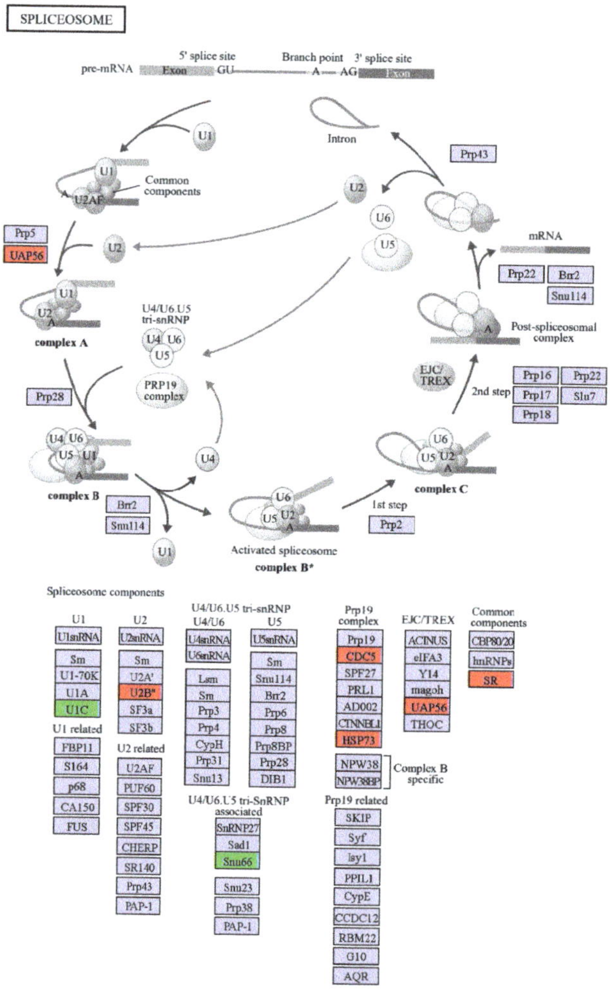

Figure 6. Graphical representation of the spliceosome pathway based on the KEGG database. The identified proteins from the spliceosome pathway in our study are color-coded: up-regulated proteins are displayed in red and down-regulated in green.

Ribosome is the core molecular machinery that governs protein synthesis and is closely linked to cell activation and proliferation. It has been well established that elevated protein synthesis and up-regulated ribosome biogenesis are characteristic hallmarks of cancer cells [44,45]. Targeting ribosome biogenesis in various cancer types has been an ongoing activity, both in improving the specificity of existing drugs and in identifying new potent compounds with enhanced pharmacological properties [46]. Moreover, there is increasing evidence about the relationship between the expression level of certain ribosomal proteins and cancer progression and differentiation in various cancer types [47]. Comparative proteomics analysis of the [Pd(*dach*)Cl_2] treatment of HeLa cells revealed altered levels of six ribosomal proteins—two up-regulated (RPL4, RPS25) and four down-regulated (RPL32, RPL36AL, RPS15A, RPS9) (Figure 7). One of these, RPL4, i.e., ribosomal protein L4, is vital for ribosome biogenesis and may play a role in transcriptional regulation [48]. Additionally, it is significantly involved in the p53 signaling pathway by direct interaction with MDM2 which is a negative regulator of p53, and suppresses MDM2-mediated p53 ubiquitination and degradation [49]. The overexpression of RPL4 also promotes the binding of MDM2 to RPL5 and RPL11, which further suppress MDM2 in cells [49]. Similarly, it has been demonstrated that ribosomal protein S25 (RPS25) also interacts with MDM2 and inhibits its E3 ligase activity, resulting in the reduction of MDM2-mediated p53 ubiquitination and subsequent stabilization and activation of p53 [50].

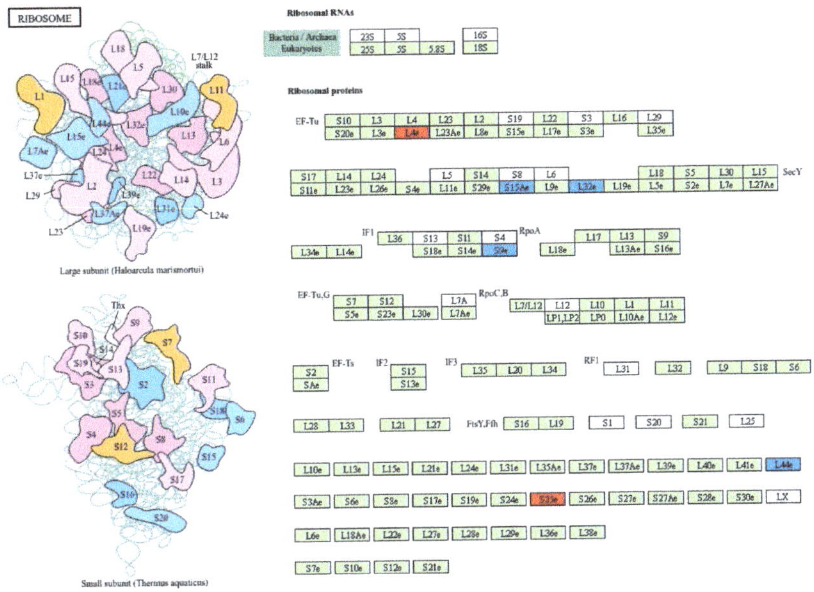

Figure 7. Graphical representation of the ribosome pathway based on the KEGG database. Proteins colored with light green are part of the human ribosome. The identified proteins from the ribosome pathway in our study are color-coded: up-regulated proteins are displayed in red and down-regulated in blue.

On the other hand, overexpression of RPL32 has been observed in lung [51], breast [52], and hepatocellular cancer [52]. High levels of RPL32 were associated with greater metastatic potential and unfavorable outcomes in patients while its silencing inhibits cancer proliferation. Similarly, overexpression of RPS15A and RPS9 has been observed in many cancer types. RPS15A has been implicated in the development and progression of liver, colorectal, and gastric cancers [53–55]. Down-regulation of RPS15A has been shown to inhibit cancer cell growth and trigger apoptosis in some cases [53,55]. Overexpression of RPS9 predicted poor prognosis of non-small cell lung cancer patients while its knockdown repressed cell proliferation, metastasis, and induced apoptosis [56]. RPS9 was also up-regulated in osteosarcoma through the activation of the MAPK signaling pathway, while its down-regulation inhibited cell growth in this cancer type [57]. The observed up-regulation of RPL4 and RPS25, which suppresses MDM2-mediated p53 degradation and leads to its activation, as well as the down-regulation of RPL32, RPS15A, and RPS9, which have confirmed role in tumor progression, strongly suggests the potential anti-tumor effect of the [Pd(*dach*)Cl_2] treatment.

Analysis using the Reactome Pathways database, in addition to showing a significant association with "Translation" and "Splicing", also indicated an alteration in NMD independent of the EJC pathway due to [Pd(*dach*)Cl_2] treatment. This pathway is part of the mRNA surveillance pathway, which represents a cellular mechanism that ensures accurate gene expression by eliminating defective mRNAs and preventing the production of potentially toxic proteins [20]. The mRNA surveillance pathway relies on the ribosome to detect and signal the presence of problematic mRNA transcripts. Overactivation or failure of these pathways could lead to either excessive degradation of functional mRNAs or accumulation of faulty ones, both of which are potentially toxic to the cell.

In our study, we detected significantly altered amounts of four proteins involved in the mRNA surveillance pathway, with DDX39B being up-regulated and PABPC1L, PPP1CA, and ETF1 being down-regulated in the treated group. PABPC1L is a protein that plays a crucial role in cancer development and progression, particularly in the immune evasion and the survival of tumor cells. This protein is being investigated as a potential therapeutic target in various cancers, including renal cell carcinoma [58], colorectal cancer [59], and gastric cancer [60]. PPP1CA, which encodes the alpha catalytic subunit of Protein Phosphatase 1 (PP1), has been implicated in various cancers, and its role can be complex, sometimes acting as an oncogene and at other times as a tumor suppressor. Research suggests it can influence tumor cell growth, migration, invasion, and even apoptosis [61]. ETF1, or eukaryotic translation termination factor 1, is a protein involved in translation termination. Although not directly classified as a cancer gene, research suggests ETF1 overexpression may play a role in cancer development and progression in breast cancer and myeloid malignancies [62,63]. The enrichment of the mRNA surveillance pathway suggests that [Pd(*dach*)Cl_2] disrupts the balance of RNA quality control. In particular, the fact that [Pd(*dach*)Cl_2]-treated HeLa cells exhibit down-regulation of PABPC1L, PPP1CA, and ETF1, which is strongly correlated with cancer progression, points to its possible anti-cancer effects.

However, we should also point out that this study has some limitations. Since this was a proof-of-concept study, the sample size was not adequate to perform multiple testing correction. Therefore, identification of DAPs was based on the unadjusted *p*-values. Despite this fact, our findings were in line with the existing literature of HeLa cells' protein and mRNA expression. Moreover, the observed down-regulation of a number of proteins highly correlated with tumor development and progression was consistent with the observed cytotoxic effect of [Pd(*dach*)Cl_2] treatment. Although this confirms the validity of the reported results, further validation in the context of statistically well-powered study is required.

4. Materials and Methods

4.1. Reagents

Human cervical carcinoma (HeLa) cells were acquired from the American TypeCulture Collection (ATCC, Manassas, VA, USA). All chemicals were purchased from Sigma-Aldrich, GmbH (Taufkirchen, Germany). The [Pd(*dach*)Cl_2] was synthesized and characterized in the lab of Prof. Biljana Petrović (Faculty of Natural Sciences, University of Kragujevac, Kragujevac, Serbia) [64].

4.2. Cell Culture

HeLa cells were cultured in DMEM medium at 37 °C and 5% CO_2. After being treated with 50 µg/mL [Pd(*dach*)Cl_2] (dissolved in 7% DMSO in PBS), the cells were incubated for 48 h. Control cells were treated with 7% DMSO in PBS for 48 h. [Pd(*dach*)Cl_2]-treated HeLa cells and controls were prepared in five replicates each.

4.3. Sample Preparation

The frozen cells from each biological replicate were resuspended in 100 µL of Lysis buffer (4% SDS, 5 mM $MgCl_2 \cdot 6H_2O$, 10 mM CHAPS, 100 mM NH_4HCO_3, 50 mM DTT), then lysed by freezing at −80 °C for 15 min, and subsequently thawing for a total of three times. The samples were then sonicated for 20 min in an ice bath to dissolve them further. The protein content was quantified using the Bradford method [65] and stored at −80 °C until use. Samples were prepared for LC-MS/MS using the RapiGest protocol as previously described [66]. The starting protein amount for tryptic digestion per sample was 25 µg protein and the final protein concentration was 200 ng/µL. Internal standard protein, yeast ADH, was added to all samples with a final concentration of 25 fmol/µL.

4.4. LC-MS/MS Acquisition

The proteomics profiling was performed using label-free, data-independent nano-LC-MS/MS acquisition on ACQUITY UPLC® M-Class (Waters Corporation, Milford, MA, USA) coupled with SYNAPT G2-Si High-Definition Mass Spectrometer (Waters Corporation). Data were obtained using the ultra-definition mass spectrometry ($UDMS^E$) mode described in detail elsewhere [67]. The protein load for the $UDMS^E$ runs was determined to be 150 ng by testing pool samples (containing an equal amount of each of the ten individual samples), starting from 0.5 to 3.0 µL of sample per run, and processing in ProteinLynx Global SERVER (PLGS, version 3.0.3, Waters Corp.). One test run for each sample was performed for quality assurance and validation of the protein concentration, followed by run at 150 ng. LC and MS parameters were as previously described in detail [68].

4.5. MS Data Processing and Identification

Test runs were processed with PLGS (Waters Corporation) with the following settings: (1) Low-energy (LE) and high-energy (HE) thresholds of 450 counts and 20 counts, respectively; (2) Precursor and fragment ion mass tolerances—auto; (3) One missed cleavages, carbamidomethyl C as a fixed modification, and oxidized M as a variable modification; (4) A minimum of two fragment ion matches per peptide identification and five fragment ion matches per protein identification, with at least one peptide match per protein identification; (5) The protein false discovery rate (FDR) < 1%. The typical RMS error for precursor and product ions were ±5 and ±10 ppm, respectively.

Comparative proteomics analysis was performed by Progenesis QIP version 4.1 (Nonlinear dynamics, Waters Corporation) with the following settings: (1) LE and HE threshold—auto; (2) reference run—auto; (3) normalization—"normalize to all proteins"; (4) digest reagent—trypsin; (5) maximum missed cleavages—one; (6) maximum protein mass—250 kDa; (7) fixed modifications—carbamidomethyl C; (8) variable modification— oxidation M; (9) peptide tolerance—auto; (10) fragment tolerance—auto; (11) FDR < 1%; (12) minimum of two fragment ion matches per peptide identification and five fragment ion matches per protein identification, with at least one peptide match per protein identification; (13) quantification based on non-conflicting peptides; (14) grouping of similar proteins. A combined target-decoy database based on the UniProtKB/Swiss-Prot database with 20,370 proteins (June 2020) was used. The obtained identifications were filtrated to remove peptides ≤ 6 amino acids and a score below 4. Data was exported as a .csv output file for subsequent analysis. The calculated FDR on the whole dataset was 3.2%.

4.6. Proteomics Data Analysis

The criteria for the selection of DAPs were identification based on ≥2 peptides and an ANOVA p-value ≤ 0.05. The enriched GO annotation terms and pathways were retrieved using the Cytoscape plug-in ClueGO [69]. ClueGo selection criteria were the following: Analysis mode—Functional analysis; Organism—Homo sapiens [9606]; Show only pathways with pV $\leq$ 0.05; GO term selection—GO tree interval—3 to 8 with a minimum of 4 genes to be associated with a term, and these genes represent at least 1% from the total number of associated genes; GO pathway selection—minimum of 3 genes to be associated with a pathway, and these genes represent at least 4% from the total number of associated genes; GO term/pathway network connectivity (Kappa Score) = 0.4; Two-sided hypergeometric test and Bonferroni step down with cutoff $p \leq 0.05$; Grouping options—YES; Leading Group Term based on—Highest Significance (Kappa score) with min 50% of genes for group merge and min 50% of terms for group merge. STRING analysis [70] settings: (1) complete STRING network; (2) evidence setting; (3) all active interaction sources; (4) confidence score = 0.600; and (5) max number of interactors to show—none (query proteins only). Cytoscape plug-in cytoHubba [71] was used to rank nodes (proteins) in the network by their network features using the MCC topological analysis method.

4.7. Computational Biology Analysis of the [Pd(dach)Cl_2] Interaction with HeLa Cellular Proteins

The ISM focused on the cross-spectral analysis of the informational spectra of the Pd(II) complex and proteins expressed in HeLa cells, serving as a baseline. The workflow consisted of the following steps: (1) Each molecule is presented as a series of numbers based on the electron–ion interaction potential (EIIP) [15], and each amino acid in the protein sequence is replaced by its own EIIP value; (2) Numerical sequence was then transformed into an IS by Fourier transform; (3) The spectra of two molecules, which interact potentially, were multiplied to obtain cross-spectrum (CS) results. The approach used for small molecules (ISM-SM) is modified and utilized to generate the IS of the Pd(II) complex [13].

The list of genes expressed in HeLa cells above the baseline (TPM > 2.0) was retrieved from the Expression Atlas [72]. Protein sequences needed for further ISM analysis were obtained from the UniProt database [73]. The potential protein interactors of the Pd (II) complex in HeLa cells were further analyzed by enrichment analysis using DAVID [16].

5. Conclusions

In conclusion, the significant enrichment of the ribosome as the core molecular machinery governing protein synthesis, along with the spliceosome and the mRNA surveillance pathways in HeLa cervical cancer cells treated with the [Pd(*dach*)Cl_2] complex, suggests a broad disruption in post-transcriptional gene regulation mechanisms. These pathways are relevant as potential targets, as they play essential roles in RNA processing and quality control. The findings presented in this work indicate that the [Pd(*dach*)Cl_2] complex may exert its effects by interfering with RNA metabolism, which might be accompanied with alterations of the expression of key regulatory proteins involved in cell survival and proliferation. Moreover, by using a combination of proteomics analysis and ISM, we predicted the involvement of specific ribosomal proteins in p53 stabilization and activation, which needs to be further experimentally documented. This hypothesis, as well as the confirmed sensitivity of cancer cells to spliceosomal stress, further underlines the therapeutic relevance of these pathways and provides mechanistic insights into the anti-cancer potential of [Pd(*dach*)Cl_2].

Supplementary Materials: The following supporting information can be downloaded at: https://www.mdpi.com/article/10.3390/inorganics13070215/s1, Table S1: List of identified proteins from [Pd(*dach*)Cl_2]-treated HeLa cells and untreated HeLa cells (control group) based on ≥2 peptides.

Author Contributions: Conceptualization, M.P. and K.D.; methodology, K.D., V.R., B.G., and M.S. (Milan Senćanski); formal analysis, K.D. and V.R.; investigation, V.R., K.D., M.D.N., and J.Ž.; writing—original draft preparation, V.R., K.D., M.P. and M.S. (Milutin Stepić); writing—review and editing, M.S. (Milutin Stepić), J.Ž., M.D.N., M.S. (Milan Senćanski), and B.G.; funding acquisition, M.P. and K.D. All authors have read and agreed to the published version of the manuscript.

Funding: The research work was funded and APC paid by the Ministry of Ministry of Science, Technological Development, and Innovation of the Republic of Serbia (Grant No. 451-03-136/2025-03/200017). Mobility scheme to support joint experiments was supported by the Western Balkans Fund Move Grants (PROMETHEAN I and PROMETHEAN II, MO-1-041 and MO-5-084, respectively).

Institutional Review Board Statement: Not applicable.

Informed Consent Statement: Not applicable.

Data Availability Statement: The original contributions presented in this study are included in the article/Supplementary Materials. Further inquiries can be directed to the corresponding authors.

Acknowledgments: The authors are thankful to Lela Korićanac (Department of Molecular Biology and Endocrinology, VINČA Institute of Nuclear Sciences, National Institute of the Republic of Serbia, University of Belgrade, Serbia) for her assistance in the cell experiments and Iva Popović (COHERENCE-Centre, Department of Atomic Physics, VINČA Institute of Nuclear Sciences, National Institute of the Republic of Serbia, University of Belgrade, Serbia) for her help in the manuscript preparation.

Conflicts of Interest: The authors declare no conflicts of interest. The funders had no role in the design of the study; in the collection, analyses, or interpretation of data; in the writing of the manuscript; or in the decision to publish the results.

Abbreviations

The following abbreviations are used in this manuscript:

DAPs	Differentially abundant proteins
DNA	Desoxyribonucleic acid
EJC	Exon Junction Complex
GO	Gene Ontology
HSP	Heat Shock Protein
ISM	Informational spectrum method
LC-MS/MS	Liquid Chromatography-Tandem mass spectrometry
MCC	Maximal Clique Centrality
NDM	Nonsense Mediated Decay
PPI	Protein–protein interaction
RNA	Ribonucleic Acid
snRNPs	Small Nuclear Ribonucleoproteins
SOSS DNA	Sensor of the single-stranded DNA
TP53	Tumor Suppressor Protein p53

References

1. Fowler, J.R.; Maani, E.V.; Dunton, C.J.; Gasalberti, D.P.; Jack, B.W. Cervical Cancer. In *StatPearls*; StatPearls Publishing: Treasure Island, FL, USA, 2025.
2. Sung, H.; Ferlay, J.; Siegel, R.L.; Laversanne, M.; Soerjomataram, I.; Jemal, A.; Bray, F. Global Cancer Statistics 2020: GLOBOCAN Estimates of Incidence and Mortality Worldwide for 36 Cancers in 185 Countries. *CA Cancer J. Clin.* **2021**, *71*, 209–249. [CrossRef]

3. Arbyn, M.; Weiderpass, E.; Bruni, L.; De Sanjosé, S.; Saraiya, M.; Ferlay, J.; Bray, F. Estimates of incidence and mortality of cervical cancer in 2018: A worldwide analysis. *Lancet Glob. Health* **2020**, *8*, e191–e203. [CrossRef] [PubMed]
4. Ralić, V.; Nešić, M.D.; Dučić, T.; Stepić, M.; Korićanac, L.; Davalieva, K.; Petković, M. Analysis of Biomolecular Changes in HeLa Cervical Cancer Cell Line Induced by Interaction with [Pd(dach)Cl_2]. *Inorganics* **2025**, *13*, 20. [CrossRef]
5. Yodoshi, M.; Okabe, N. Structures and Interaction with DNA of Ternary Palladium(II) Complexes: [Pd(Gly)(X)] (Gly=Glycine; X=2,2′-Bipyridine, 1,10-Phenanthroline and 2,2′-Bi-pyridylamine). *Chem. Pharm. Bull.* **2008**, *56*, 908–914. [CrossRef]
6. Gao, E.; Liu, F.; Zhu, M.; Wang, L.; Huang, Y.; Liu, H.; Ma, S.; Shi, Q.; Wang, N. Synthesis, characterization, DNA interaction, and cytotoxicity of novel Pd(II) and Pt(II) complexes. *J. Enzyme Inhib. Med. Chem.* **2010**, *25*, 557–564. [CrossRef]
7. Hobart, D.B.; Berg, M.A.G.; Merola, J.S. Bis-glycinato complexes of palladium(II): Synthesis, structural determination, and hydrogen bonding interactions. *Inorganica Chim. Acta* **2014**, *423*, 21–30. [CrossRef]
8. Hobart, D.B.; Berg, M.A.G.; Rogers, H.M.; Merola, J.S. Synthesis, Characterization, and Non-Covalent Interactions of Palladium(II)-Amino Acid Complexes. *Molecules* **2021**, *26*, 4331. [CrossRef]
9. Hobart, D.B.; Merola, J.S.; Rogers, H.M.; Sahgal, S.; Mitchell, J.; Florio, J.; Merola, J.W. Synthesis, Structure, and Catalytic Reactivity of Pd(II) Complexes of Proline and Proline Homologs. *Catalysts* **2019**, *9*, 515. [CrossRef]
10. Wang, Y.; Chiu, J.-F. Proteomic Approaches in Understanding Action Mechanisms of Metal-Based Anticancer Drugs. *Met.-Based Drugs* **2008**, *2008*, 716329. [CrossRef]
11. Fotis, C.; Antoranz, A.; Hatziavramidis, D.; Sakellaropoulos, T.; Alexopoulos, L.G. Network-based technologies for early drug discovery. *Drug Discov. Today* **2018**, *23*, 626–635. [CrossRef]
12. Nešić, M.D.; Dučić, T.; Gonçalves, M.; Stepić, M.; Algarra, M.; Soto, J.; Gemović, B.; Bandosz, T.J.; Petković, M. Biochemical changes in cancer cells induced by photoactive nanosystem based on carbon dots loaded with Ru-complex. *Chem. Biol. Interact.* **2022**, *360*, 109950. [CrossRef]
13. Sencanski, M.; Perovic, V.; Pajovic, S.B.; Adzic, M.; Paessler, S.; Glisic, S. Drug Repurposing for Candidate SARS-CoV-2 Main Protease Inhibitors by a Novel In Silico Method. *Molecules* **2020**, *25*, 3830. [CrossRef] [PubMed]
14. Veljkovic, N.; Glisic, S.; Prljic, J.; Perovic, V.; Botta, M.; Veljkovic, V. Discovery of New Therapeutic Targets by the Informational Spectrum Method. *Curr. Protein Pept. Sci.* **2008**, *9*, 493–506. [CrossRef] [PubMed]
15. Veljkovic, V. *A Theoretical Approach to the Preselection of Carcinogens and Chemical Carcinogenesis*; Gordon and Breach Science Publishers: New York, NY, USA, 1980.
16. Sherman, B.T.; Hao, M.; Qiu, J.; Jiao, X.; Baseler, M.W.; Lane, H.C.; Imamichi, T.; Chang, W. DAVID: A web server for functional enrichment analysis and functional annotation of gene lists (2021 update). *Nucleic Acids Res.* **2022**, *50*, W216–W221. [CrossRef]
17. Huang, J.; Gong, Z.; Ghosal, G.; Chen, J. SOSS Complexes Participate in the Maintenance of Genomic Stability. *Mol. Cell* **2009**, *35*, 384–393. [CrossRef]
18. Ren, W.; Chen, H.; Sun, Q.; Tang, X.; Lim, S.C.; Huang, J.; Song, H. Structural Basis of SOSS1 Complex Assembly and Recognition of ssDNA. *Cell Rep.* **2014**, *6*, 982–991. [CrossRef]
19. Li, Y.H.; Gao, Z.Q.; Dong, Y.H. Crystal structure of SSB from homo sapiens: 5d8e. 2016. Available online: https://www.rcsb.org/structure/5D8E (accessed on 1 June 2025).
20. Zidovska, A. The self-stirred genome: Large-scale chromatin dynamics, its biophysical origins and implications. *Curr. Opin. Genet. Dev.* **2020**, *61*, 83–90. [CrossRef]
21. Hanel, W.; Moll, U.M. Links between mutant p53 and genomic instability. *J. Cell Biochem.* **2012**, *113*, 433–439. [CrossRef] [PubMed]
22. Yang, H.; Salz, T.; Zajac-Kaye, M.; Liao, D.; Huang, S.; Qiu, Y. Overexpression of histone deacetylases in cancer cells is controlled by interplay of transcription factors and epigenetic modulators. *FASEB J.* **2014**, *28*, 4265–4279. [CrossRef]
23. Levine, A.J.; Oren, M. The first 30 years of p53: Growing ever more complex. *Nat. Rev. Cancer* **2009**, *9*, 749–758. [CrossRef]
24. Bi, Y.; Kong, P.; Zhang, L.; Cui, H.; Xu, X.; Chang, F.; Yan, T.; Li, J.; Cheng, C.; Song, B.; et al. EP300 as an oncogene correlates with poor prognosis in esophageal squamous carcinoma. *J. Cancer* **2019**, *10*, 5413–5426. [CrossRef]
25. Richard, D.J.; Bolderson, E.; Cubeddu, L.; Wadsworth, R.I.M.; Savage, K.; Sharma, G.G.; Nicolette, M.L.; Tsvetanov, S.; McIlwraith, M.J.; Pandita, R.K.; et al. Single-stranded DNA-binding protein hSSB1 is critical for genomic stability. *Nature* **2008**, *453*, 677–681. [CrossRef]
26. Long, Q.; Sebesta, M.; Sedova, K.; Haluza, V.; Alagia, A.; Liu, Z.; Stefl, R.; Gullerova, M. The phosphorylated trimeric SOSS1 complex and RNA polymerase II trigger liquid-liquid phase separation at double-strand breaks. *Cell Rep.* **2023**, *42*, 113489. [CrossRef] [PubMed]
27. Venables, J.P.; Klinck, R.; Koh, C.; Gervais-Bird, J.; Bramard, A.; Inkel, L.; Durand, M.; Couture, S.; Froehlich, U.; Lapointe, E.; et al. Cancer-associated regulation of alternative splicing. *Nat. Struct. Mol. Biol.* **2009**, *16*, 670–676. [CrossRef] [PubMed]
28. Hsu, T.Y.-T.; Simon, L.M.; Neill, N.J.; Marcotte, R.; Sayad, A.; Bland, C.S.; Echeverria, G.V.; Sun, T.; Kurley, S.J.; Tyagi, S.; et al. The spliceosome is a therapeutic vulnerability in MYC-driven cancer. *Nature* **2015**, *525*, 384–388. [CrossRef]

29. Higareda-Almaraz, J.C.; Valtierra-Gutiérrez, I.A.; Hernandez-Ortiz, M.; Contreras, S.; Hernandez, E.; Encarnacion, S. Analysis and Prediction of Pathways in HeLa Cells by Integrating Biological Levels of Organization with Systems-Biology Approaches. *PLoS ONE* **2013**, *8*, e65433. [CrossRef]
30. Zhang, Y.; Shan, N.; Zhou, W.; Zhang, S. Identification of HSPA8 as a candidate biomarker for endometrial carcinoma by using iTRAQ-based proteomic analysis. *OncoTargets Ther.* **2016**, *2016*, 2169–2179. [CrossRef]
31. Li, B.; Ming, H.; Qin, S.; Zhou, L.; Huang, Z.; Jin, P.; Peng, L.; Luo, M.; Zhang, T.; Wang, K.; et al. HSPA8 Activates Wnt/β-Catenin Signaling to Facilitate BRAF V600E Colorectal Cancer Progression by CMA-Mediated CAV1 Degradation. *Adv. Sci.* **2024**, *11*, 2306535. [CrossRef]
32. Wu, J.; Lu, F.; Yu, B.; Wang, W.; Ye, X. The oncogenic role of SNRPB in human tumors: A pan-cancer analysis. *Front. Mol. Biosci.* **2022**, *9*, 994440. [CrossRef]
33. Szymura, S.J.; Bernal, G.M.; Wu, L.; Zhang, Z.; Crawley, C.D.; Voce, D.J.; Campbell, P.-A.; Ranoa, D.E.; Weichselbaum, R.R.; Yamini, B. DDX39B interacts with the pattern recognition receptor pathway to inhibit NF-κB and sensitize to alkylating chemotherapy. *BMC Biol.* **2020**, *18*, 32. [CrossRef]
34. Wang, L.; Wang, Y.; Su, B.; Yu, P.; He, J.; Meng, L.; Xiao, Q.; Sun, J.; Zhou, K.; Xue, Y.; et al. Transcriptome-wide analysis and modelling of prognostic alternative splicing signatures in invasive breast cancer: A prospective clinical study. *Sci. Rep.* **2020**, *10*, 16504. [CrossRef] [PubMed]
35. Zhang, H.; He, C.; Guo, X.; Fang, Y.; Lai, Q.; Wang, X.; Pan, X.; Li, H.; Qin, K.; Li, A.; et al. DDX39B contributes to the proliferation of colorectal cancer through direct binding to CDK6/CCND1. *Cell Death Discov.* **2022**, *8*, 30. [CrossRef] [PubMed]
36. Nakata, D.; Nakao, S.; Nakayama, K.; Araki, S.; Nakayama, Y.; Aparicio, S.; Hara, T.; Nakanishi, A. The RNA helicase DDX39B and its paralog DDX39A regulate androgen receptor splice variant AR-V7 generation. *Biochem. Biophys. Res. Commun.* **2017**, *483*, 271–276. [CrossRef] [PubMed]
37. Meng, T.; Huang, R.; Zeng, Z.; Huang, Z.; Yin, H.; Jiao, C.; Yan, P.; Hu, P.; Zhu, X.; Li, Z.; et al. Identification of Prognostic and Metastatic Alternative Splicing Signatures in Kidney Renal Clear Cell Carcinoma. *Front. Bioeng. Biotechnol.* **2019**, *7*, 270. [CrossRef]
38. Walbrecq, G.; Lecha, O.; Gaigneaux, A.; Fougeras, M.R.; Philippidou, D.; Margue, C.; Tetsi Nomigni, M.; Bernardin, F.; Dittmar, G.; Behrmann, I.; et al. Hypoxia-Induced Adaptations of miRNomes and Proteomes in Melanoma Cells and Their Secreted Extracellular Vesicles. *Cancers* **2020**, *12*, 692. [CrossRef]
39. Turunen, J.J.; Niemelä, E.H.; Verma, B.; Frilander, M.J. The significant other: Splicing by the minor spliceosome. *WIREs RNA* **2013**, *4*, 61–76. [CrossRef]
40. Alioto, T.S. U12DB: A database of orthologous U12-type spliceosomal introns. *Nucleic Acids Res.* **2007**, *35*, D110–D115. [CrossRef]
41. Ye, Z.; Bing, A.; Zhao, S.; Yi, S.; Zhan, X. Comprehensive analysis of spliceosome genes and their mutants across 27 cancer types in 9070 patients: Clinically relevant outcomes in the context of 3P medicine. *EPMA J.* **2022**, *13*, 335–350. [CrossRef]
42. Peng, Q.; Zhou, Y.; Oyang, L.; Wu, N.; Tang, Y.; Su, M.; Luo, X.; Wang, Y.; Sheng, X.; Ma, J.; et al. Impacts and mechanisms of alternative mRNA splicing in cancer metabolism, immune response, and therapeutics. *Mol. Ther.* **2022**, *30*, 1018–1035. [CrossRef]
43. Kitamura, K.; Nimura, K. Regulation of RNA Splicing: Aberrant Splicing Regulation and Therapeutic Targets in Cancer. *Cells* **2021**, *10*, 923. [CrossRef]
44. Ruggero, D.; Pandolfi, P.P. Does the ribosome translate cancer? *Nat. Rev. Cancer* **2003**, *3*, 179–192. [CrossRef] [PubMed]
45. Van Riggelen, J.; Yetil, A.; Felsher, D.W. MYC as a regulator of ribosome biogenesis and protein synthesis. *Nat. Rev. Cancer* **2010**, *10*, 301–309. [CrossRef] [PubMed]
46. Zisi, A.; Bartek, J.; Lindström, M.S. Targeting Ribosome Biogenesis in Cancer: Lessons Learned and Way Forward. *Cancers* **2022**, *14*, 2126. [CrossRef]
47. Ramalho, S.; Dopler, A.; Faller, W.J. Ribosome specialization in cancer: A spotlight on ribosomal proteins. *NAR Cancer* **2024**, *6*, zcae029. [CrossRef] [PubMed]
48. Trifa, Y.; Privat, I.; Gagnon, J.; Baeza, L.; Lerbs-Mache, S. The Nuclear RPL4 Gene Encodes a Chloroplast Protein That Co-purifies with the T7-like Transcription Complex as Well as Plastid Ribosomes. *J. Biol. Chem.* **1998**, *273*, 3980–3985. [CrossRef]
49. He, X.; Li, Y.; Dai, M.-S.; Sun, X.-X. Ribosomal protein L4 is a novel regulator of the MDM2-p53 loop. *Oncotarget* **2016**, *7*, 16217–16226. [CrossRef]
50. Zhang, X.; Wang, W.; Wang, H.; Wang, M.-H.; Xu, W.; Zhang, R. Identification of ribosomal protein S25 (RPS25)–MDM2–p53 regulatory feedback loop. *Oncogene* **2013**, *32*, 2782–2791. [CrossRef]
51. Xie, J.; Zhang, W.; Liang, X.; Shuai, C.; Zhou, Y.; Pan, H.; Yang, Y.; Han, W. RPL32 Promotes Lung Cancer Progression by Facilitating p53 Degradation. *Mol. Ther.—Nucleic Acids* **2020**, *21*, 75–85. [CrossRef]
52. Xu, L.; Wang, L.; Jiang, C.; Zhu, Q.; Chen, R.; Wang, J.; Wang, S. Biological effect of ribosomal protein L32 on human breast cancer cell behavior. *Mol. Med. Rep.* **2020**, *22*, 2478–2486. [CrossRef]

53. Xu, M.; Wang, Y.; Chen, L.; Pan, B.; Chen, F.; Fang, Y.; Yu, Z.; Chen, G. Down-regulation of ribosomal protein S15A mRNA with a short hairpin RNA inhibits human hepatic cancer cell growth in vitro. *Gene* **2014**, *536*, 84–89. [CrossRef]
54. Chen, J.; Wei, Y.; Feng, Q.; Ren, L.; He, G.; Chang, W.; Zhu, D.; Yi, T.; Lin, Q.; Tang, W.; et al. Ribosomal protein S15A promotes malignant transformation and predicts poor outcome in colorectal cancer through misregulation of p53 signaling pathway. *Int. J. Oncol.* **2016**, *48*, 1628–1638. [CrossRef] [PubMed]
55. Shi, D.; Liu, J. RPS15a Silencing Suppresses Cell Proliferation and Migration of Gastric Cancer. *Yonsei Med. J.* **2018**, *59*, 1166. [CrossRef] [PubMed]
56. Kong, Y.; Shuangshuang, D.; Liang, X.; Zhou, X. RPS9 promotes the progression of NSCLC via activation Stat3 and Erk signaling pathways. *J. Cancer* **2022**, *13*, 1346–1355. [CrossRef]
57. Cheng, D.; Zhu, B.; Li, S.; Yuan, T.; Yang, Q.; Fan, C. Down-regulation of RPS9 Inhibits Osteosarcoma Cell Growth through Inactivation of MAPK Signaling Pathway. *J. Cancer* **2017**, *8*, 2720–2728. [CrossRef]
58. Shu, G.; Chen, M.; Liao, W.; Fu, L.; Lin, M.; Gui, C.; Cen, J.; Lu, J.; Chen, Z.; Wei, J.; et al. PABPC1L Induces IDO1 to Promote Tryptophan Metabolism and Immune Suppression in Renal Cell Carcinoma. *Cancer Res.* **2024**, *84*, 1659–1679. [CrossRef] [PubMed]
59. Wu, Y.-Q.; Ju, C.-L.J.; Wang, B.-J.; Wang, R.-G. PABPC1L depletion inhibits proliferation and migration via blockage of AKT pathway in human colorectal cancer cells. *Oncol. Lett.* **2019**, *17*, 3439–3445. [CrossRef]
60. An, T.; Deng, L.; Wang, Y.; Yang, Z.; Chai, C.; Ouyang, J.; Lu, X.; Zhang, C. The Prognostic Impacts of PABPC1 Expression on Gastric Cancer Patients. *Future Oncol.* **2021**, *17*, 4471–4479. [CrossRef]
61. Castro, M.E.; Ferrer, I.; Cascon, A.; Guijarro, M.V.; Lleonart, M.; Cajal, S.R.Y.; Leal, J.F.M.; Robledo, M.; Carnero, A. PPP1CA contributes to the senescence program induced by oncogenic Ras. *Carcinogenesis* **2007**, *29*, 491–499. [CrossRef]
62. Hu, A.; Hong, F.; Li, D.; Xie, Q.; Chen, K.; Zhu, L.; He, H. KDM3B-ETF1 fusion gene downregulates LMO2 via the WNT/β-catenin signaling pathway, promoting metastasis of invasive ductal carcinoma. *Cancer Gene Ther.* **2022**, *29*, 215–224. [CrossRef]
63. Stoddart, A.; Qian, Z.; Fernald, A.A.; Bergerson, R.J.; Wang, J.; Karrison, T.; Anastasi, J.; Bartom, E.T.; Sarver, A.L.; McNerney, M.E.; et al. Retroviral insertional mutagenesis identifies the del(5q) genes, CXXC5, TIFAB and ETF1, as well as the Wnt pathway, as potential targets in del(5q) myeloid neoplasms. *Haematologica* **2016**, *101*, e232–e236. [CrossRef]
64. Petrović, B.; Bugarčić, Ž.D.; Van Eldik, R. Kinetic studies on the reactions of [Pd(dach)(X–Y)] complexes with some DNA constituents. *Dalton Trans.* **2008**, 807–813. [CrossRef] [PubMed]
65. Bradford, M.M. A rapid and sensitive method for the quantitation of microgram quantities of protein utilizing the principle of protein-dye binding. *Anal. Biochem.* **1976**, *72*, 248–254. [CrossRef] [PubMed]
66. Davalieva, K.; Rusevski, A.; Velkov, M.; Noveski, P.; Kubelka-Sabit, K.; Filipovski, V.; Plaseski, T.; Dimovski, A.; Plaseska-Karanfilska, D. Comparative proteomics analysis of human FFPE testicular tissues reveals new candidate biomarkers for distinction among azoospermia types and subtypes. *J. Proteomics* **2022**, *267*, 104686. [CrossRef] [PubMed]
67. Distler, U.; Kuharev, J.; Navarro, P.; Levin, Y.; Schild, H.; Tenzer, S. Drift time-specific collision energies enable deep-coverage data-independent acquisition proteomics. *Nat. Methods* **2014**, *11*, 167–170. [CrossRef]
68. Davalieva, K.; Kiprijanovska, S.; Dimovski, A.; Rosoklija, G.; Dwork, A.J. Comparative evaluation of two methods for LC-MS/MS proteomic analysis of formalin fixed and paraffin embedded tissues. *J. Proteomics* **2021**, *235*, 104117. [CrossRef]
69. Bindea, G.; Mlecnik, B.; Hackl, H.; Charoentong, P.; Tosolini, M.; Kirilovsky, A.; Fridman, W.-H.; Pagès, F.; Trajanoski, Z.; Galon, J. ClueGO: A Cytoscape plug-in to decipher functionally grouped gene ontology and pathway annotation networks. *Bioinformatics* **2009**, *25*, 1091–1093. [CrossRef]
70. Szklarczyk, D.; Gable, A.L.; Lyon, D.; Junge, A.; Wyder, S.; Huerta-Cepas, J.; Simonovic, M.; Doncheva, N.T.; Morris, J.H.; Bork, P.; et al. STRING v11: Protein–protein association networks with increased coverage, supporting functional discovery in genome-wide experimental datasets. *Nucleic Acids Res.* **2019**, *47*, D607–D613. [CrossRef]
71. Chin, C.-H.; Chen, S.-H.; Wu, H.-H.; Ho, C.-W.; Ko, M.-T.; Lin, C.-Y. cytoHubba: Identifying hub objects and sub-networks from complex interactome. *BMC Syst. Biol.* **2014**, *8*, S11. [CrossRef]
72. Papatheodorou, I.; Moreno, P.; Manning, J.; Fuentes, A.M.-P.; George, N.; Fexova, S.; Fonseca, N.A.; Füllgrabe, A.; Green, M.; Huang, N.; et al. Expression Atlas update: From tissues to single cells. *Nucleic Acids Res.* **2019**, *48*, D77–D83. [CrossRef]
73. The UniProt Consortium. UniProt: A worldwide hub of protein knowledge. *Nucleic Acids Res.* **2019**, *47*, D506–D515. [CrossRef]

inorganics

MDPI

Article

Theoretical Study of the Effects of Different Coordination Atoms (O/S/N) on Crystal Structure, Stability, and Protein/DNA Binding of Ni(II) Complexes with Pyridoxal-Semi, Thiosemi, and Isothiosemicarbazone Ligand Systems

Violeta Jevtovic [1,†], Aleksandra Rakić [2,†], Odeh A. O. Alshammari [1], Munirah Sulaiman Alhar [1], Tahani Alenezi [1], Violeta Rakic [3] and Dušan Dimić [2,*]

1 Department of Chemistry, College of Science, University of Ha'il, Ha'il 81451, Saudi Arabia
2 Faculty of Physical Chemistry, University of Belgrade, Studentski trg 12–16, 11000 Belgrade, Serbia
3 Department of Agriculture and Food Technology Studies Prokuplje, Toplica Academy of Applied Studies, Ćirila i Metodija 1, 18400 Prokuplje, Serbia
* Correspondence: ddimic@ffh.bg.ac.rs
† These authors contributed equally to this work.

Abstract: Nickel transition metal complexes have shown various biological activities that depend on the ligands and geometry. In this contribution, six Ni(II) nitrate complexes with pyridoxal-semi, thiosemi, and isothiosemicarbazone ligands were examined using theoretical chemistry methods. The structures of three previously reported complexes ($[Ni(PLSC)(H_2O)_3]{\cdot}2NO_3^-$, $[Ni(PLTSC)_2]{\cdot}2NO_3^-{\cdot}H_2O$, and $[Ni(PLITSC)(H_2O)_3]{\cdot}2NO_3^-$) were investigated based on Hirshfeld surface analysis, and the most important stabilization interactions in the crystal structures were outlined. These structures were optimized at the B3LYP/6-311++G(d,p)(H,C,N,O,(S))/LanL2DZ(Ni) level of theory, and the applicability was checked by comparing theoretical and experimental bond lengths and angles. The same level of theory was applied for the optimization of three additional structures, ($[Ni(PLSC)_2]^{2+}$, $[Ni(PLTSC)(H_2O)_3]^{2+}$, and $[Ni(PLITSC)_2]^{2+}$). The interactions between selected ligands and Ni(II) were examined using the Natural Bond Orbital (NBO) and Quantum Theory of Atoms in Molecules (QTAIM) approaches. Particular emphasis was placed on interactions between oxygen, sulfur, and nitrogen donor atoms and Ni(II). Human Serum Albumin (HSA) and the DNA-binding properties of these complex cations were assessed using molecular docking simulations. The presence of water molecules and various substituents in the thermodynamics of the processes was demonstrated. The results showed significant effects of structural parameters on the stability and reactivity towards important biomolecules.

Keywords: nickel(II) complexes; DFT; QTAIM; HSA; DNA; molecular docking

1. Introduction

Transition metal complexes are an important group of compounds with pronounced biological activity. They are commonly used in medicine as contrast agents in Magnetic Resonance Imaging (MRI), radiopharmaceuticals, chemotherapeutics, and compounds for arthritis treatment [1]. The success of platinum compounds in suppressing cell division led to the introduction of transition metal complexes in modern medical research. The primary mechanism of cisplatin action includes binding to DNA molecules and disrupting the transcription process [2,3]. The limited selectivity and toxicity of cisplatin and the development of resistance led to research on other transition metal complexes. Platinum and ruthenium ions and their compounds have been the most widely investigated [4]. However, cheaper first-row transition metals are being examined as potential alternatives for efficient cytotoxic agents [5]. Nickel is one of the essential elements in biological systems, and it constitutes several metalloproteins.

Nickel complexes are being investigated as antibacterial, leishmanicidal, antifungal, anticancer, antioxidant, and antiretroviral agents [6–10]. The nature of coordinating ligands influences these compounds' thermodynamic and kinetic stability [1]. It has been shown that the size and shape of octahedral nickel complexes directly affect DNA affinity [11].

Complexes containing semi- and thiosemicarbazone ligands have attracted the attention of scientists due to their interesting structural characteristics and the possibility of complexation to different transition metals [12]. The biological activity of these compounds includes potential use as radioprotectors, anti-protozoa agents, chelating ligands, and antimicrobials [13]. The thiosemicarbazone complexes have shown a broader spectrum of biological activities, as discussed in [14]. Thiosemicarbazones ($R^1R^2C^2$=N^3-N^2(H)-C^1(=S)$N^1R^3R^4$) are a diverse group of ligands containing both nitrogen and sulfur donor atoms. They are prepared via condensation between ketone or aldehyde and thiosemicarbazide. Depending on protonation, these ligands form differently charged complexes [15]. Pyridoxal–thiosemicarbazone (PLTSC) is formed by reacting pyrodixal, a vitamin B6 analog, and thiosemicarbazone [12]. A comprehensive review of the transition metal complexes with PLTSC is presented in [13]. The pyridoxal–semicarazone (PLSC) ligand is an oxygen analog of PLTSC, with ONO donor atoms [12]. The third ligand of interest for this study is pyridoxal-S-methylisothiosemicarbazone (PLITSC), formed in the reaction between pyridoxal and S-methyl-isothiosemicarbazone. In this ligand, the sulfur atom is protected by the presence of a methyl substituent, which leads to an ONN donor system. PLITSC can also be found in neutral, mono, and dianionic forms [16]. The structures of neutral PLSC, PLTSC, and PLITSC ligands are presented in Figure 1.

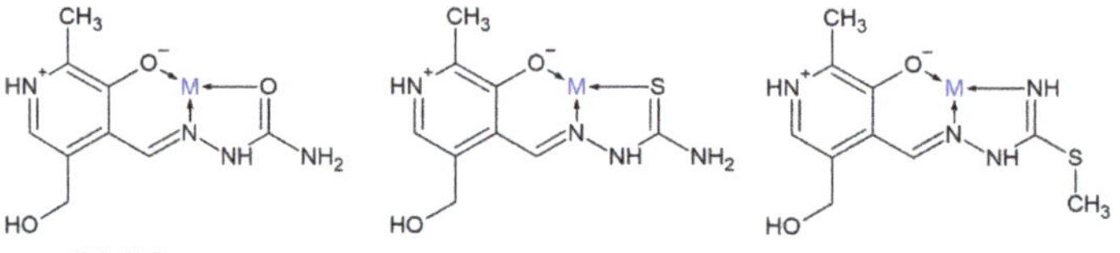

Figure 1. Complexation modes of neutral PLSC, PLTSC, and PLITSC ligands.

Several Ni(II) complexes containing the mentioned ligands have been described in the literature. Jevtović et al. investigated the effects of metal ions (Fe, Co, Ni, and Cu) in complexes with PLTSC ligands on structural properties, protein binding, and cytotoxicity [17]. The crystal structure of $[Ni(PLTSC)_2]\cdot 2NO_3^-\cdot H_2O$ was similar to the previously obtained one by Leovac et al. [12]. Density functional theory (DFT) optimization of structure at the B3LYP/6-311++G(d,p)(H,C,N,O,(S))/LanL2DZ(Ni) level of theory gave bond lengths and angles comparable to the experimental ones, and this structure was used for further theoretical analyses. The experimental transport protein binding affinity of the mentioned complex was further verified by molecular docking. Synthesis, X-ray structures, and spectra of two complexes containing PLSC and PLITSC neutral ligands along with three molecules of water ($[Ni(PLSC)(H_2O)_3]\cdot 2NO_3^-$ and $[Ni(PLITSC)(H_2O)_3]\cdot 2NO_3^-\cdot H_2O$) were described in references [18,19]. Other Ni(II) complexes with three ligands from Figure 1 include [Ni(PLSC-H)2]·H_2O [20], $[Ni(PLSC)Cl_2]\cdot 3.5H_2O$, $[Ni(PLSC)(NCS)_2]\cdot 4H_2O$, $[Ni(PLSC\text{-}2H)NH_3]\cdot 1.5H_2O$, [Ni(PLTSC-H]NCS], [Ni(PLTSC-H)py]·$2NO_3^-$ (py-piridine) [18], $[Ni(PLTSC\text{-}H)(NO_3)]\cdot 2H_2O$, and [Ni(PLTSC-2H]·$2H_2O$ [21], and their structural and spectral properties are described in a review paper by Leovac et al. [13]. Structures with other counterions are also known [22].

This paper aims to investigate the differences in the crystal structure, stability, and protein/DNA affinity of Ni(II) nitrate complexes with PLSC, PLTSC, and PLITSC ligands. The crystal structures of known complexes, namely $[Ni(PLSC)(H_2O)_3]\cdot 2NO_3^-$, $[Ni(PLTSC)_2]\cdot 2NO_3^-$, and $[Ni(PLITSC)(H_2O)_3]\cdot 2NO_3^-$, were examined using Hirshfeld surface analysis, and the percentages of different stabilization interactions were compared. The structures of complex cations from these crystal structures, together with three theoretical cations ($[Ni(PLSC)_2]^{2+}$, $[Ni(PLTSC)(H_2O)_3]^{2+}$, and $[Ni(PLITSC)_2]^{2+}$), were optimized at the B3LYP/6-311++G(d,p)(H,C,N,O,(S))/LanL2DZ(Ni) level of theory. The changes in bond lengths and overall geometry were discussed based on the present donor atoms. The Quantum Theory of Atoms in Molecules (QTAIM) was applied to examine the strength of coordination bonds between various donor atoms and nickel(II). The molecular docking study was used to determine the affinity of these complex cations towards Human Serum Albumin (HSA), a significant transport protein, and DNA. The effects of ligand structure and the presence of water molecules on the interaction with biomolecules were outlined.

2. Results and Discussion

2.1. Hirshfeld Surface Analysis

Selected crystal structures of nickel(II) nitrate complexes with PLSC, PLTSC, and PLITSC were examined using Hirshfeld surface analysis. These three structures were chosen because all contain nickel in +2 state, a ligand in a neutral form, a charge of the complex cation of +2, and surrounding nitrate counterions. Therefore, these structures allow for the elucidation of the most important contacts in the crystal structures, as three ligands differ in the donor atoms attached to central metal ion. Figure 2 presents the Hirshfeld surfaces of these complex compounds, while the most important contacts are listed in Table 1. The fingerprint plots of the most numerous contacts are given in Figures S1–S3. It is important to outline that only the complex cation was included when examining the stabilization interactions, as the positions of nitrate counterions are significantly different in the examined structures.

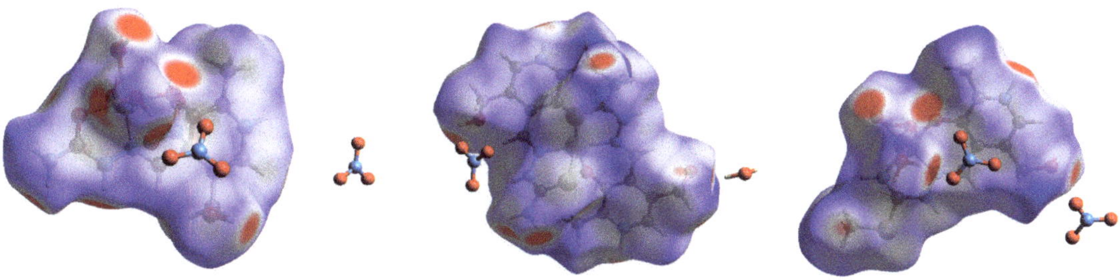

$[Ni(PLSC)(H_2O)_3]{\cdot}2NO_3^-$ $[Ni(PLTSC)_2]{\cdot}2NO_3^-{\cdot}H_2O$ $[Ni(PLITSC)(H_2O)_3]{\cdot}2NO_3^-$

Figure 2. Hirshfeld surfaces of different nickel(II) nitrate complexes included in this study.

Table 1. Percentages of the most important contacts in the crystallographic structures determined based on Hirshfeld surface analysis.

Contact	$[Ni(PLSC)(H_2O)_3]{\cdot}2NO_3^-$	$[Ni(PLTSC)_2]{\cdot}2NO_3^-{\cdot}H_2O$	$[Ni(PLITSC)(H_2O)_3]{\cdot}2NO_3^-$
O···H	49.7	36.3	43.8
H···H	31.5	27.4	36.1
N···H	3.4	4.7	2.4
C···H	5.5	9.7	3.9
O···N	1.9	1.8	0.5
O···O	1.7	0.6	/
O···C	4.1	1.3	4.0
N···C	1.4	0.5	/
S···H	/	10.4	2.2
S···C	/	2.8	2.0
S···O	/	0.3	1.8
S···N	/	0.9	1.0

The octahedral geometry of three complexes excludes the possibility of interactions involving the central metal ion, as previously discussed for similar compounds [17,23]. The most numerous contacts include oxygen and hydrogen atoms. The percentages of O···H contacts are 49.1, 36.3, and 43.8% for the three complexes (Table 1). This value is higher when water molecules are present. The highest percentage in the case of $[Ni(PLSC)(H_2O)_3]{\cdot}2NO_3^-$ is expected, since PLSC contains one oxygen atom more than the PLITSC ligand. The red spots on the Hirshfeld surfaces denote the positions of hydrogen bonds, where most are located around water molecules. These complexes act mainly as hydrogen atom donors through water molecules and the OH group attached to the pyridine ring. In the crystal structure of the Ni(II) complex with a deprotonated PLSC ligand, the amount of O···H interactions is 23.3% due to the decrease in the number of hydrogen bonds [20]. The second-most numerous contacts are formed between hydrogen atoms, ranging from 27.4 to 36.1%. These interactions are usually weaker and include interactions between hydrogen atoms attached to different atoms. The relative abundance of hydrogen atoms in ring structures, water molecules, and methyl group of PLITSC is responsible for forming these interactions. Strong hydrogen bonds also include N···H contacts. The protonated nitrogen atoms of the pyridine ring, aliphatic chain, and ending amino group of PLSC and PLTSC ligands are suitable for the formation of stabilization interactions. The percentage of N···H (2.4%) is the lowest in the third complex due to the exchange of amino group with –S(CH_3). The interactions between carbon and hydrogen atoms, denoted as C···H, have percentages between 3.9 and 9.7%, consistent with previously investigated dioxovanadium(V) complexes with PLSC, PLTSC, and PLITSC ligands [24]. The existence of two ligand molecules in the second complex increases the amount of carbon atoms and the abundance of the interactions. Again, the presence of methyl group in PLITSC, compared to PLSC, leads to an increase in the amount of C···H contacts. The interactions between positively charged hydrogen atoms and negatively charged π-electron clouds are also part of this group [25]. The position of the sulfur atom in the structure of PLTSC and PLITSC significantly influences the percentage of weaker hydrogen bonds (S···H). The methyl group attached to the sulfur atom limits the interactions, leading to a much lower percentage of 2.2% (Table 1).

Interactions between electronegative atoms also depend on the present substituents and relative positions of groups. In the first structure, the most numerous are N···O (1.9%) contacts due to the present groups. Once the oxygen atom is exchanged with sulfur, these interactions are lowered to 1.8% in the second structure. The O···O interactions are present only in the first two structures, proving that there is no direct contact between the water molecules of the two complex cations in the third structure. The abundance of oxygen atoms in the first structure is responsible for the formation of O···C interactions. Structures with sulfur also have a higher percentage of S···C interactions when compared to interactions with nitrogen and oxygen atoms. The effects of geometry and stabilization interactions will be investigated in the following structures.

2.2. DFT Optimization of Structures

The structures of three experimental and three theoretical structures were optimized at the B3LYP/6-311++G(d,p)(H,C,N,O,(S))/LanL2DZ(Ni) level of theory, as shown in Figure 3. The triplet state of the complexes was considered, as it was shown to be more stable than the singlet one [26]. The applicability of the selected level of theory was examined by comparing the crystallographic and optimized bond lengths and angles for ($[Ni(PLSC)(H_2O)_3]^{2+}$, $[Ni(PLITSC)(H_2O)_3]^{2+}$, and $[Ni(PLTSC)_2]^{2+}$ complex cations. These structures were extracted from previously discussed crystal structures. The quantitative parameters used for comparison were the correlation coefficient and mean absolute error (MAE). The second parameter calculates the average absolute difference between the two datasets. The experimental and theoretical structural parameters are listed in Tables S1–S3, while the R and MAE values are presented in Table 2. The optimization of theoretical structures was needed to further analyze the coordination of ligands to the central metal ions, as well as for the molecular docking simulations of the binding to transport proteins and DNA.

The optimized bond lengths coincide well with the crystallographic ones. The correlation coefficients are 0.96 for $[Ni(PLITSC)(H_2O)_3]^{2+}$ and 0.99 for the other two structures, with MAE values between 0.017 ($[Ni(PLSC)(H_2O)_3]^{2+}$) and 0.041 Å ($[Ni(PLTSC)_2]^{2+}$). These values were of the order of experimental error. In the first structure, the bond lengths between nickel(II) and water oxygen atoms are between 2.054 and 2.115 Å in the experimental structures and between 2.086 and 2.153 Å in the optimized structures. The distances between the central metal ion and carbonyl group attached to the pyridine ring are 1.983 and 1.992 Å in the crystallographic and optimized structures, respectively. A larger distance is found between the second carbonyl group and Ni(II) (2.068/2.073 Å), while the Ni–N bond lengths are 2.016 (exp.) and 2.064 Å (theor.). A similar range of bond lengths between the central metal ion and coordinated water is found in the structure of $[Ni(PLITSC)(H_2O)_3]^{2+}$, between 2.041 and 2.131 Å, which proves the assumption that these bonds are not significantly influenced by the ligand structure. The change in the ligand structure of PLITSC in comparison to PLSC leads to lower bond distances between Ni(II) and the carbonyl group oxygen atom (1.880/1.999 Å) and the nitrogen atom of the azomethine group (1.938/2.078 Å). It is evident that due to the optimization and relaxation of the structure, the bond lengths in $[Ni(PLITSC)(H_2O)_3]^{2+}$ increase and become similar to the first complex. The distances between Ni(II) and the nitrogen donor atoms are 1.938 and 2.078 Å in the experimental and theoretical structures, which are comparable to when an oxygen atom is present. In the experimental structure of $[Ni(PLTSC)_2]^{2+}$, the bond lengths between the oxygen/nitrogen atoms and the central metal ion are 1.906/1.907 Å, which are lower than values found in the two complexes with coordinated water molecules. The bond distance between Ni(II) and sulfur is 2.277 in the experimental structure and 2.483 Å in the theoretical structure. Based on these values, it can be assumed that the weakest interactions include the sulfur atoms due to their low electronegativity and electron-donation ability. These assumptions are examined in detail in the following section. The bond lengths of atoms within PLSC, PLTSC, and PLITSC are within the expected range, as explained in references [16,24]. The extended delocalization within the ligand structures prevents significant changes in bond lengths upon complexation [17].

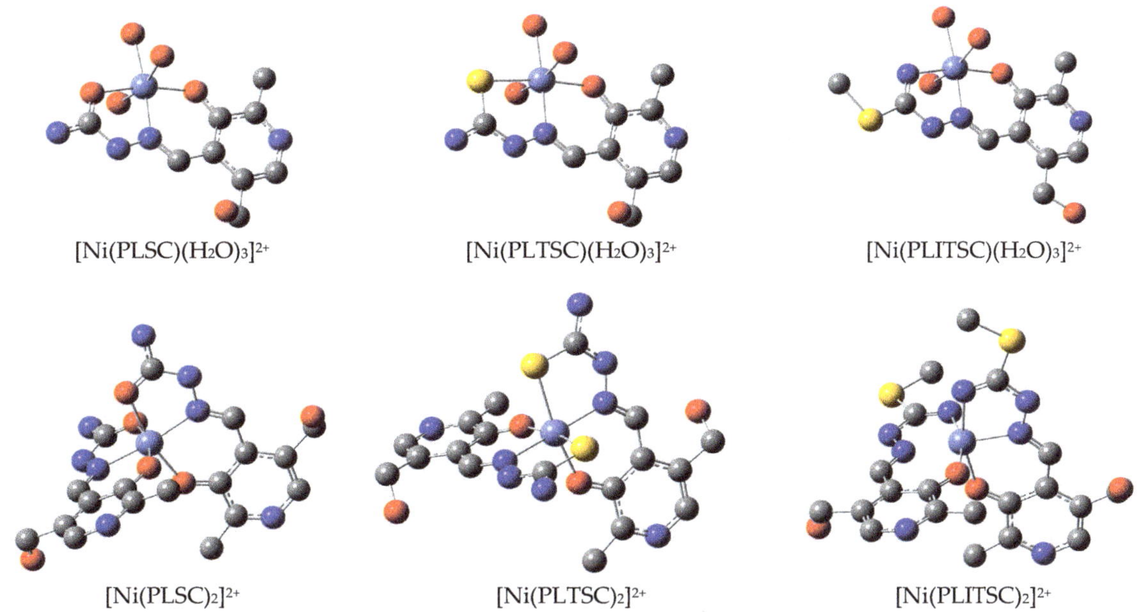

Figure 3. Optimized structures (at the B3LYP/6-311++G(d,p)(H,C,N,O,(S)/LanL2DZ(Ni) level of theory) of selected octahedral Ni(II) complexes. Carbon—gray; nitrogen—blue; oxygen—red; sulfur—yellow; nickel—teal; hydrogen atoms have been omitted for clarity.

Table 2. Bonding energy (BE), electron configuration of Ni, and charges on selected atoms in the optimized structures of complex cations at the B3LYP/6-311++G(d,p)(H,C,N,O,(S)/LanL2DZ(Ni) level of theory.

Complex Cation	BE [kJ mol^{-1}]	Electron Configuration Ni	Charge Ni	Charge O_{arom}	Charge $N_{azomethine}$	Charge X	Charge O_{water}
$[Ni(PLSC)(H_2O)_3]^{2+}$	−981	$4s^{0.23}$ $3d^{8.27}$ $4p^{0.40}$	1.096	−0.710	−0.300	−0.668 (O)	−0.906/−0.891/−0.906
$[Ni(PLTSC)(H_2O)_3]^{2+}$	−854	$4s^{0.27}$ $3d^{8.32}$ $4p^{0.51}$	0.899	−0.697	−0.287	−0.053 (S)	−0.900/−0.890/−0.900
$[Ni(PLITSC)(H_2O)_3]^{2+}$	−850	$4s^{0.24}$ $3d^{8.28}$ $4p^{0.43}$	1.045	−0.718	−0.285	−0.794 (N)	−0.889/−0.898/−0.898
$[Ni(PLSC)_2]^{2+}$	−2122	$4s^{0.26}$ $3d^{8.29}$ $4p^{0.44}$	1.003	−0.676/−0.675	−0.280/−0.279	−0.637/−0.637 (O)	/
$[Ni(PLTSC)_2]^{2+}$	−2082	$4s^{0.33}$ $3d^{8.39}$ $4p^{0.64}$	0.639	−0.656/0.654	−0.264/−0.261	−0.024/−0.024 (S)	/
$[Ni(PLITSC)_2]^{2+}$	−2150	$4s^{0.28}$ $3d^{8.31}$ $4p^{0.49}$	0.916	−0.667/−0.667	−0.267/−0.267	−0.736/−0.736 (N)	/

The bond angles are more prone to change upon optimization due to the system's relaxation, which leads to octahedral geometry. Nevertheless, the correlation coefficients are 0.99 for all three structures, with MAE values between 1.50 ($[Ni(PLITSC)(H_2O)_3]^{2+}$) and 1.94° ($[Ni(PLTSC)_2]^{2+}$). The optimization of the first crystal structure led to significant changes in bond angles that include water molecules. For example, the experimental value of the O2-Ni1-O3 is 86.94°, while the theoretical value is 82.30°. These changes of several degrees can be explained by the system's relaxation and the absence of interactions with surrounding units. The optimization was performed for isolated complexes in a vacuum, and certain differences were expected. The discrepancies between the two sets of data are less pronounced for angles including two donor atoms of the PLSC ligand, such as between the carbonyl group oxygen attached to the pyridine ring, Ni(II), and the nitrogen atom (89.69 (exp.) and 87.98° (theor.), Table S2). Again, the rigidity of the ligand prevents significant changes in the angle values. A similar result was found for the structure of the second examined complex cation, $[Ni(PLITSC)(H_2O)_3]^{2+}$. The third experimentally obtained complex contains two PLTSC ligands with three different donor atoms. The optimized structure is characterized by the change in angles; for example, the S–Ni–O angles, in which S and O belong to the same ligand, are equilibrated to 169°, although their experimental values were 179.29 and 164.36°. The increase in local symmetry is a consequence of the system's relaxation. The changes in the other angles are less pronounced, usually less than two degrees. It is important to mention that the crystal structure of $[Ni(PLTSC)_2]\cdot 2NO_3^-\cdot H_2O$ contains two counterions and a solvent molecule that influence the overall geometry through intermolecular interactions. Based on these results, it can be concluded that the selected level of theory optimized the structure of the examined complex cations well, and it could be applied to the structural examination of other theoretical structures and stabilization interactions.

The theoretical structures of the other three complex cations were optimized at the same level of theory. In the optimized structure of $[Ni(PLTSC)(H_2O)_3]^{2+}$, the bond length of Ni–S is 2.419 Å, which is higher than other bonds in corresponding complexes with PLSC and PLITSC ligands. A certain elongation of Ni–O bonds (carbonyl groups attached to aromatic ring/aliphatic chain, 2.123/2.028 Å) is also found in $[Ni(PLSC)_2]^{2+}$ compared to the structure with three molecules of water (2.073/1.992 Å). The presence of another ligand changes the positional preferences due to its size. The same can be observed in the structure of $[Ni(PLITSC)_2]^{2+}$, especially for the Ni–O (amino substituent). Therefore, these interactions were subjected to the QTAIM analysis to determine their type and strength.

The binding energy of ligands to the central metal ion was estimated using the following equations for mono and bisligand systems for the optimizations in the gas phase [27]:

$$BE = E(\text{monoligand complex}) - [(E(Ni(H_2O)_3)^{2+} - E(\text{ligand})] \quad (1)$$

$$BE = E(\text{bisligand complex}) - [(E(Ni)^{2+} - 2E(\text{ligand})] \quad (2)$$

As explained in the Methodology section, the obtained values were corrected for the basis set superposition error. The binding energies, calculated at the B3LYP/6-311++G(d,p)

(H,C,N,O,(S)/LanL2DZ(Ni) level of theory, are shown in Table 2, and they confirm the stability of the examined species. Caramori et al. discussed that several parameters influence the binding energy, such as the ligand size, hydrogen bonds, and electrostatic interactions between positively charged central metal ion and negatively charged ligands [28]. The highest binding energy for complex cations with three water molecules is calculated for the PLSC ligand (941 kJ mol^{-1}) due to an electronegative oxygen atom attached to an aliphatic chain. When a sulfur atom is present in the structure, the binding energies are lowered to 854 and 850 kJ mol^{-1}. These values are comparable to those found for Ni(II) complexes with chalcone-based Schiff bases [29]. Complex cations with two ligands show even higher binding energies of −2112 (PLSC), −2082 (PLTSC), and −2150 kJ mol^{-1} (PLITSC). It is important to observe that these energies are more than two times higher than the previously discussed ones, which leads to the conclusion that interactions with PLSC/PLTSC/PLITSC are stronger than those with three water molecules. The presence of four nitrogen atoms in the structure of $[Ni(PLITSC)_2]^{2+}$ and their electron-donating ability are the most important parameters for structure stability.

The natural atomic charges of nickel, oxygen, nitrogen, and sulfur atoms that form the coordination sphere are listed in Table 2, together with the electron configurations of Ni ions. The NBO analysis is an appropriate method for studying the charge transfer in compounds [30]. The electron configuration of the free metal ion is $4s^0$ $3d^8$. The investigated complexes have an equal donation from each ligand to d and s orbitals, as the occupations are between 0.23 and 0.33 for the 4s orbital and 8.27 and 8.39 for the 3d orbital. A significant donation is observed for the 4p orbital (0.40–0.64). A much lower donation was found in square-planar Ni(II) complexes with Schiff bases synthesized from 2-hydroxy-5-methylbenzaldehyde, although nitrogen and oxygen atoms were also involved [31]. The octahedral Ni(II) complex with phenylacetic acid and 1,10′-phenanthroline contained significant donations to the 4p orbital [32]. The charges on Ni, O_{arom}, $N_{azomethine}$, and O_{aliph} are 1.096, −0.710, −0.300, and −0.668 *e*, respectively. The expected values are 2, −2, −1, and −2 *e*, which proves the assumption that electron donation occurs from oxygen and nitrogen atoms' orbitals to the orbitals of the central metal ion. The charge on the water oxygen atoms is higher than on oxygen atoms attached to the pyridine ring and aliphatic chain. Interestingly, the electron donation of a water molecule in the same plane as the PLSC ligand is much higher than that of two other molecules. A lower charge on the nickel ion is calculated for $[Ni(PLTSC)(H_2O)_3]^{2+}$ due to the significant donation from the sulfur atom, especially to the 4p orbital. Once the sulfur atom is exchanged with nitrogen, the charge on Ni(II) is 1.045 *e* (Table 2). The electron density donation is much more pronounced in complexes with two ligands, leading to nickel ion charges of 1.045 (PLSC), 0.639 (PLTSC), and 0.916 *e* (PLITSC). In these complexes, the lowest charge is obtained for nitrogen atoms of the PLITSC ligand (−0.736 *e*), followed by oxygen (−0.637 *e*) and sulfur (−0.024 *e*). These charges further influence the complexes' bond lengths and stability, as discussed in the following section.

2.3. QTAIM Analysis

The QTAIM approach is based on Bader's theory of interacting atoms in molecules, and it is useful for examining metal–ligand interactions [26]. Within this approach, the topological properties of electron density and Laplacian in Bond Critical Points (BCPs) are calculated. In this section, the stabilization interactions between donor atoms and central metal ions are examined, along with the interactions formed between ligand molecules. The following parameters are included in the discussion: the electron density (ρ(r)), Laplacian ($\nabla^2\rho(r)$), Lagrangian kinetic electron density (*G*(r)), potential electron density (*V*(r)), density of total electron energy (*H*(r) = G(r) + *V*(r)), and interatomic bond energy (E_{bond} = *V*(r)/2), as presented in [33]. Based on the classification proposed by Bader and Essen, there are two types of interactions. Shared (covalent) interactions are characterized by a high electron density (>0.1 a.u.), while closed-shell regions (ionic bonds, van der Waals interactions, and hydrogen bonds) commonly have an electron density of around 0.01 a.u. [34]. These parameters are shown in Table 2 for the selected bonds, while the complete list is given in Table S4.

All Laplacian values are positive in the examined complexes, allowing for their characterization as electron-shared interactions [35]. Bianchi et al. proposed a more detailed bond regiment based on the ratio of G(r) and V(r) into three categories: shared shell region of covalent bonds ($-G(r)/V(r) > 2$), intermediate (transit) region with dative bonds and ionic bonds of weak covalent degree ($1 > -G(r)/V(r) > 2$), and closed-shell region of ionic bonds and van der Walls interactions ($-G(r)/V(r) < 1$) [36]. Additionally, the bond degree is defined as the ratio between total electron energy and electron density ($BD = H(r)/\rho(r)$), or the total energy per electron [37]. The meaning of this parameter depends on the total energy density. If $H > 0$, the BD determines the softening degree of the non-covalent interaction. When $H < 0$, the parameter can be denoted as a covalent degree (CD), and a higher value indicates a stronger covalent feature [37,38]. The interactions between nitrogen, oxygen, sulfur atoms, and central metal ions have Laplacian values between 0.323 and 0.503 a.u. ($[Ni(PLSC)(H_2O)_3]^{2+}$), while the values of $-G(r)/V(r)$ depend on the chosen atom. Three water molecules interact through bonds with weak covalent characters ($-G(r)/V(r) = 1.1$). On the other side, Ni–O_{aliph} falls within the same category, while Ni–O_{arom} and Ni–$N_{azomethine}$ can be classified as dative bonds with $-G(r)/V(r) \leq 1$. This character is additionally proven by the negative total electron density value of the Ni–$N_{azomethine}$ bond. The degree of covalence character can be estimated by the Espinosa et al. approach of calculating the $H(r)/\rho(r)$ parameter [39]. The value of this parameter for water molecules is 0.13 and 0.20, depending on its position. A water molecule in the same plane as the PLSC ligand has a lower $H(r)/\rho(r)$ value. Where electronegative atoms of PLSC are concerned, the values are much lower, between 0.03 (Ni–O_{arom}) and 0.09 (Ni–O_{aliph}), proving a lower covalent character of these bonds than in the case of water molecules. The bond between nitrogen and Ni(II) has negative total energy, and the covalent degree is only -0.04. The interatomic bond energy nicely follows this discussion, as the strongest bond is formed with the aromatic oxygen atom (-160.3 kJ mol^{-1}), followed by azomethine nitrogen (-143.4 kJ mol^{-1}) and aliphatic oxygen (-118.2 kJ mol^{-1}) (Table 2 and Table S4). The structure of the ligand also contains a weak intermolecular hydrogen bond (C–H···O) with $-G(r)/V(r) = 1.2$ and bond energy of -16.9 kJ mol^{-1}.

Once the oxygen atom is exchanged with the sulfur atom, the interaction between sulfur and the central metal ion is characterized by electron density and Laplacian values of 0.054 and 0.190 a.u. (Table 3). These values, together with $-G(r)/V(r) = 1$ and negative total electron energy (Table 2), classify this interaction as dative or ionic. The covalent degree of Ni–S is higher than that of Ni–$N_{azomethine}$ (-0.07 vs. 0.01, Table S4). The interaction energy is lower (-72.4 kJ mol^{-1}) compared to Ni–O_{aliph}. This change is reflected in the increased bond distance, as previously discussed. Coordinated water molecules in $[Ni(PLTSC)(H_2O)_3]^{2+}$ have bonding energies of -80.4 and -103.4 kJ mol^{-1}, again depending on their position. It is important to observe that the parameters of Ni–O_{arom} and Ni–O_{aliph} are very similar to the first complex cation, although the interaction energies are lower for several kJ mol^{-1}. The interaction with the azomethine nitrogen atom still has negative total electron energy. The presence of the second donor nitrogen atom in the structure of $[Ni(PLITSC)(H_2O)_3]^{2+}$ affects the formation of a novel Ni–N_{amino} bond, with negative total energy (-10.2 kJ mol^{-1}) and an interaction energy of -141.0 kJ mol^{-1}. The CD parameter is lower when the amino group nitrogen atom is included. The bond energy in this structure is higher than that between azomethine nitrogen and Ni(II). When three different donor atoms (O_{aliph}, S, and N_{amino}) in complexes with three water molecules are compared, it can be concluded that the strongest interaction is formed with nitrogen, followed by oxygen and sulfur atoms. This is an expected order, bearing in mind the donating abilities of these atoms. These bonds have a pure dative or ionic character, with a slight contribution from covalent bonds in the case of oxygen atoms. These parameters influence the bond lengths from previous sections.

The structures of complex cations with two tridentate ligands were also analyzed. The parameters of interactions between donor atoms and the central metal ion in $[Ni(PLSC)_2]^{2+}$ are equal for two ligands, proving that a certain degree of symmetry exists in the structure. The interatomic bond energies of Ni–O_{arom} and Ni–O_{aliph} were lowered (-138.9 and -96.2 kJ mol^{-1}) compared to complex cation with one PLSC ligand, while the bond energy increased for

Ni–$N_{azomethine}$ bond. The CD parameter of both bonds with azomethine nitrogen was the same (−0.04). This is a consequence of the bulkiness of ligands and their accommodation around the central metal ion. When two PLTSC ligands are present, the bond energy of Ni–S is even lower (−59.6 kJ mol^{-1}), as expected for a structure with the longest donor atom–central metal ion bond. It should be outlined that the total electron density of Ni–$N_{azomethine}$ bonds is positive, with a low CD value of (0.002) compared to bonds with oxygen atoms (0.09). The same is found for the third bisligand complex, with the bond energy of Ni–N_{amino} (−111.1 kJ mol^{-1}). The −G(r)/V(r) ratio is the highest for oxygen atoms, signifying a slight covalent character, while only Ni–$N_{azomethine}$ has a negative total electron energy and the lowest covalent degree.

Table 3. The calculated Bond Critical Point (BCP) properties of selected bonds at the DFT/B3LYP-D3BJ/6-311+G(d,p)(H,C,N,O,S)/LanL2DZ(Ni) level of theory: the electron density (ρ(r)) and its Laplacian ($\nabla^2\rho(r)$); the Lagrangian kinetic electron density (G(r)) and the potential electron density (V(r)); the density of the total energy of electrons (H(r))–Cremer–Kraka electronic energy density; the interatomic bond energy, E_{bond}.

Bond	ρ(r) [a.u.]	$\nabla^2\rho(r)$ [a.u.]	G(r) [kJ mol^{-1}]	V(r) [kJ mol^{-1}]	H(r) [kJ mol^{-1}]	−G(r)/V(r)	E_{bond} [kJ mol^{-1}]	H(r) [a.u.]/ρ(r) [a.u.]
			$[Ni(PLSC)(H_2O)_3]^{2+}$					
Ni–O_{aliph}	0.059	0.404	250.8	−236.4	14.4	1.1	−118.2	0.09
			$[Ni(PLTSC)(H_2O)_3]^{2+}$					
Ni–S	0.054	0.190	134.9	−144.7	−9.9	0.9	−72.4	−0.07
			$[Ni(PLITSC)(H_2O)_3]^{2+}$					
Ni–N_{amino}	0.071	0.398	271.7	−282.0	−10.2	1.0	−141.0	−0.05
			$[Ni(PLSC)_2]^{2+}$					
Ni–O_{aliph} 1	0.052	0.347	210.1	−192.4	17.7	1.1	−96.2	0.13
Ni–O_{aliph} 2	0.052	0.347	210.2	−192.5	17.7	1.1	−96.2	0.13
			$[Ni(PLTSC)_2]^{2+}$					
Ni–S 1	0.047	0.168	114.8	−119.2	−4.4	1.0	−590.5	−0.04
Ni–S 2	0.047	0.167	114.3	−118.7	−4.4	1.0	−59.3	−0.04
			$[Ni(PLITSC)_2]^{2+}$					
Ni–N_{amino} 1	0.060	0.344	224.1	−222.3	1.9	1.0	−111.1	0.01
Ni–N_{amino} 2	0.060	0.344	224.1	−222.2	1.9	1.0	−111.1	0.01

Another parameter for investigation of Ni–X (X=O, S, N) stability is the ellipticity index (ε). According to Bader's theory, the high value of this parameter signifies unstable bonds [40]. The values of ε from Table S4 are relatively low and in a narrow region (0.044–0.149), indicating the dynamic stability of these bonds [27]. The increase in ε value is a consequence of the electron delocalization and the increase in the π character of bonding [37]. When water-containing complexes are compared, the bond ellipticities are 0.140 (Ni–O_{aliph}), 0.101 (Ni–S), and 0.149 (Ni–N_{amino}) (Table S4). Higher ε values indicate deviations from cylindrical symmetry, which are less pronounced in bonds containing sulfur, which is a larger and less electronegative atom than nitrogen and oxygen. The reasons for the deviation from cylindrical symmetry include π-bonding or multiple bonding characters. The same trend was observed for complexes with two PLSC/PLTSC/PLITSC ligands. These results prove the assumption that the presence of different donor atoms influences the stability of compounds and that theoretical methods can be applied to predict compounds, even before their experimental preparation. This is an important finding, showing that that structures of known compounds can be used to estimate the stability and reactivity of novel ones, as discussed in [35].

2.4. Protein and DNA Binding Properties of Complexes

The examination of interactions between transport proteins/DNA and metal complexes is one of the common methods for predicting activity. Transport proteins are essential for distributing important compounds, such as fatty acids, metal ions, drugs, and toxins [41,42]. Their interactions with DNA are considered one of the main pathways for the cytotoxic activity of compounds [43]. Several experimental and theoretical methodologies have been applied to examine similar compounds in the literature [16,17,23]. In this section, molecular docking simulations were performed to determine the affinity of selected complex cations towards protein and DNA binding. The most stable structures optimized at the B3LYP-D3BJ/6-311+G(d,p)(H,C,N,O,S)/LanL2DZ(Ni) level of theory were selected as flexible ligands, while the structures of HSA and DNA, obtained from the Cambridge Crystallographic Data Centre, were kept rigid.

The structure of HSA includes three domains (I, II, and III), further divided into two subdomains (A and B), leading to six subdomains, as shown in Figure 4. Peptide chains connect two subdomains within each domain, represented in light gray. The cavities of the subdomains include the fatty acid binding sites (FA) that are important for transporting fatty acids and other compounds. The fluorescence emission of HSA originates from Trp214 amino acid, located between domains IIA and IIB. The selected HSA structure contains six myristic acid molecules that mimic different fatty acids bound to transport proteins. Only the most stable conformations are discussed in the main text, while the complete list of molecular docking results is given in Table S5. Previous findings have shown that one of the chosen complex cations ($[Ni(PLTSC)_2]^{2+}$) bonded spontaneously to HSA and induced a decrease in the fluorescence intensity [17]. This result and the molecular docking simulation in the same reference verified that the binding position of compounds should be around Trp2014.

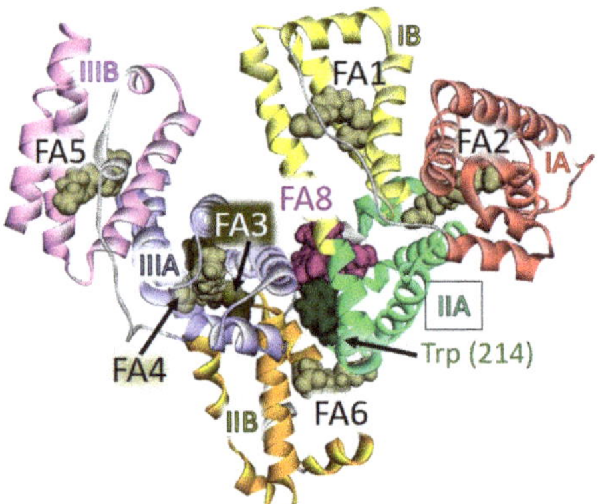

Figure 4. Structure of HSA with subdomains: IA in reddish, IB in yellow, IIA in green, IIB in orange, IIIA in lilac, and IIIB in pink. The active positions FA1–FA8 are denoted with the representation of myristic acids (yellow CPK model), commonly bound in the structures. The position of a fluorescent amino acid, Trp214, is depicted as dark green CPK balls, while $[Ni(PLSC)_2]^{2+}$ is presented as purple CPK balls.

Where interactions with HSA are concerned, examining the effect of ligands on the binding affinity is important. In the case of complex cations with three molecules of water, the changes in Gibbs free energy of binding are −24.6 ($[Ni(PLSC)(H_2O)_3]^{2+}$), −33.6 ($[Ni(PLTSC)(H_2O)_3]^{2+}$), and −26.2 kJ mol^{-1} ($[Ni(PLITSC)(H_2O)_3]^{2+}$). The most stable structures are formed within the FA8 site containing Trp214, as given in Table S5. A narrow range of values suggests that similar interactions were formed between amino acids and various ligands. Figure 5 (left) depicts the most important interactions with $[Ni(PLSC)(H_2O)_3]^{2+}$. There are classic hydrogen bonds with Glu153 and Glu291 in which both amino acids act as hydrogen atom acceptors in interactions with amino groups and water molecules. A weak hydrogen bond includes a hydrogen atom of an aliphatic chain of complex cations and Ser192/Glu188. Hydrophobic interactions between His288/Ala291 and the methyl group attached to the pyridine ring are present. A higher number of classic hydrogen bonds is found for $[Ni(PLTSC)(H_2O)_3]^{2+}$ (Figure S7), leading to the lowest ΔG_b value, comparable to the native ligand, warfarin (-34.0 kJ mol^{-1}). These interactions are formed with Ser454, Ser202, Glu450, and electronegative groups (hydroxyl and amino) and water molecules. The presence of a sulfur atom in the ligand structure of PLTSC leads to the sulfur–π interaction with Trp214. Other weak interactions are hydrophobic (Val344 and Arg485) and carbon–hydrogen bonds (Ser454). Five hydrogen bonds with Asp451, Glu450, and Ser202 are included through interactions with water molecules and hydrogen attached to the pyridine nitrogen atom of $[Ni(PLITSC)(H_2O)_3]^{2+}$. The methylthio group is flexible and can form sulfur–π interactions, in this case, with Trp214 (Figure S7). The hydrophobic and carbon/π-donor hydrogen bonds additionally stabilize the investigated system. Based on these results, it can be concluded that the water molecules in the structures of complex cations are important for the formation of interactions with surrounding amino acids. At the same time, the sulfur atom introduces a different type of interaction that lowers the change in Gibbs free energy of binding.

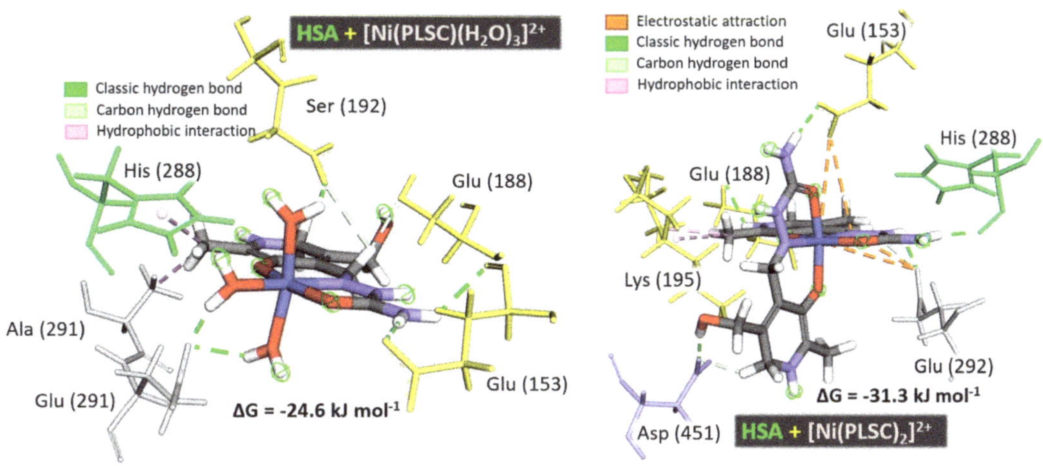

Figure 5. The most important interactions between HSA and $[Ni(PLSC)(H_2O)_3]^{2+}$ (**left**) and $[Ni(PLSC)_2]^{2+}$ (**right**) complexes, as obtained in the molecular docking simulations.

When two PLSC ligands are present in the structure, the value of ΔG_b is lowered to -31.3 kJ mol^{-1} due to the increased number of hydrogen bonds with His288, Glu292, Glu188, and Asp451. The electrostatic attraction is between the central metal ion and Glu153 and Glu292. The increase in the size of $[Ni(PLSC)_2]^{2+}$ compared to $[Ni(PLSC)(H_2O)_3]^{2+}$ allows for a higher number of strong interactions, especially hydrogen bonds. Additionally, alkyl interactions with Ala191 and Lys195 are shown in Figure 5. The binding energies of the other two complexes are lower than those of complex cations with three molecules of water, -25.9 and -26.1 kJ mol^{-1} for $[Ni(PLTSC)_2]^{2+}$ and $[Ni(PLITSC)_2]^{2+}$, respectively (Table S5 and Figure S8). The results from Table S5 also show that complex cations form interactions with amino acids within FA8. The multitude of amino groups is responsible for a high number of hydrogen bonds between amino acid residues and $[Ni(PLTSC)_2]^{2+}$. The lower binding affinities of these two complex cations can be explained by the existence of unfavorable interactions with His440 ($[Ni(PLTSC)_2]^{2+}$), Lys195, and Arg218 ($[Ni(PLITSC)_2]^{2+}$). His440 is also engaged in sulfur–π interactions with PLTSC ligands. In the case of PLITSC, two sulfur–π interactions include Asp451 and Arg222. These findings support the assumption that the selection of ending groups of the aliphatic chain has a profound effect on the binding affinity, along with the presence of sulfur atoms and aromatic structures.

The molecular docking simulations were performed for isolated PLSC, PLTSC, and PLITSC ligands to examine their binding positions and affinity towards HSA protein. The binding energies of these ligands were −26.7, −26.4, and −28.1 kJ mol^{-1} (Table S5), respectively. These values were comparable to the values obtained for complexes with two tridentate ligands, thus proving the importance of ligand structures for the binding of complexes. An important difference in these cases was the binding site. Namely, all three ligands were bound to the active sites IB and IIA due to their size (Figure S9). The elongated structure of ligands with several polar groups allowed for the formation of several hydrogen bonds, all presented in Figure S10. All three ligands were positioned further from the Trp214 amino acid than the respective complexes.

During molecular docking simulations, several positions were obtained for the interactions between complex cations and DNA. For example, 13 different orientations were found for $[Ni(PLSC)(H_2O)_3]^{2+}$, with changes in Gibbs free energy of binding of between −35.9 and −26.8 kJ mol^{-1}. Some of these positions are depicted in Figure S11, denoted by the changes in Gibbs free energies ΔG_1, ΔG_6, ΔG_7, and ΔG_{10}. It should be mentioned that each of these positions included several orientations of the complex cation within the DNA structure. The most stable conformation is presented in Figure 6. There are several hydrogen bonds between water molecules and the oxygen and nitrogen atoms of a phosphate group, sugar, and aromatic rings. PLSC ligands also interact with the oxygen atoms of aromatic rings and sugars through hydroxymethyl oxygen, amino, and azomethine nitrogens. The change in Gibbs energy of binding is the lowest among the investigated complex cations (−35.9 kJ mol^{-1}, Table S6) due to the positions of electronegative groups in the minor groove of DNA. The number of hydrogen bonds is lowered in the complex cations with three molecules of water and PLTSC/PLITSC ligands, leading to binding energies of −31.9 and −31.3 kJ mol^{-1}. As presented in Figure S12, eight and seven classic hydrogen bonds exist between water molecules and the hydroxyl group of ligands on one side and oxygen atoms of sugars and aromatic rings. These hydrophobic interactions include the methyl groups of PLTSC/PLITSC and the aromatic rings of guanine and cytosine in positions 10, 11, 14, and 15.

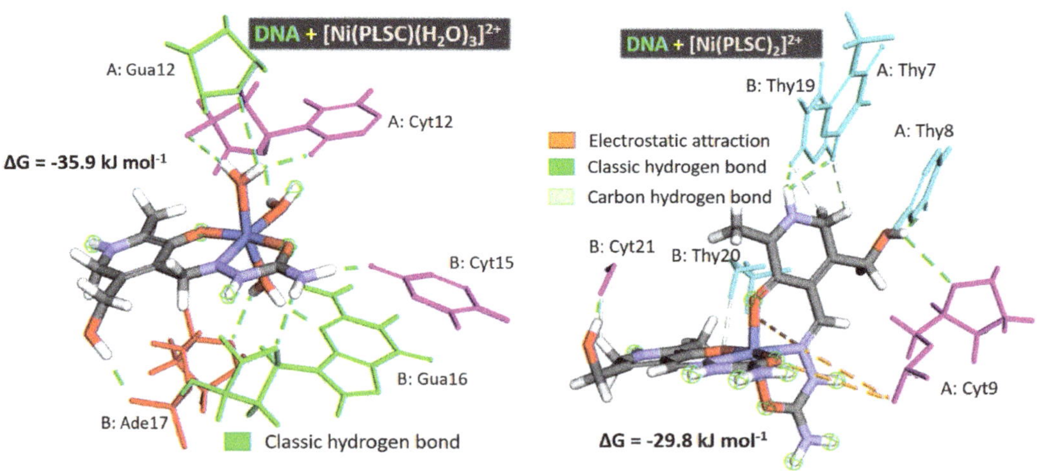

Figure 6. The most important interactions between DNA and $[Ni(PLSC)(H_2O)_3]^{2+}$ (**left**) and $[Ni(PLSC)_2]^{2+}$ (**right**) complexes, as obtained via the molecular docking simulations.

Removing water molecules from complex cations lowers the change in Gibbs free binding energy for several kJ mol^{-1}, although novel interactions are introduced. In the case of $[Ni(PLSC)_2]^{2+}$, the number of classic hydrogen bonds is reduced to five (Figure 6), and the change in Gibbs free energy of binding is −29.8 kJ mol^{-1}. Electrostatic interactions between the positively charged pyridine nitrogen and azomethine nitrogen of PLSC and the negatively charged oxygen atoms of the phosphate groups of cytosine and thymine can be observed. Carbon hydrogen bonds also exist with thymine residues in positions 7 and 19 and cytosine residues in position 21. The binding of $[Ni(PLTSC)_2]^{2+}$ and $[Ni(PLITSC)_2]^{2+}$ is even less spontaneous: −26.8 and −28.7 kJ mol^{-1}. The same types of interactions are presented in Figure S13, as previously discussed. Therefore, the binding to DNA is guided by water molecules, leading to a much more spontaneous process. When three ligands are compared, the ending amino group and oxygen atom in the PLSC ligand are responsible for the highest number of hydrogen bonds. The sulfur atom in the structure is excluded from these interactions due to low electronegativity and protection by the methyl group in the PLITSC ligand. The most important result is that all the analyzed complex cations bind in the minor groove of DNA, as the octahedral geometry limits other mechanisms.

The interactions between DNA and free ligands were also investigated, and the binding energies are presented in Table S6. These energies were −26.7 (PLSC), −26.4 (PLITSC), and −28.6 kJ mol^{-1} (PLITSC). All free ligands were located in the minor groove of the DNA, which coincides with their respective complexes. The binding energies were lower than for the complexes with three water molecules, concluding that they were important for the system's stability. Complexes with two tridentate ligands showed affinity almost equal to the free ligands. The most important interactions are depicted in Figure S14. Due to the presence of electronegative groups, the number of classic hydrogen bonds was five (PLTSC and PLITSC) and eight (PLTSC), similar to the binding of complex cations. Free PLSC ligands and two complexes showed the highest binding affinity towards DNA. The experimental examination of DNA affinity and interactions in vivo is recommended to verify these theoretical findings, as additional ions and solvent molecules might influence these processes.

3. Materials and Methods

3.1. Hirshfeld Surface Analysis

The Hirshfeld surface analysis was applied to investigate the intramolecular interactions responsible for the stabilization of the crystal structure. The CrystalExplorer program [44] was utilized for this analysis. The results are represented by a graph connecting two distances: one between the two nearest nuclei (d_e) and the second between the nuclei and the external surface (d_i) [45,46]. The distance values are normalized and colored depending on the sum of the van der Waals radii of the interacting atoms. Red, white, and blue colors are used for the shorter, equal, and longer separations than the respective van der Walls radii. The specific contacts are shown in fingerprint plots that allow for the determination of their percentages in the total number of contacts. The crystallographic structures of four Ni(II) nitrate complexes with the mentioned ligands are taken from the Cambridge Crystallographic Data Centre (CCDC), as given in the following section. A similar analysis was previously discussed for PLSC, PLTSC, and PLITSC ligands with various transition metals [17,47].

3.2. Quantum-Chemical Analysis

The structures of the complex cations $[Ni(PLSC)(H_2O)_3]^{2+}$, $[Ni(PLISC)_2]^{2+}$, $[Ni(PLTSC)(H_2O)_3]^{2+}$, and $[Ni(PLSC)2]^{2+}$ were taken from crystallographic structures of their nitrate salts from references [12,18,19]. Additionally, two structures, $[Ni(PLTSC)_2]^{2+}$ and $[Ni(PLITSC)(H_2O)_3]^{2+}$, were prepared based on the structural parameters of similar compounds. The structures were optimized using the Gaussian 09 Program package [48] at the B3LYP/6-311++G(d,p)(H,C,N,O,S)/LanL2DZ(Ni) level of theory [49–52]. The same level of theory was previously applied for the optimization, spectral prediction, and reactivity description of other Ni(II) complexes [53–55]. The geometry optimization was performed without any geometrical constraints for the octahedral geometry, with a charge of +2 and a multiplicity of 3, corresponding to the common multiplicity of nickel-containing compounds. The absence of imaginary frequencies proved that the minimum on the potential energy surface was found. Counterpose (CP) corrections were applied to all binding energy values to minimize the basis set superposition error [56]. The natural bond orbital charges were calculated using the NBO analysis approach [57], as implemented in the

Gaussian 09. The Quantum Theory of Atoms in Molecules (QTAIM) analysis, as proposed by Bader [58,59], was performed using the AIMAll program (aim.tkgristmill.com) package [60]. Different parameters calculated from the total electron energy contributions are described in the main text.

3.3. Molecular Docking

Molecular-level investigations into the binding interactions between nickel(II) complexes, human serum albumin (HSA), and deoxyribonucleic acid (DNA) were performed using molecular docking simulations with AutoDock 4.2.6 [61], facilitated by the AMDock program (version 1.5.2) [62]. The structures of all six nickel(II) coordination complexes were optimized using the Gaussian 09 Program package at the B3LYP-D3BJ/6-311+G(d,p)(H,C,N,O,S)/LanL2DZ(Ni) level of theory. The structures of HSA (PDB ID: 1H9Z [63]) and DNA (PDB ID: 1BNA [64]) were retrieved from the Protein Data Bank (PDB). The number of poses was set to 20, the number of runs was set to 100, and the energy evaluations were set to 10,000,000. BIOVIA Discovery Studio was employed for molecular preparation and the docking setup and for visualizing and analyzing the docking results.

4. Conclusions

The most numerous contacts in the crystal structures of $[Ni(PLSC)(H_2O)_3]{\cdot}2NO_3^-$, $[Ni(PLTSC)_2]{\cdot}2NO_3^-{\cdot}H_2O$, and $[Ni(PLITSC)(H_2O)_3]{\cdot}2NO_3^-$ are denoted as O···H, with the following percentages: 49.1, 36.3, and 43.8. A significant part of these interactions is due to three molecules of water. When PLTSC and PLITSC are part of the structure, the sulfur atom is included in S···H, S···C, S···O, and S···N contacts. Important stabilization interactions also exist between protonated nitrogen atoms and surrounding electronegative groups. The optimized structures of the three mentioned complexes were performed at the B3LYP/6-311++G(d,p)(H,C,N,O,(S))/LanL2DZ(Ni) level of theory. The correlation coefficients for bond length comparisons are between 0.96 and 0.99, with MAE values between 0.017 and 0.041 Å. The bond angles are also well reproduced, with correlation coefficients of 0.99 and MAE values between 1.50 and 1.94°. Upon optimization, the structures adopted almost perfect octahedral geometry. The structures of the other three complex cations were optimized at the same level of theory. The highest binding energy (941 kJ mol^{-1}) was calculated for the PLSC ligand due to the existence of oxygen and nitrogen donor atoms. Where complex cations with two PLSC/PLTSC/PLITSC ligands are concerned, the highest binding energy was obtained for $[Ni(PLITSC)_2]^{2+}$, in which two to three nitrogen atoms of each ligand surround the central metal ion. The NBO charge analysis proved that significant donations from ligands to the Ni(II) orbitals existed. The interactions between donor atoms and Ni(II) were further examined using the QTAIM approach. The strongest interaction, as determined by the interatomic bond energy, was between the nitrogen atom of the amino group of PLITSC (−140.0 kJ mol^{-1}), followed by the oxygen atom of PLSC (−118.2 kJ mol^{-1}) and the sulfur atom of PLTSC (−72.4 kJ mol^{-1}). All the bonds showed a partial covalent character, especially with oxygen atoms. The same trend was found in the complexes with two ligands. The HSA binding affinity was also dependent on the ligands. All complex cations were positioned close to the fluorescent amino acid Trp214 in the most stable structures. The most important interactions included classic hydrogen bonds, carbon–hydrogen bonds, and hydrophobic interactions. The structures of PLTSC and PLITSC allowed for the formation of sulfur–π interactions. The binding energy of $[Ni(PLTSC)(H_2O)_3]^{2+}$ was comparable to warfarin, a native ligand in the examined crystal structure. The most stable position for the interactions with DNA was in the minor groove, which was expected due to the size and geometry of the complexes. The highest affinity towards DNA was calculated for two complex cations with the PLSC ligand due to the spatial distribution of electronegative groups interacting with nucleobases. Further experimental studies are recommended to obtain the binding constants and to examine the influence of solvent molecules and types of interactions between biomolecules and nickel(II) complexes.

Supplementary Materials: The following supporting information can be downloaded at: https://www.mdpi.com/article/10.3390/inorganics12090251/s1, Figure S1: Fingerprint plots for the most numerous contacts in [Ni(PLSC)(H_2O)$_3$]·2NO_3^- structure; Figure S2: Fingerprint plots for the most numerous contacts in [Ni(PLTSC)$_2$]·2NO_3^-·H_2O structure; Figure S3: Fingerprint plots for the most numerous contacts in [Ni(PLITSC)(H_2O)$_3$]·2NO_3^- structure; Figure S4: Optimized structure of $[Ni(PLTSC)(H_2O)_3]^{2+}$ with atom numbering scheme; Table S1: Experimental and theoretical (at the B3LYP/6-311++G(d,p)(H,C,N,O)/LanL2DZ(Ni) level of theory) bond lengths (Å) and angles (°) in the structure of $[Ni(PLSC)(H_2O)_3]^{2+}$; Figure S5: Optimized structure of $[Ni(PLITSC)(H_2O)_3]^{2+}$ with atom numbering scheme; Table S2: Experimental and theoretical (at the B3LYP/6-311++G(d,p)(H,C,N,O,S)/ LanL2DZ(Ni) level of theory) bond lengths (Å) and angles (°) in the structure of $[Ni(PLITSC)(H_2O)_3]^{2+}$; Figure S6: Optimized structure of $[Ni(PLTSC)_2]^{2+}$ with atom numbering scheme; Table S3: Experimental and theoretical (at the B3LYP/6-311++G(d,p)(H,C,N,O,S)/LanL2DZ(Ni) level of theory) bond lengths (Å) and angles (°) in the structure of $[Ni(PLTSC)_2]^{2+}$; Table S4: The calculated Bond Critical Point (BCP) properties at the DFT/B3LYP-D3BJ/6-311+G(d,p)/def2-TZVP level of theory: the electron density (ρ(r)) and its Laplacian ($\nabla^2\rho(r)$); the Lagrangian kinetic electron density (G(r)) and the potential electron density (V(r)); the density of the total energy of electrons (H(r))–Cremer–Kraka electronic energy density; the interatomic bond energy, E_{bond}, ε–ellipticity parameter; Table S5: Binding energies, sites, and number of runs for the selected Ni(II) complexes, warfarin, and HAS, as obtained via the molecular docking simulations; Figure S7: The most important interactions between HSA and $[Ni(PLTSC)(H_2O)_3]^{2+}$ (left) and $[Ni(PLITSC)(H_2O)_3]^{2+}$ (right) complexes, as obtained via the molecular docking simulations; Figure S8: The most important interactions between HSA and $[Ni(PLTSC)_2]^{2+}$ (left) and $[Ni(PLITSC)_2]^{2+}$ (right) complexes, as obtained via the molecular docking simulations; Figure S9: The binding positions of PLITSC, $[Ni(PLITSC)(H_2O)_3]^{2+}$, and $[Ni(PLITSC)_2]^{2+}$ within the structure of HAS; Figure S10: The most important interactions between HSA and PLSC, PLTSC, and PLITSC ligands, as obtained via the molecular docking simulations; Table S6: Binding energies, sites, and number of runs for the selected Ni(II) complexes, warfarin, and HAS, as obtained via the molecular docking simulations; Figure S11: Different docking positions of $[Ni(PLSC)(H_2O)_3]^{2+}$ in the structure of DNA; Figure S12: The most important interactions between DNA and $[Ni(PLTSC)(H_2O)_3]^{2+}$ (left) and $[Ni(PLITSC)(H_2O)_3]^{2+}$ (right) complexes, as obtained via the molecular docking simulations; Figure S13: The most important interactions between DNA and $[Ni(PLTSC)_2]^{2+}$ (left) and $[Ni(PLITSC)_2]^{2+}$ (right) complexes, as obtained via the molecular docking simulations; Figure S14: The most important interactions between HSA and PLSC, PLTSC, and PLITSC ligands, as obtained via the molecular docking simulations.

Author Contributions: Conceptualization, V.J. and D.D.; methodology, A.R. and O.A.O.A.; software, A.R. and M.S.A.; validation, T.A., V.R. and M.S.A.; formal analysis, O.A.O.A., M.S.A. and T.A.; investigation, O.A.O.A., T.A. and V.R.; resources, V.J. and A.R.; data curation, O.A.O.A., M.S.A. and V.R.; writing—original draft preparation, V.J. and A.R.; writing—review and editing, D.D.; visualization, T.A. and V.R.; supervision, D.D.; project administration, D.D.; funding acquisition, V.J. All authors have read and agreed to the published version of the manuscript.

Funding: This research was funded by the Scientific Research Deanship at the University of Ha'il, Kingdom of Saudi Arabia, grant number RG-23080.

Data Availability Statement: The data are contained in this article.

Acknowledgments: The authors are thankful to the University of Ha'il, Kingdom of Saudi Arabia. This research was funded by the Scientific Research Deanship at the University of Ha'il, Kingdom of Saudi Arabia, through Project number RG-23080.

Conflicts of Interest: The authors declare no conflicts of interest. The funders had no role in the design of the study; in the collection, analyses, or interpretation of data; in the writing of the manuscript; or in the decision to publish the results.

References

1. Barone, G.; Terenzi, A.; Lauria, A.; Almerico, A.M.; Leal, J.M.; Busto, N.; García, B. DNA-binding of nickel(II), copper(II) and zinc(II) complexes: Structure–affinity relationships. *Coord. Chem. Rev.* **2013**, *257*, 2848–2862. [CrossRef]
2. Jamieson, E.R.; Lippard, S.J. Structure, Recognition, and Processing of Cisplatin–DNA Adducts. *Chem. Rev.* **1999**, *99*, 2467–2498. [CrossRef] [PubMed]
3. Todd, R.C.; Lippard, S.J. Inhibition of transcription by platinum antitumor compounds. *Metallomics* **2009**, *1*, 280. [CrossRef] [PubMed]

4. Busto, N.; Valladolid, J.; Martínez-Alonso, M.; Lozano, H.J.; Jalón, F.A.; Manzano, B.R.; Rodríguez, A.M.; Carrión, M.C.; Biver, T.; Leal, J.M.; et al. Anticancer Activity and DNA Binding of a Bifunctional Ru(II) Arene Aqua-Complex with the 2,4-Diamino-6-(2-pyridyl)-1,3,5-triazine Ligand. *Inorg. Chem.* **2013**, *52*, 9962–9974. [CrossRef]
5. Sigel, A.; Sigel, H.; Sigel, R.K.O. (Eds.) *Nickel and Its Surprising Impact in Nature*; Wiley: Hoboken, NJ, USA, 2007; ISBN 9780470016718.
6. Alper, P.; Erkisa, M.; Genckal, H.M.; Sahin, S.; Ulukaya, E.; Ari, F. Synthesis, characterization, anticancer and antioxidant activity of new nickel(II) and copper(II) flavonoid complexes. *J. Mol. Struct.* **2019**, *1196*, 783–792. [CrossRef]
7. Wu, H.; Yuan, J.; Bai, Y.; Pan, G.; Wang, H.; Shu, X. Synthesis, structure, DNA-binding properties and antioxidant activity of a nickel(II) complex with bis(N-allylbenzimidazol-2-ylmethyl)benzylamine. *J. Photochem. Photobiol. B Biol.* **2012**, *107*, 65–72. [CrossRef]
8. Maia, D.O.; Santos, V.F.; Barbosa, C.R.S.; Fróes, Y.N.; Muniz, D.F.; Santos, A.L.E.; Santos, M.H.C.; Silva, R.R.S.; Silva, C.G.L.; Souza, R.O.S.; et al. Nickel(II) chloride schiff base complex: Synthesis, characterization, toxicity, antibacterial and leishmanicidal activity. *Chem. Biol. Interact.* **2022**, *351*, 109714. [CrossRef]
9. Alomar, K.; Landreau, A.; Allain, M.; Bouet, G.; Larcher, G. Synthesis, structure and antifungal activity of thiophene-2,3-dicarboxaldehyde bis(thiosemicarbazone) and nickel(II), copper(II) and cadmium(II) complexes: Unsymmetrical coordination mode of nickel complex. *J. Inorg. Biochem.* **2013**, *126*, 76–83. [CrossRef] [PubMed]
10. Pelosi, G.; Bisceglie, F.; Bignami, F.; Ronzi, P.; Schiavone, P.; Re, M.C.; Casoli, C.; Pilotti, E. Antiretroviral Activity of Thiosemicarbazone Metal Complexes. *J. Med. Chem.* **2010**, *53*, 8765–8769. [CrossRef]
11. Talib, J.; Harman, D.G.; Dillon, C.T.; Aldrich-Wright, J.; Beck, J.L.; Ralph, S.F. Does the metal influence non-covalent binding of complexes to DNA? *Dalt. Trans.* **2009**, *3*, 504–513. [CrossRef]
12. Leovac, V.M.; Jovanović, L.S.; Jevtović, V.S.; Pelosi, G.; Bisceglie, F. Transition metal complexes with thiosemicarbazide-based ligand—Part LV: Synthesis and X-ray structural study of novel Ni(II) complexes with pyridoxal semicarbazone and pyridoxal thiosemicarbazone. *Polyhedron* **2007**, *26*, 2971–2978. [CrossRef]
13. Leovac, V.M.; Jevtović, V.S.; Jovanović, L.S.; Bogdanović, G.A. Metal complexes with schiff-base ligands—Pyridoxal and semicarbazide-based derivatives. *J. Serbian Chem. Soc.* **2005**, *70*, 393–422. [CrossRef]
14. West, D.X.; Liberta, A.E.; Padhye, S.B.; Chikate, R.C.; Sonawane, P.B.; Kumbhar, A.S.; Yerande, R.G. Thiosemicarbazone complexes of copper(II): Structural and biological studies. *Coord. Chem. Rev.* **1993**, *123*, 49–71. [CrossRef]
15. Lobana, T.S.; Butcher, R.J. Metal—Thiosemicarbazone interactions. Synthesis of an iodo-bridged dinuclear [diiodobis(pyrrole-2-carbaldehydethiosemicarbazone)dicopper(I)] complex. *Transit. Met. Chem.* **2004**, *29*, 291–295.
16. Jevtovic, V.; Golubović, L.; Alshammari, B.; Alshammari, M.R.; Rajeh, S.Y.; Alreshidi, M.A.; Alshammari, O.A.O.; Rakić, A.; Dimić, D. Crystal Structure, Theoretical Analysis, and Protein/DNA Binding Activity of Iron(III) Complex Containing Differently Protonated Pyridoxal–S-Methyl-Isothiosemicarbazone Ligands. *Int. J. Mol. Sci.* **2024**, *25*, 7058. [CrossRef]
17. Jevtovic, V.; Alshamari, A.K.; Milenković, D.; Dimitrić Marković, J.; Marković, Z.; Dimić, D. The Effect of Metal Ions (Fe, Co, Ni, and Cu) on the Molecular-Structural, Protein Binding, and Cytotoxic Properties of Metal Pyridoxal-Thiosemicarbazone Complexes. *Int. J. Mol. Sci.* **2023**, *24*, 11910. [CrossRef]
18. Leovac, V.M.; Jovanović, L.S.; Divjaković, V.; Pevec, A.; Leban, I.; Armbruster, T. Transition metal complexes with thiosemicarbazide-based ligands. Part LIV. Nickel(II) complexes with pyridoxal semi- (PLSC) and thiosemicarbazone (PLTSC). Crystal and molecular structure of $[Ni(PLSC)(H_2O)_3](NO_3)_2$ and $[Ni(PLTSC\text{-}H)py]NO_3$. *Polyhedron* **2007**, *26*, 49–58. [CrossRef]
19. Jevtovic, V.; Vidovic, D. Synthesis, characterization and X-Ray crystal structure of the tri aqua (3-Hydroxy-5-Hydroxymethyl-2-Methylpyridine-4-Carboxaldehyde-3-Methylisotiosemicarbazone: K3, O3, N7, N10) Ni(II) nitrate. *J. Chem. Crystallogr.* **2010**, *40*, 794–798. [CrossRef]
20. Jevtovic, V.; Alshammari, N.; Latif, S.; Alsukaibi, A.K.D.; Humaidi, J.; Alanazi, T.Y.A.; Abdulaziz, F.; Matalka, S.I.; Pantelić, N.Đ.; Marković, M.; et al. Synthesis, Crystal Structure, Theoretical Calculations, Antibacterial Activity, Electrochemical Behavior, and Molecular Docking of Ni(II) and Cu(II) Complexes with Pyridoxal-Semicarbazone. *Molecules* **2022**, *27*, 6322. [CrossRef]
21. Ferrari Belicchi, M.; Fava Gasparri, G.; Leporati, E.; Pelizzi, C.; Tarasconi, P.; Tosi, G. Thiosemicarbazones as co-ordinating agents. Solution chemistry and X-ray structure of pyridoxal thiosemicarbazone trihydrate and spectroscopic properties of its metal complexes. *J. Chem. Soc. Dalt. Trans.* **1986**, *3*, 2455–2461.
22. Leovac, V.M.; Marković, S.; Divjaković, V.; Szécsényi, K.M.; Joksović, M.D.; Lebanc, I. Structural and DFT studies on molecular structure of Ni(II) chloride complex with pyridoxal semicarbazone (PLSC). Unusual coordination mode of PLSC. *Acta Chim. Slov.* **2008**, *55*, 850–860.
23. Jevtovic, V.; Alhar, M.S.O.; Milenković, D.; Marković, Z.; Dimitrić Marković, J.; Dimić, D. Synthesis, Structural Characterization, Cytotoxicity, and Protein/DNA Binding Properties of Pyridoxylidene-Aminoguanidine-Metal (Fe, Co, Zn, Cu) Complexes. *Int. J. Mol. Sci.* **2023**, *24*, 14745. [CrossRef] [PubMed]
24. Alshammari, O.A.O.; Maisara, S.; Alshammari, B.; Alshammari, M.R.; Rakic, V.; Dimitrić Marković, J.; Jevtovic, V.; Dimić, D. Theoretical Insights into Different Complexation Modes of Dioxovanadium(V) Compounds with Pyridoxal Semicarbazone/Thiosemicarbazone/S-Methyl-iso-thiosemicarbazone Ligands. *Molecules* **2024**, *29*, 1213. [CrossRef] [PubMed]
25. Gak Simić, K.; Đorđević, I.; Lazić, A.; Radovanović, L.; Petković-Benazzouz, M.; Rogan, J.; Trišović, N.; Janjić, G. On the supramolecular outcomes of fluorination of cyclohexane-5-spirohydantoin derivatives. *CrystEngComm* **2021**, *23*, 2606–2622. [CrossRef]
26. Gall, M.; Breza, M. QTAIM study of transition metal complexes with cyclophosphazene-based multisite ligands I: Zinc(II) and nickel(II) complexes. *Polyhedron* **2009**, *28*, 521–524. [CrossRef]

27. Karaush, N.N.; Baryshnikov, G.V.; Minaeva, V.A.; Minaev, B.F. A DFT and QTAIM study of the novel d-block metal complexes with tetraoxa[8]circulene-based ligands. *New J. Chem.* **2015**, *39*, 7815–7821. [CrossRef]
28. Caramori, G.F.; Parreira, R.L.T.; Ferreira, A.M.D.C. Isatin-Schiff base copper(II) complexes—A DFT study of the metal-ligand bonding situation. *Int. J. Quantum Chem.* **2012**, *112*, 625–646. [CrossRef]
29. Pitchumani Violet Mary, C.; Shankar, R.; Vijayakumar, S. Theoretical insights into the metal chelating and antimicrobial properties of the chalcone based Schiff bases. *Mol. Simul.* **2019**, *45*, 636–645. [CrossRef]
30. Kargar, H.; Ashfaq, M.; Fallah-Mehrjardi, M.; Behjatmanesh-Ardakani, R.; Munawar, K.S.; Tahir, M.N. Unsymmetrical Ni(II) Schiff base complex: Synthesis, spectral characterization, crystal structure analysis, Hirshfeld surface investigation, theoretical studies, and antibacterial activity. *J. Mol. Struct.* **2022**, *1265*, 133381. [CrossRef]
31. Guelai, A.; Brahim, H.; Guendouzi, A.; Boumediene, M.; Brahim, S. Structure, electronic properties, and NBO and TD-DFT analyses of nickel(II), zinc(II), and palladium(II) complexes based on Schiff-base ligands. *J. Mol. Model.* **2018**, *24*, 301. [CrossRef]
32. Dikmen, G.; Kani, İ. Synthesis, spectroscopic characterization (FT-IR, Raman, UV-VIS, XRD), DFT studies and DNA binding properties of $[Ni(C_6H_5CH_2COO)(C_{12}H_8N_2)_2](ClO_4)(CH_3OH)$ compound. *J. Mol. Struct.* **2020**, *1209*, 127955. [CrossRef]
33. Kasalović, M.P.; Jelača, S.; Milanović, Ž.; Maksimović-Ivanić, D.; Mijatović, S.; Lađarević, J.; Božić, B.; Marković, Z.; Dunđerović, D.; Rüffer, T.; et al. Novel triphenyltin(iv) compounds with carboxylato *N* -functionalized 2-quinolones as promising potential anticancer drug candidates: In vitro and in vivo evaluation. *Dalt. Trans.* **2024**, *53*, 8298–8314. [CrossRef]
34. Soliman, S.M.; Albering, J.; Abu-Youssef, M.A.M. Structural analyses of two new highly distorted octahedral copper(II) complexes with quinoline-type ligands; Hirshfeld, AIM and NBO studies. *Polyhedron* **2017**, *127*, 36–50. [CrossRef]
35. Lepetit, C.; Vabre, B.; Canac, Y.; Alikhani, M.E.; Zargarian, D. Pentacoordinated, square pyramidal cationic PCP Ni(II) pincer complexes: ELF and QTAIM topological analyses of nickel–triflate interactions. *Theor. Chem. Acc.* **2018**, *137*, 141. [CrossRef]
36. Bianchi, R.; Gervasio, G.; Marabello, D. Experimental Electron Density Analysis of Mn 2 (CO) 10: Metal–Metal and Metal–Ligand Bond Characterization. *Inorg. Chem.* **2000**, *39*, 2360–2366. [CrossRef]
37. Sánchez-Coronilla, A.; Sánchez-Márquez, J.; Zorrilla, D.; Martín, E.I.; de los Santos, D.M.; Navas, J.; Fernández-Lorenzo, C.; Alcántara, R.; Martín-Calleja, J. Convergent study of Ru–ligand interactions through QTAIM, ELF, NBO molecular descriptors and TDDFT analysis of organometallic dyes. *Mol. Phys.* **2014**, *112*, 2063–2077. [CrossRef]
38. Uribe, E.A.; Daza, M.C.; Villaveces, J.L.; Delgado, S.A. On the nature of copper–hydrogen bonding: AIM and NBO analysis of CuH_n ($1 \leq n \leq 6$) complexes. *Int. J. Quantum Chem.* **2010**, *110*, 524–531. [CrossRef]
39. Espinosa, E.; Alkorta, I.; Elguero, J.; Molins, E. From weak to strong interactions: A comprehensive analysis of the topological and energetic properties of the electron density distribution involving X–H···F–Y systems. *J. Chem. Phys.* **2002**, *117*, 5529–5542. [CrossRef]
40. Bader, R.F.W.; Slee, T.S.; Cremer, D.; Kraka, E. Description of conjugation and hyperconjugation in terms of electron distributions. *J. Am. Chem. Soc.* **1983**, *105*, 5061–5068. [CrossRef]
41. Fasano, M.; Curry, S.; Terreno, E.; Galliano, M.; Fanali, G.; Narciso, P.; Notari, S.; Ascenzi, P. The extraordinary ligand binding properties of human serum albumin. *IUBMB Life (Int. Union Biochem. Mol. Biol. Life)* **2005**, *57*, 787–796. [CrossRef]
42. Fanali, G.; di Masi, A.; Trezza, V.; Marino, M.; Fasano, M.; Ascenzi, P. Human serum albumin: From bench to bedside. *Mol. Asp. Med.* **2012**, *33*, 209–290. [CrossRef]
43. Palchaudhuri, R.; Hergenrother, P.J. DNA as a target for anticancer compounds: Methods to determine the mode of binding and the mechanism of action. *Curr. Opin. Biotechnol.* **2007**, *18*, 497–503. [CrossRef]
44. Turner, M.J.; McKinnon, J.J.; Wolff, S.K.; Grimwood, D.J.; Spackman, P.R.; Jayatilaka, D.; Spackman, M.A. *CrystalExplorer17*; University of Western Australia: Perth, Australia, 2017.
45. Spackman, M.A.; Jayatilaka, D. Hirshfeld surface analysis. *CrystEngComm* **2009**, *11*, 19–32. [CrossRef]
46. Spackman, M.A.; Byrom, P.G. A novel definition of a molecule in a crystal. *Chem. Phys. Lett.* **1997**, *267*, 215–220. [CrossRef]
47. Abdulaziz, F.; Alabbosh, K.F.; Alshammari, O.A.O.; Bin Tuwalah, W.M.; Alanazi, T.Y.A.; Rakić, A.; Barić, M.; Marković, M.; Jevtovic, V.; Dimić, D. Crystallographic Structure and Quantum-Chemical Analysis of Biologically Active Co(III)-Pyridoxal–Isothiosemicarbazone Complex. *Inorganics* **2023**, *11*, 466. [CrossRef]
48. Frisch, M.J.; Trucks, G.W.; Schlegel, H.B.; Scuseria, G.E.; Robb, M.A.; Cheeseman, J.R.; Scalmani, G.; Barone, V.; Mennucci, B.; Petersson, G.A.; et al. *Gaussian 09*; Gaussian Inc.: Wallingford, CT, USA, 2009.
49. Becke, A.D. Density-functional thermochemistry. III. The role of exact exchange. *J. Chem. Phys.* **1993**, *98*, 5648–5652. [CrossRef]
50. Dunning, T.H. Gaussian basis sets for use in correlated molecular calculations. I. The atoms boron through neon and hydrogen. *J. Chem. Phys.* **1989**, *90*, 1007. [CrossRef]
51. Hay, P.J.; Wadt, W.R. Ab initio effective core potentials for molecular calculations. Potentials for K to Au including the outermost core orbitale. *J. Chem. Phys.* **1985**, *82*, 299–310. [CrossRef]
52. Hay, P.J.; Wadt, W.R. Ab initio effective core potentials for molecular calculations. Potentials for the transition metal atoms Sc to Hg. *J. Chem. Phys.* **1985**, *82*, 270–283. [CrossRef]
53. Macit, M.; Tanak, H.; Orbay, M.; Özdemir, N. Synthesis, crystal structure, spectroscopic characterization and DFT studies of bis[(1Z,2E)-N-(2,6-diethylphenyl)-N′-hydroxy-2-(hydroxyimino)acetimidamidato]nickel(II). *Inorg. Chim. Acta* **2017**, *459*, 36–44. [CrossRef]
54. Genc, Z.K.; Tekin, S.; Sandal, S.; Sekerci, M.; Genc, M. Synthesis and DFT studies of structural and some spectral parameters of nickel(II) complex with 2-(2-hydroxybenzoyl)-N-(1-adamantyl) hydrazine carbothioamide. *Res. Chem. Intermed.* **2015**, *41*, 4477–4488. [CrossRef]
55. Rahnamaye Aliabad, H.A.; Chahkandi, M. Optoelectronic and structural studies of a Ni(II) complex including bicyclic guanidine ligands: DFT calculations. *Comput. Theor. Chem.* **2017**, *1122*, 53–61. [CrossRef]

56. Boys, S.F.; Bernardi, F. The calculation of small molecular interactions by the differences of separate total energies. Some procedures with reduced errors. *Mol. Phys.* **1970**, *19*, 553–566. [CrossRef]
57. Reed, A.E.; Weinstock, R.B.; Weinhold, F. Natural population analysis. *J. Chem. Phys.* **1985**, *83*, 735–746. [CrossRef]
58. Bader, R.F.W. Atoms in molecules. *Acc. Chem. Res.* **1985**, *18*, 9–15. [CrossRef]
59. Bader, R.F.W. *Atoms in Molecules: A Quantum Theory;* Oxford University Press: Oxford, UK, 1990.
60. Keith, T.A. TK Gristmill Software. *AIMAll (Version 19.10.12)*, Overland Park, KS, USA, 2019.
61. Morris, G.M.; Huey, R.; Lindstrom, W.; Sanner, M.F.; Belew, R.K.; Goodsell, D.S.; Olson, A.J. AutoDock4 and AutoDockTools4: Automated docking with selective receptor flexibility. *J. Comput. Chem.* **2009**, *30*, 2785–2791. [CrossRef]
62. Valdés-Tresanco, M.S.; Valdés-Tresanco, M.E.; Valiente, P.A.; Moreno, E. AMDock: A versatile graphical tool for assisting molecular docking with Autodock Vina and Autodock4. *Biol. Direct* **2020**, *15*, 12. [CrossRef]
63. Petitpas, I.; Bhattacharya, A.A.; Twine, S.; East, M.; Curry, S. Crystal structure analysis of warfarin binding to human serum albumin. Anatomy of drug site I. *J. Biol. Chem.* **2001**, *276*, 22804–22809. [CrossRef]
64. Drewt, H.R.; Wingtt, R.M.; Takanot, T.; Brokat, C.; Tanakat, S.; Itakuraii, K.; Dickersont, R.E. Structure of a B-DNA dodecamer: Conformation and dynamics. *Proc. Natl. Acad. Sci. USA* **1981**, *78*, 2179–2183.

inorganics

MDPI

Article

A New Azide-Bridged Polymeric Manganese (III) Schiff Base Complex with an Allylamine-Derived Ligand: Structural Characterization and Activity Spectra

Aynaz Talebi [1], Mehdi Salehi [1], A. J. Lopes Jesus [2,*], Maciej Kubicki [3], Rui Fausto [4,5] and Reza Golbedaghi [6]

1 Department of Chemistry, College of Science, Semnan University, Semnan 3513119111, Iran; talebi.aynaz@yahoo.com (A.T.); msalehi@semnan.ac.ir (M.S.)
2 University of Coimbra, CQC-IMS, Faculty of Pharmacy, 3004-295 Coimbra, Portugal
3 Faculty of Chemistry, Adam Mickiewicz University, 61-614 Poznan, Poland; maciej.kubicki@amu.edu.pl
4 University of Coimbra, CQC-IMS, Department of Chemistry, 3004-535 Coimbra, Portugal; rfausto@ci.uc.pt
5 Istanbul Kultur University, Faculty Sciences and Letters, Department of Physics, 34158 Bakirkoy, Istanbul, Turkey
6 Chemistry Department, Payame Noor University, Tehran 19395-4697, Iran; golbedaghi82@gmail.com
* Correspondence: ajorge@ff.uc.pt

Abstract: This paper reports the synthesis and structural characterization of a novel azide-bridged polymeric manganese (III) Schiff base complex, using 2-((allylimino)methyl)-6-ethoxyphenol as a ligand. The crystal structure of the synthesized compound, elucidated by single-crystal X-ray diffraction analysis, indicates that it crystallizes in the monoclinic space group $P2_1/c$. The complex is found to display an octahedral geometry in which the central manganese Mn(III) coordinates with two bidentate donor Schiff base ligands via oxygen and nitrogen atoms. In addition, the metallic centers are linked together to form a one-dimensional chain bridged by end-to-end azide ligands. To offer a more thorough characterization of the synthesized compound, the study incorporates experimental data from FT-IR, UV-Vis, and cyclic voltammetry, alongside computational results from Hirshfeld surface analysis and DFT calculations conducted for both the ligand and complex. The computational analyses provided valuable insights into the intrachain and interchain interactions within the crystal structure, clarified the conformational characteristics of the isolated ligand molecule, and aided in the interpretation of the experimental IR spectra. Furthermore, an assessment of the compound's drug-like properties was conducted using activity spectra for substances (PASS) predictions, revealing potential pharmacological activities.

Keywords: Schiff base complex; crystal structure; spectroscopy; Hirshfeld surface analysis; intermolecular interactions; DFT calculations; activity spectra

1. Introduction

Schiff bases constitute an extensive group of versatile compounds characterized by the presence of a carbon–nitrogen double bond, denoted by the general structure $R_1R_2C=NR_3$, where R_1, R_2, and R_3 are typically composed of alkyl or aryl groups ($R_3 \neq H$) [1–3]. These compounds are known for their capacity to form ligands that interact effectively with most transition metals through their azomethine nitrogen atoms [4,5]. Schiff bases and their metal complexes have considerable potential as pharmaceutical agents, displaying a diverse spectrum of biological activities. These include anti-inflammatory, antitumor, and antimicrobial effects [5–10].

Some Schiff base complexes with allylamine derivatives as ligands have been reported in the literature [11–15]. For instance, Behnam et al. synthesized and characterized complexes of Zn(II), Ni(II), Co(III), Cu(II), and V(IV) with 5-bromo-2-((allylimino)methyl) phenol [12,13]. Identical studies were conducted by Kazemi et al. for complexes of Ni(II), Cu(II), Pd(II), and V(IV) with 2-hydroxy-1-allyliminomethyl-naphthalen [14]. In another study by Salehi et al., complexes of Zn(II), Mn(III), and Co(III) with the same Schiff base ligand were synthesized and investigated using a variety of theoretical (DFT) and experimental approaches [15]. Besides the synthesis and structural characterization, these novel compounds were also evaluated for their potential biological applications and molecular docking properties.

The objective of our current research is to synthesize and conduct a thorough characterization of a manganese (III) complex utilizing a Schiff base ligand derived from 3-ethoxy-hydroxybenzaldehyde and allylamine [2-((allylimino)methyl)-6-ethoxyphenol, Scheme 1]. The selection of this Schiff base ligand was mainly guided by its structural features, which offer the potential for different coordination types, such as chelation through the phenolate oxygen and imine nitrogen atoms, thereby facilitating the formation of stable and well-defined metal–ligand complexes. In addition, this choice of ligand builds on previous studies performed by some of the authors in analogous coordination systems [12,13,15]. This continuity of research thus enables the exploration of new ligand–metal combinations and their associated structural and biological properties. As we delve deeper into our investigation, this complex was found to represent an interesting example of an Mn(III)-based coordination polymer where the metallic centers are connected through an end-to-end μ-(1,3) azide linkage [16–19]. The incorporation of the allylamine-derived Schiff base ligand introduces distinct structural features and electronic properties, which may significantly influence the overall behavior of the complex. This ligand design has not been extensively explored in the context of Mn(III) coordination polymers [15], making our study a valuable addition to the existing reports on the structure of such systems [16–22]. Metal–organic coordination polymers have attracted significant interest in the scientific community due to their potential applications in diverse fields, including magnetism, catalysis, and sensing [23–27].

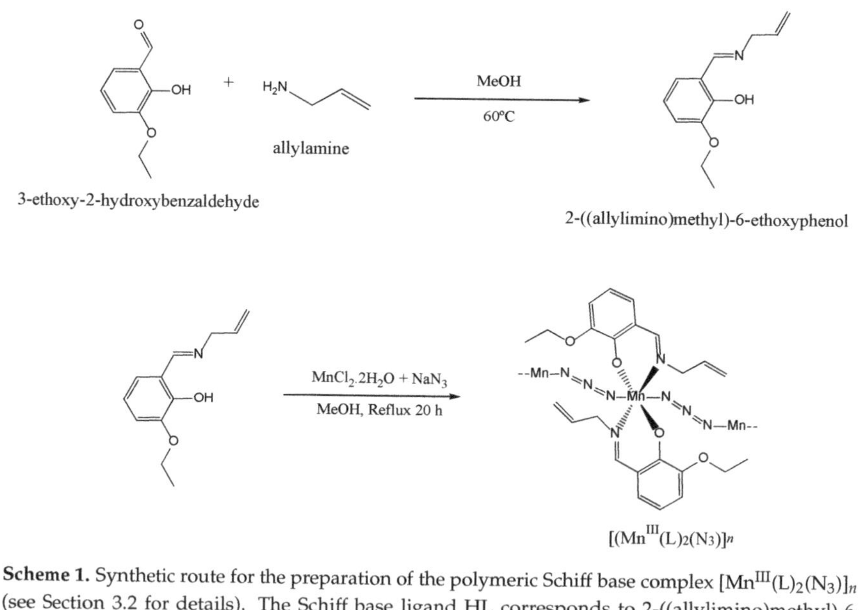

Scheme 1. Synthetic route for the preparation of the polymeric Schiff base complex $[Mn^{III}(L)_2(N_3)]_n$ (see Section 3.2 for details). The Schiff base ligand HL corresponds to 2-((allylimino)methyl)-6-ethoxyphenol.

In summary, the main structural features of the synthesized polymeric Mn(III) Schiff base complex have been thoroughly investigated by analyzing the most relevant geometrical parameters around the metallic center provided by X-ray diffraction data. In addition, Hirshfeld surfaces were utilized to study intermolecular interactions, providing valuable insights into the compound's molecular arrangement in the crystal. Further investigations involving infrared and UV-Vis spectroscopy, cyclic voltammetry, and theoretical computations at the DFT level in both the ligand and the complex were also performed. These combined approaches provided a comprehensive insight into the properties and behavior of the compound. Furthermore, the synthetized complex was subjected to activity spectra for substances (PASS) analysis, which allowed the evaluation of its potential for combating pathogenic viruses.

2. Results and Discussion

2.1. Synthesis and Spectroscopic Characterization of the SCHIFF Base and Complex

The synthetic route for preparing the polymeric Schiff base complex $[Mn^{III}(L)_2(N_3)]_n$ is outlined in Scheme 1. The bidentate Schiff base ligand (HL) was synthesized by reacting allylamine with 3-ethoxy-2-hydroxybenzaldehyde (1:1 molar ratio) in a methanol solution. Subsequently, the mixture was treated with $MnCl_2$ and NaN_3 methanolic solutions. For a detailed description of the experimental procedure used for synthesizing both the ligand and complex, please refer to Section 3.2.

The IR spectra of the Schiff base ligand and complex are provided in Figure S1a,b, and the identified bands are characterized in Table S1. The vibrational calculations performed at the B3LYP/6-311++G(d,p) level on the fully optimized geometries of both species were used to help in the assignment of the most important experimental bands (the computed vibrational data are provided in Tables S2 and S3). In terms of what the IR spectrum of HL is concerned with, it is worth noting the presence of a strong feature at 1659/1647 cm^{-1}, which is undoubtedly ascribed to the C=N stretching vibration of the azomethine group [13,28,29]. This band is theoretically predicted at 1652.8 (I = 198.3 km mol^{-1}). The observation of this band, together with the absence of bands due to the antisymmetric and symmetric NH_2 stretching vibrations, which have been observed in the IR spectra of allylamine within the 3390–3260 cm^{-1} range [30] (refer to Figure S1 for more details), unequivocally confirms the condensation of all primary amino groups during the synthesis of the ligand. The other prominent bands observed in the fingerprint region of the HL spectrum are located at 1464, 1254, and 739 cm^{-1}. In the spectrum calculated for the isolated molecule (Table S2), these bands are predicted at 1463.4 (I = 246.1 km mol^{-1}), 1252.7 (I = 437.2 km mol^{-1}), and 728.8 cm^{-1} (I = 53.2 km mol^{-1}). The first is assigned to the aromatic C–C and C1–O1 stretching vibrations, the second to the C2–O2 stretching vibration, and the third to ring C–H out-of-plane deformations (atom numbering is presented in Figure 1). In the high-frequency region of the IR spectrum, the broad band centered at around 3440 cm^{-1} is most likely assigned to the stretching vibration of the OH groups involved in intermolecular H-bonds. It is important to note that the spectrum calculated for the most stable conformer of HL predicts a very intense absorption at 3038 cm^{-1} (I = 547.8 km mol^{-1}), which is ascribed to the stretching vibration of the OH group engaged in an intramolecular hydrogen bond with the azomethine nitrogen atom (OH$\cdots$N=C) (see Section 2.5 for details). However, no prominent experimental band is evident in the vicinity of this spectral position, confirming the absence of such a hydrogen bond in the synthetized Schiff base ligand. The weak band observed at 2982 cm^{-1} (as shown in Figure S1) is likely associated with CH stretching vibrations (see Table S2).

Regarding the IR spectrum of the complex, the band corresponding to the C=N stretching vibration is shifted to lower frequencies as compared to the free ligand, being observed at 1614 cm^{-1} (calculated at 1662.0 cm^{-1}, I = 607.7 km mol^{-1}). This experimentally observed red shift is most likely attributed to the electron donation from the nitrogen to the empty d-orbitals of the metallic center [31]. The strongest absorption band observed at 2046 cm^{-1} (calculated at 2172 cm^{-1}, I = 1878 km mol^{-1}), which is not present in the spectrum of the ligand, is assigned to the antisymmetric stretching vibration of the azide group (N_3), in consonance with the IR spectroscopic data obtained for other azide-bridged polymeric Mn(III) Schiff base complexes [15,32]. In the lower-frequency region of the IR spectrum, it is noteworthy to observe the bands at 624 and 495 cm^{-1}, which also do not have a corresponding presence in the IR spectrum of the ligand. These bands can be attributed to the antisymmetric stretching vibrations of the Mn–O1 and Mn–N8 bonds. In the computed IR spectrum, the corresponding absorptions are predicted at 582.8 cm^{-1} (I = 66.4 km mol^{-1}) and 461.7 cm^{-1} (I = 66.1 km mol^{-1}). An intense absorption corresponding to the Mn–N1 (or N3) stretching vibration is predicted to occur at 411.9 cm^{-1} (I = 159.7 km mol^{-1}). The non-observation of this band in the experimental spectrum may be attributed to the fact that it is most likely situated below 400 cm^{-1}, placing it outside the detection range of the used IR spectrometer.

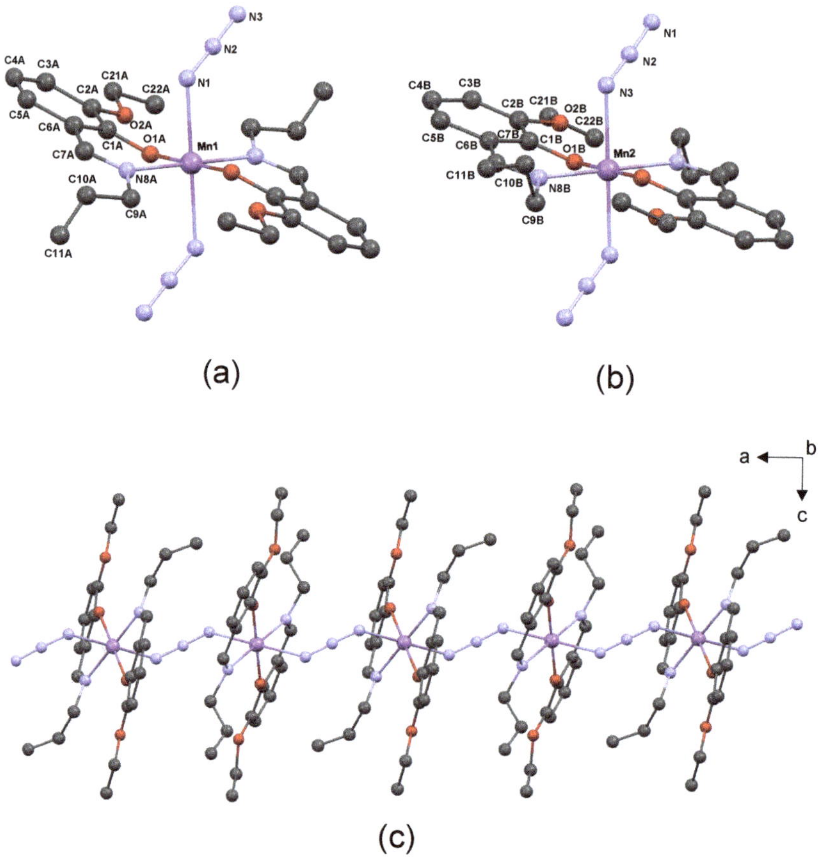

Figure 1. (**a**,**b**) Coordination environment around the Mn1 and Mn2 cations in the Mn(III) complex. (**c**) Segment of the one-dimensional zigzag chain along the a-axis connecting individual monomeric $Mn^{III}(L_2)$ units through the azide ligand.

The UV-vis absorption spectrum of the Mn(III) complex recorded in a DMF solution is displayed in Figure S2. This spectrum follows the general pattern of that previously reported for other similar Mn(III) complexes [12,13,15]. In the UV region, the absorption bands at 235 and 274 nm correspond to π-π^* transitions involving molecular orbitals located on phenolic chromophores, while that at 384 nm is most likely associated with intra-ligand (C=N) transitions. In the visible region, the complex exhibits a characteristic broad and low-intensity band at 624 nm, indicative of d $\rightarrow$ d transitions occurring within its complex.

2.2. Electrochemical Studies

The cyclic voltammogram of the synthesized complex in a DMF solution, recorded between −2.3 and 0.2 V, is shown in Figure S3. In this voltammogram, the cathodic peaks at −0.13 and −1.77 V correspond, to the Mn(III) → Mn(II) and Mn(II) → Mn(I) reductions. No anodic peaks were observed during the reverse scan, thus revealing the irreversibility of these redox processes. This behavior is in line with that previously observed for other structurally similar manganese (III) Schiff base complexes [15,33]. The observed irreversibility can be attributed to the inherent instability of the reduced manganese species in the given experimental conditions. Additionally, the electronic properties of the Schiff base ligand could further influence redox behavior, potentially affecting the ease with which these reduced forms revert to their higher oxidation state. This insight underscores the importance of both the ligand environment and the stability of the metal centers in determining the overall redox characteristics of the complex.

2.3. Crystal Structure

The X-ray diffraction analysis of the synthesized Mn(III) complex reveals that it crystallizes in the monoclinic system with the space group $P2_1/c$. The asymmetric unit of the crystal comprises two Mn(III) cations, identified as Mn1 and Mn2, along with two Schiff base ligands and one azide ion (N_3^-). Both cations are positioned at distinct centers of inversion within the space group $P2_1/c$, located at coordinates ($\frac{1}{2}$, $\frac{1}{2}$, $\frac{1}{2}$) and (0, $\frac{1}{2}$, $\frac{1}{2}$). As shown in Figure 1, the coordination environment around each manganese cation involves two deprotonated Schiff base ligands (Ls), where each ligand is connected to the metallic center through a nitrogen (N8) and oxygen (O1) atom, thus acting as a bidentate NO-donor (Figure 1a,b). In the complex, these two pairs of atoms are arranged in the equatorial plane of an octahedral structure, enveloping the manganese cation. The azide ion (N_3^-) serves as a bridging ligand, linking the two metallic centers (see Figure 1c for details). The most relevant geometric characteristics of the Mn(III) complex, obtained from the X-ray data, are provided in Table 1.

Table 1. Selected parameters extracted from the X-ray data bond lengths (Å) and bond angles (deg.) around the metallic center (Mn1 or Mn2) of the studied polymeric Mn(III) complex.

Mn1		Mn2	
	Bond length/Å		
Mn1–N8A	2.035	Mn2–N8B	2.044
Mn1–O1A	1.885	Mn2–O1B	1.866
Mn1–N1	2.272	Mn2–N3	2.256
	Bond angle/deg.		
O1A–Mn1–N8A	89.2–90.8	O1B-Mn2–N8B	89.6–90.4
O1A–Mn1–N1	89.7–90.3	O1B–Mn2–N3	88.6–91.3
N8A–Mn1–N1	87.5–92.6	N8B–Mn2–N3	88.1–91.9
O1A–Mn1–O1A	180.0	O1B–Mn2–O1B	180.0
N8A–Mn1–N8A	180.0	N8B–Mn2–N8B	180.0
N1–Mn1–N1	180.0	N3–Mn2–N3	180.0

Analyzing the provided geometrical parameters obtained from the X-ray data, several observations can be made. Firstly, the axial Mn–N1 (or N3) bonds are approximately 0.237–0.213 Å longer than the equatorial Mn–N8 and Mn–O1 bonds, indicating a deviation from an ideal octahedral geometry. This observed deviation is consistent with the patterns identified in previously reported Mn(III) Schiff base complexes that share similar structural features with our complex [15,17–19,34], and is typical in high-spin Mn (III) octahedral complexes that undergo Jahn–Teller distortions [35,36]. Secondly, within the equatorial plane, the Mn–O1 bonds are shorter than the Mn–N8 bonds, which is expected due to the higher electron density around the oxygen atom. This geometrical feature is consistent with the IR spectroscopic data presented in Section 2.1, where the Mn–O1 stretching vibration was observed at a higher frequency than the Mn–N8 stretching vibration. Lastly, a comparison between the two metal centers, Mn1 and Mn2, reveals variations in bond lengths. In Mn2, both the axial Mn–N1(or N3) and equatorial Mn–O1 bonds are shorter by approximately 0.015 and 0.019 Å, respectively, compared to Mn1. Conversely, the equatorial Mn–N8 bond in Mn1 is only 0.009 Å shorter than in Mn2. These differences suggest that, globally, the arrangement of ligands around the Mn2 center is more compact than that around the Mn1 center. Examining the bond angles around the central metal, it can be observed that the O1–Mn–N8, O–Mn–O, and O1–Mn–N1 (or N3) and N8–Mn–N1 (or N3) angles fall within the range of 88–93°, while the O1–Mn–O1, N8–Mn–N8, and N1–Mn–N1 (or N3–Mn–N3) angles are equal to 180°.

The crystal structure of the complex reveals the presence of a coordination polymer that extends along the [100] direction, as depicted in Figure 1c. This supramolecular structure is obtained by linking individual monomeric $Mn^{III}(L_2)$ units through the azide ligand, involving an end-to-end μ-(1,3) bridging mode [15,18,34,37]. The bridging arrangement gives rise to a chain-like structure, exhibiting a one-dimensional zigzag pattern. Within this chain, the Mn1 and Mn2 metal centers are arranged in an alternating manner. The distance between adjacent Mn1 and Mn2 centers within the chain is 5.763 Å, while the shortest distance between Mn atoms in different chains is 10.291 Å. These data agree with the those reported for other azide-bridged polymeric Mn(III) Schiff base complexes [17,19,34].

Apart from the coordination bonds, the crystal structure of the compound is characterized by the formation of short C–H···N, C–H···O, and C–H···C intermolecular interactions. These interactions are observed within the polymeric chains, as well as between different chains. Although these contacts are associated with the formation of weak stabilizing interactions, they play a significant role in the arrangement of molecules in the crystal lattice. To better comprehend these intermolecular interactions and their influence on the crystal structure, an analysis of Hirshfeld surfaces is presented below.

2.4. Hirshfeld Surface Analysis

The Hirshfeld surfaces generated for the Schiff base complex are presented in Figure 2a. The larger red spots observed on these surfaces correspond to the axial Mn–N1/N3 coordination bonds, which have been already characterized in Section 2.3 and are therefore excluded from this analysis. The majority of the remaining red spots is attributed to the formation of weak stabilizing C–H··· A interactions (A = N, O, or C) [38], which are denoted in Figure 2 by letters a–h. Red spots labeled as i, j are associated with the formation of C···C repulsive interactions. The geometrical parameters characterizing all these interactions are presented in Table 2, with the most relevant ones highlighted in Figure 3.

Table 2. Geometrical characterization of the most relevant intermolecular contacts found in the crystalline structure of the polymeric Mn(III) complex [a].

Interaction	Label	H···A/Å	C···A/Å	C–H···A/°
Interchain				
C22B–H···N1/N2	a/a′	2.742/2.562	3.631/3.345	150.9/136.8
C21B–H···N3	b	2.634	3.167	114.0
C22A–H···C22B	c	2.697	3.607	154.6
C21B–H···C2A	d	2.789	3.574	136.6
Intrachain				
C9B–H···O1A	e	2.549	3.499	160.9
C9A–H···N2/N3	f/g	2.603/2.668	3.347/3.573	131.9/152.1
C11A–H···O1B	h	2.699	3.359	127.1
C11A···C2B	i		3.219	
C11A···C1B	j		3.320	

[a] A = N, O, or C.

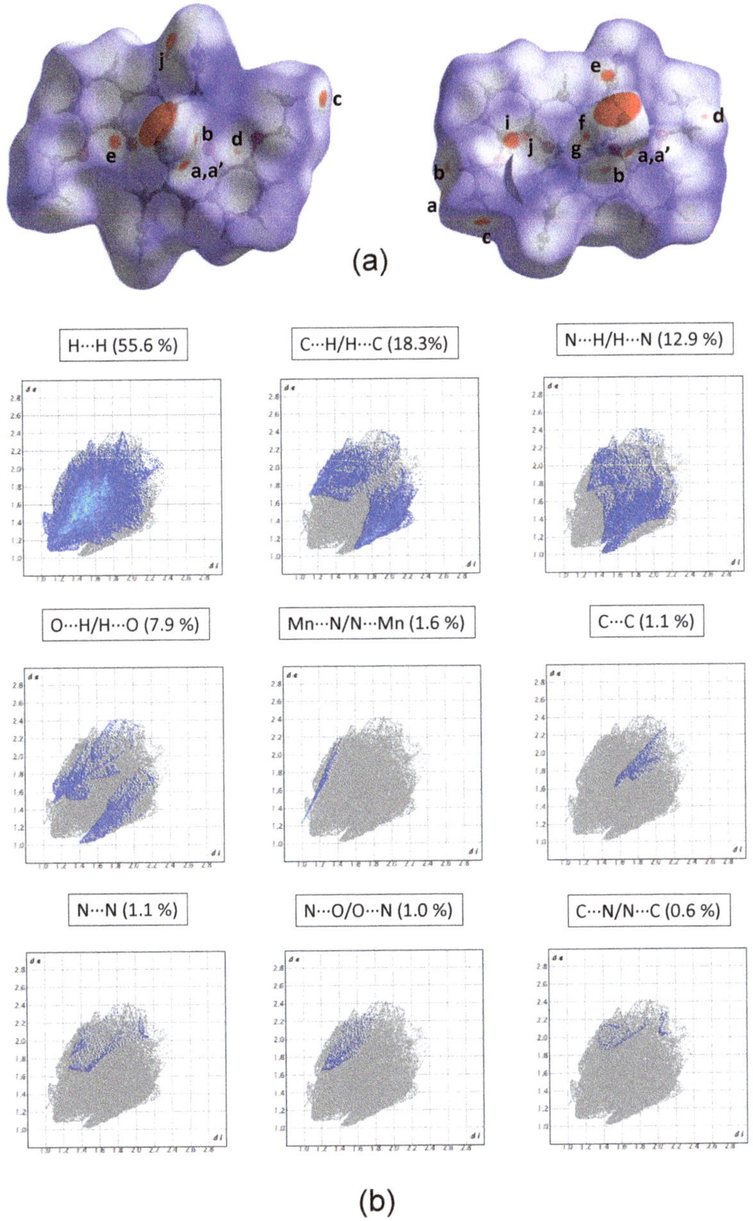

Figure 2. (**a**) Three-dimensional Hirshfeld surfaces generated for the $Mn^{III}(L_2)(N_3)_2^-$ fragment. For the left surface, the metallic center is Mn1, while for the right surface, the metallic center is Mn2. The surfaces are mapped with the normalized contact distance (d_{norm}) and the relevant intermolecular contacts are denoted by small letters. (**b**) Fingerprint plots calculated for the Mn (III) complex, decomposed into different contacts. The d_i and d_e values correspond, to the closest internal and external distances (Å) from given points on the Hirshfeld surface contacts. The contributions (in %) of the different contacts to the total Hirshfeld surface area are included.

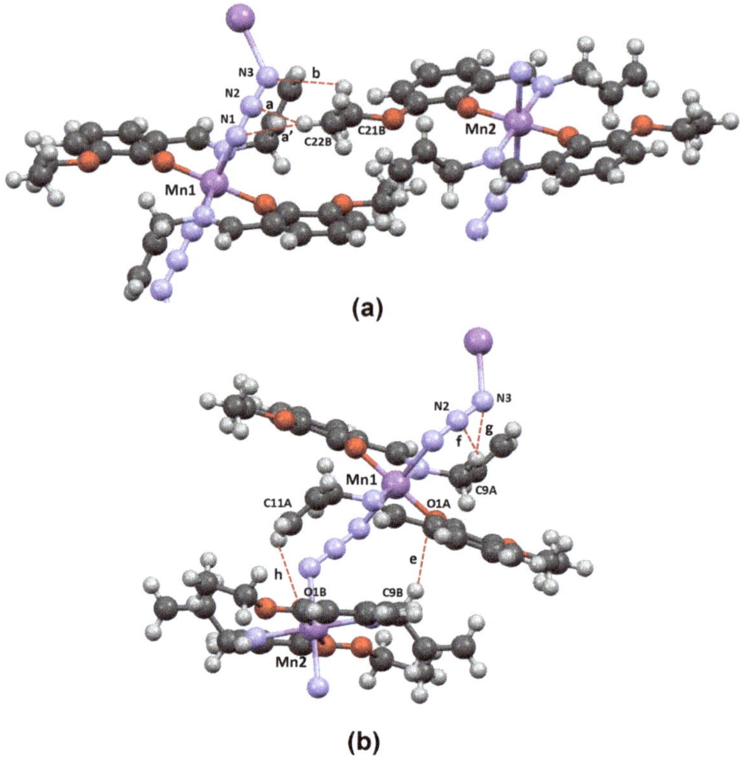

Figure 3. Most relevant interchain (**a**) and intrachain (**b**) hydrogen bond-like interactions formed in the crystalline structure of the polymeric Schiff base complex $[Mn^{III}(L)_2(N_3)]_n$.

The short contacts labeled in the Hirshfeld surfaces as a, a′, and b correspond to the formation of C–H$\cdots$N hydrogen-bond like interactions that occur between adjacent polymeric chains (interchain). A common characteristic of these interactions is that the carbon atoms serving as donors are part of the ethoxy fragment, while the nitrogen azide atoms function as acceptors. In the case of the first two interactions (a, a′), C22B (from the methyl group of the ethoxy fragment) acts as the donor atom, while the azide nitrogen atoms N1 or N2 act as acceptors. The third interaction (b) involves one of the hydrogens connected to C21B (from the methylene group of the ethoxy fragment) and the other azide N3 atom. The geometrical parameters presented in Table 2 indicate that interaction b is weaker when compared to interactions a and a′. This difference in strength is mainly attributed to the significantly smaller C–H$\cdots$N angle observed in interaction b (114°), in contrast to the wider angles found in the other interactions (137–151°).

The remaining stabilizing interchain contacts (labeled as c and d) are characterized as very weak C–H$\cdots$C interactions and are not shown in Figure 3. Interactions e-h take place within the same chain (intrachain). The positioning of the carbon atoms acting as donors is an important difference from the interchain interactions. Unlike the former atoms, the donor carbon atoms form part of the allyl fragment. Two of the intrachain interactions (e and h) are of the C–H$\cdots$O type, where C9B or C11A serve as donors while O1A or O1B act as acceptors, respectively. Based on the geometrical parameters, interaction e (C9B–H$\cdots$O1A) emerges as the strongest of all the ones present in the crystal structure of the complex. The other two intrachain interactions (f, g) are classified as C–H$\cdots$N, involving C9A as a donor and the azide N2 or N3 atoms as acceptors. It is worth highlighting that the formation of the C11A–H$\cdots$O1B interaction (h) has two significant structural consequences: (i) the conformation of the allyl fragment (–CH_2–CH=CH_2) in the Schiff base ligands connected to Mn2 differs from that connected to Mn1, as illustrated in Figure 1; and (ii) C11A approaches the C1B and C2B atoms of the phenyl ring of a Schiff base situated in adjacent intrachain complexes, resulting in the formation of C11A· · · C1B and C11A· · · C2B contacts, which are also characterized in Table 2.

The 2D-fingerprint plots, extracted from the Hirshfeld surfaces, divided into contributions stemming from key atomic interactions, are included in Figure 2b. Given the substantial number of hydrogen atoms within the molecular structure of the analyzed compound, it is not surprising to observe a predominance of the H· · · H contacts, representing 55.6% of the Hirshfeld surface. In line with our expectations from the discussion above, the subsequent noteworthy contributions arise from the H· · · C/C· · · H, H· · · N/N· · · H, and H· · · O/O· · · H contacts, accounting for approximately 18.3%, 12.9%, and 7.9% of the total surface area, respectively. On the other hand, less prominent contributions are attributed to Mn· · · N, C· · · C, N· · · N, N· · · O/O· · · N, and C· · · N/N· · · C contacts, which together represent only 5.4% of the surface.

2.5. Conformational Analysis of the Schiff Base Ligand

In addition to the structure of the complex, it is also instructive to investigate the structure of the Schiff base ligand (HL), with particular emphasis on exploring its conformational behavior. The potential energy surfaces around the C7=N8-C9-C10 (θ_1) and N8-C9-C10=C11 (θ_2) torsions (see Figure S4a,b) led to the identification of eight conformers. The B3LYP/6-311++G(d,p) fully optimized geometries of two representative conformers are shown in Figure 4, while the geometries of all eight identified conformers, along with their energetic characterizations, are presented in Figure S5 and Table S4, respectively. The conformers are labeled as HL1 or HL2, depending on whether the O1H group is directed toward the N8 or O2 atoms, respectively, followed by the labels "c" (cis, ~0°) or "sk±" (skew±, ~±120°), defining the orientation around the θ_1 and θ_2 dihedrals, which define the conformation of the allyl fragment. According to the calculations, the HL1 conformers are more stable than the HL2 ones. This difference in stability is attributed to the formation of an O1–H· · · N8 intramolecular hydrogen bond in HL1 [*d*(H· · · N) ~1.7Å; ∠(O–H· · · 8) ~147–148°], which is much stronger than the O1–H· · · O2 intramolecular bond formed in the HL2 conformers [*d* (H· · · N) ~2.0Å; ∠(O–H· · · N) ~116–117°]. For a given orientation of the θ_1 and θ_2 torsions, the energy difference between these two sets of conformers is ~30 kJ mol^{-1}. Among the HL1 conformers, the most stable one is characterized by sk^- and sk^+ (or sk^+ and sk^-) orientations around the θ_1 and θ_2 dihedrals (θ_1 = ±121.7°; θ_2 = ±128.5°) respectively (see Figure 4, left). It is worth noting that the arrangement of the allyl fragment in this conformer is not significantly different from its arrangement in the crystalline structure of the Schiff base ligand coordinated to the Mn2 cation (θ_1 = ±74.1°; θ_2 = ±126.5°). This is consistent with the fact that, in the crystalline structure of the complex, the allyl fragment of the ligand is not involved in intrachain or interchain intermolecular interactions allowing its arrangement to closely resemble that of the most stable form of the free molecule. On the other hand, the conformation of the allyl group in the Schiff base coordinated to the Mn1 cation (θ_1 = ±8.9°; θ_2 = ±122.9°) is very close to the fifth most stable conformer (θ_1 = +18.3°; θ_2 = −120.0°), and this conformational variation is attributed to the fact that the terminal CH_2 group of the allyl fragment of the ligand is involved in the formation of an intrachain C–H· · · O interaction, as described in Section 2.4.

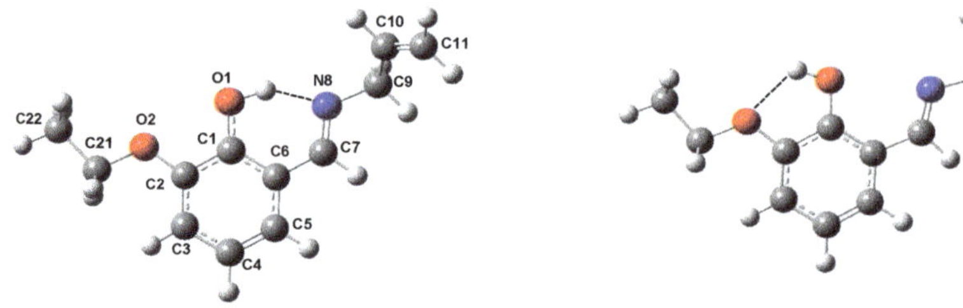

HL1(sk⁻,sk⁺) HL2 (sk⁻,sk⁺)

Figure 4. Geometries of the most stable HL1- and HL2-type conformers of the Schiff base ligand (HL), fully optimized at the B3LYP/6-311++G(d,p) level of theory.

2.6. In Silico PASS Analysis of Biological Activity

In this study, we employed PASS (Prediction of Activity Spectra for Substances), a software tool designed to evaluate the broad biological potential of organic drugs (see Section 3.7 for details). PASS utilizes structural information from chemical compounds to predict multiple types of biological activities simultaneously. We harnessed the capabilities of PASS to generate virtual molecular biological activity profiles for our synthesized complex. The "to be active" probability estimates the likelihood that our compound belongs to the sub-class of active compounds, sharing structural similarities with the most common molecular structures found in a sub-set of "actives" from the PASS training dataset. Table 3 summarizes the results of the PASS analysis, which offers valuable insights into the potential pharmacological actions of our synthesized polymeric Mn(III) complex. This analysis is crucial for understanding the diverse biological activities that this compound may exhibit based on its structural characteristics.

Table 3. Predicted biological activity spectrum of polymeric Mn(III) complex by PASS analysis.

Activity	Pa	Pi
Antiviral (Rhinovirus)	0.444	0.053
Antiviral (Influenza)	0.350	0.064
Antiviral (Picornavirus)	0.352	0.156
Antiviral	0.211	0.086
Antitussive	0.199	0.087
Antiviral (CMV)	0.224	0.122
Antiviral (Influenza A)	0.227	0.145

As one can see from Table 3, the polymeric Mn(III) complex shows promising predictions for a wide spectrum of pharmacological functions. Notably, the "Pa" values (probability of being active) consistently surpass the "Pi" values (probability of being inactive) for various biological activities. Among the most notable predictions, the compound is found to exhibit substantial "Pa" values for antiviral activities against Rhinovirus (0.444), Influenza (0.350) and Picornavirus (0.352). This indicates a strong potential for the complex to exert specific effects in these areas (Pa > 0.3). Overall, the results of this study suggest that these Schiff base complex may be considered for further investigation as a possible therapeutic agent for antiviral properties.

3. Experimental and Computational Methodologies

3.1. Materials and Instrumentation

All reagents used in the experiments were obtained from Merck (Rahway, NJ, USA) and Sigma-Aldrich (St. Louis, MI, USA) and were used without further purification. IR spectra of the synthesized solid compounds were recorded on a SHIMADZU FT-IR spectrometer (4000–400 cm^{-1}) in KBr pellets, while UV-Vis absorbance spectra in a dimethylformamide

(DMF) solution were measured using a UV-1650 PC SHIMADZU spectrophotometer. A Metrohm 757 VA Computrace instrument was utilized to conduct electrochemical analyses in a DMF solution.

3.2. Synthesis of the Polymeric Schiff Base Complex $[Mn^{III}(L)_2(N_3)]_n$

The methodology employed for synthesizing the Mn(III) complex followed that described elsewhere for an analogous complex [15]. Initially, for ligand (HL) preparation, 3-ethoxy-2-hydroxybenzaldehyde (0.185 g) was dissolved in a methanolic solution containing allylamine (0.058 g in 10 mL of methanol), and the resulting mixture was stirred for 2 h at 60 °C. To prepare the complex, 0.5 mmol of $MnCl_2.2H_2O$ was dissolved in 10 mL of methanol and added dropwise to the ligand solution. The color of the solution changed from yellow to red upon addition. Subsequently, 0.065 g of NaN_3 was dissolved in 2 mL of methanol and added to the solution, leading to a color change to black. The resulting mixture was allowed to stir magnetically under reflux for 20 h. After a few days, dark-brown single crystals were obtained, subsequently filtered, washed with a small quantity of cold *n*-hexane, and dried under vacuum. The resulting product proved suitable for X-ray analysis. Yield: 63%. FT-IR data (KBr, cm^{-1}): 2975–2923 (νC–H), 2046 [ν_{as} (N_3)]; 1616 (νC=N); 624 (νMn–O); 495 [ν(Mn–N)]. UV–Vis: λ_{max} (nm) (ε, $M^{-1}cm^{-1}$) (DMF): 235 (62500), 274 (37500), 384 (12500), 625 (3000). In the Supplementary Materials, we provide the IR and UV-Vis spectra (Figures S1 and S2, respectively).

3.3. X-ray Data Collection and Refinement Parameters

Diffraction data were collected by the ω-scan technique, at 100(1) K, on a Rigaku XCalibur four-circle diffractometer with an EOS CCD detector, equipped with a graphite-monochromatized MoK_α radiation source (λ = 0.71073 Å). The data were corrected for Lorentz polarization as well as for absorption effects [39]. The structures were solved with SHELXT [40] and refined with the full-matrix least-squares procedure on F^2 by SHELXL-2013 [13]. All non-hydrogen atoms were refined anisotropically, while hydrogen atoms were placed in idealized positions and refined as 'riding model' with isotropic displacement parameters set at 1.2 (1.5 for methyl groups) times U_{eq} of appropriate carrier atoms. Crystal data: $C_{24}H_{28}MnN_5O_4$, M_r = 505.45, monoclinic, $P2_1/c$, a = 11.5257(5) Å, b = 13.9571(6) Å, c = 15.1254(5) Å, β = 104.333(4)°, V = 2357.42(17) $Å^3$, Z = 4, d_x = 1.424 g cm^{-3}, F(000) = 1056, μ = 0.601 mm^{-1}, 18,193 reflection collected, 4736 symmetry independent (R_{int} = 4.02%), and 3458 with I > 2σ(I). Final R[I > 2σ(I)] = 0.0463, wR2[I > 2σ(I)] = 0.1104, R [all reflections] = 0.0728, wR2 [all reflections] = 0.1293, S = 1.064, and ($\Delta\rho_{max}/\Delta\rho_{min}$) = 0.60/−0.59 e·$Å^{-3}$. The CIF file containing the supplementary crystallographic data for the synthesized compound was deposited at the Cambridge Crystallographic Data Centre, with reference CCDC 2172659.

3.4. Cyclic Voltammetry

Cyclic voltammetry experiments were carried out for the complex at 298 K in a 10^{-3} M DMF solution, utilizing 0.1 M of tetrabutylammonium hexafluorophosphate (TBAH) as the supporting electrolyte. The experimental setup involved a three-electrode system, which consisted of an Ag/AgCl reference electrode, a platinum auxiliary electrode, and a glassy carbon working electrode. The voltammograms were recorded with a scan rate of 100 mV/s. Anodic and cathodic potentials were measured with reference to the Ferrocenium (Fc^+)/Ferrocene (Fc) redox couple (E^0 = 0.45 V).

3.5. Hirshfeld Surface Analysis

The CrystalExplorer software package (version 17.5) was used to perform a Hirshfeld surface (HS) analysis of the crystal structure of the Mn(III) complex [41–45]. This analysis serves as a valuable graphical tool, allowing for a comprehensive examination of the intermolecular interactions and close contacts within the crystal packing. The CIF file, obtained from the X-ray data, was used as the input for the software. The Hirshfeld surfaces were generated with high resolution and were mapped with the normalized contact distance (d_{norm}), calculated from the following equation [44,46]:

$$d_{norm} = \frac{\left(d_i - r_i^{vdW}\right)}{r_i^{vdW}} + \frac{\left(d_e - r_e^{vdW}\right)}{r_e^{vdW}} \quad (1)$$

In this equation, d_i and d_e are, respectively, the distances from the point of the surface to the nearest atoms inside and outside the surface, while r_i^{vdW} and r_e^{vdW} are the van der Waals radii of the respective atoms. Intermolecular contacts shorter than the sum of the van der Waals radii ($d_{norm} < 0$) are depicted as red spots on the surface. Contacts around the van der Waals separation ($d_{norm} = 0$) are shown as white, while contacts longer than the van der Waals distance ($d_{norm} > 0$) are represented by blue regions. Furthermore, as part of the Hirshfeld surface analysis, fingerprint plots were also generated. These plots provide a two-dimensional representation of the relationship between d_i and d_e at each surface point, assisting in understanding the nature and relative contribution of these contacts to the total Hirshfeld surface area, thereby aiding in the characterization of the complex's supramolecular architecture.

3.6. Theoretical Calculations

All theoretical calculations were performed with the Gaussian 16 software package (Revision A.03) [47]. The HL conformers were identified by calculating the relaxed potential energy surfaces around the C7=N8-C9-C10 (θ_1) and N8-C9-C10=C11 (θ_2) dihedrals. These calculations were performed with the B3LYP (DFT) functional [48–50] and the 6-311++G(d,p) basis set [51,52]. The located energy minima were then fully optimized followed by Harmonic vibrational calculations at the same B3LYP/6-311++G(d,p) level of theory. For the Mn(III) complex, an $Mn^{III}(L)_2(N_3)_2^-$ fragment extracted from the crystalline structure was fully optimized, followed by a vibrational calculation by using the same DFT functional combined with the def2-svp basis set [53]. These computations were conducted for both the high-spin (HS, quintet) and low-spin (LS, triplet) electron configurations. At this level of theory, the Gibbs energy difference at 298.15 K between the two configurations was computed to be 43.5 kJ mol^{-1} in favor of the HS state. Consequently, the vibrational data employed in the interpretation of the experimental IR spectrum of the Mn(III) complex were extracted for the fully optimized HS $Mn^{III}(L)_2(N_3)_2^-$ fragment. From the atomic spin densities, a magnetic moment of 3.9 μB was estimated for this configuration, which is consistent with the experimentally determined magnetic moments reported for HS Mn(III) complexes [54], including azide-bridged manganese(III) complexes [16,20,21]. For all optimized structures of the ligand and complex, no imaginary frequencies were observed in the vibrational calculations, confirming that they are all minima on the respective potential energy surfaces. The B3LYP/6-311++G(d,p) wavenumbers computed for the ligand were multiplied by 0.980 for the region below 2000 cm^{-1}, while for the vibrations above 2000 cm^{-1}, which are typically more anharmonic, a scale factor of 0.950 was used [55,56]. Concerning the wavenumbers calculated for the complex using the B3LYP/def2-SVP model chemistry, a single scaling factor of 0.967 was applied [57].

3.7. Prediction of Probable Activity Spectra of Substances (PASS Online Software)

Prediction of Probable Activity Spectra of Substances (PASS, version 11) analysis is an online software (https://www.way2drug.com/passonline, accessed on 20 November 2023) [58,59] that uses computer-based algorithms to assess the biological and pharmacological potential of a compound. It provides insights into the compound's interactions with various biological entities, enabling the evaluation of its utility in biological and pharmacological contexts. To perform this analysis, the software compares the 2D structure of a molecule to a vast training set consisting of over 46,000 drugs, drug candidates, and lead compounds. The molecular structure can be stored as a Mol file (*.mol) and directly uploaded to the PASS prediction website to forecast the biological activity spectrum. The molecule's activity is predicted by comparing the new compound's structure to that of a well-known biological active substrate stored in the database. The algorithm employs a Bayesian technique. The PASS tool provides a Pa:Pi ratio (active to inactive ratio) at prediction thresholds of Pa > 30%, Pa > 50%, and Pa > 70%. These thresholds indicate the predicted likelihood of a compound being active. According to leave-one-out cross-validation (LOO CV) estimates, the average prediction accuracy is approximately 95%. The accuracy of the PASS prediction relies on detailed information about the biological activity spectrum for each molecule in the PASS training set, making the biological activity estimate more accurate [59].

4. Conclusions

In this manuscript, we successfully synthesized a novel Mn(III) Schiff base complex. Its characterization was accomplished through FT-IR and UV–Vis spectroscopy. Furthermore, the molecular structure of the compound was elucidated using single-crystal X-ray diffraction analysis. The crystallographic results reveal a distinctive 1D-coordination polymer complex, where two nitrogen and oxygen atoms of the Schiff base ligand are coordinated to the central cation. The sodium azide was found to act as a bridging component, connecting the $Mn^{III}(L)_2$ complex units, originating a chain-like structure with a one-dimensional zigzag pattern. The Hirshfeld surface analysis revealed diverse C–H···A (A = N, O, or C) interactions both within the same chain or among chains. Interchain interactions prominently featured ethoxy carbon atoms (C21 and C22) as hydrogen donors and bridging azide nitrogen atoms (N1, N2, and N3) as acceptors. On the other hand, intrachain interactions were primarily driven by allyl carbon atoms (C9 and C11) as donors, involving azide nitrogens (N2 and N3) or phenoxyl oxygen (O1) as acceptors. This comprehensive analysis sheds light on the complex's structural characteristics, providing valuable insights into potential implications for its behavior. Electrochemical studies revealed an irreversible reduction process for the Mn (III) complex, providing a greater understanding of its redox behavior. Besides the characterization of the complex, we also investigated the conformational features of the isolated HL Schiff base ligand, identifying its most stable conformation as one stabilized by an intramolecular O–H···N hydrogen bond. Finally, Activity Spectra for Substances (PASS) predictions indicated potential antiviral activity for the synthesized complex. These findings can inspire and guide future studies, particularly those aimed at gaining deeper insights into the structure and behavior of the complex in biologically relevant environments. Such research will be crucial for fully exploring its therapeutic potential.

Supplementary Materials: The following supporting information can be downloaded at: https://www.mdpi.com/article/10.3390/inorganics12090234/s1, Figure S1: IR spectra of the Schiff base ligand and the polymeric Mn (III) complex; Figure S2: UV-Vis spectrum of the complex; Figure S3: Cyclic voltammogram registered for a complex solution; Figure S4: Relaxed potential energy scans computed for the Schiff base ligand (HL) at the B3LYP/6-311++G(d,p) level; Figure S5: Geometries of the conformers of HL fully optimized at the B3LYP/6-311++G(d,p) level of theory. Tables: Table S1: Experimental wavenumbers and intensity of the bands observed in the IR spectra of HL and polymeric Mn(III) complex; Table S2: B3LYP/6-311++G(d,p) wavenumbers and IR intensities calculated for the HL ligand; Table S3: B3LYP/def2-SVP wavenumbers and IR intensities calculated for the $Mn^{III}(L)_2(N_3)_2^-$ fragment of the polymeric manganese (III) Schiff base fully optimized at the same level; Table S4: Geometric data computed for the conformers of HL fully optimized at the B3LYP/6-311++G(d,p) level of theory.

Author Contributions: A.T.: investigation and writing—original draft preparation. M.S.: validation, data curation, and project administration. A.J.L.J.: formal analysis, writing—review and editing, and visualization. M.K.: methodology. R.F.: validation. R.G.: resources. All authors have read and agreed to the published version of the manuscript.

Funding: This work was supported by Semnan University, Adam Mickiewicz University, and Payame Noor University. CQC-IMS is supported by FCT through projects UI0313B/QUI/2020, UI0313P/QUI/2020, and LA/P/0056/2020 (national funds).

Data Availability Statement: All original contributions presented in this study are included within the article. For further information, inquiries can be directed to the corresponding authors.

Acknowledgments: This work results from a collaborative effort between the scientific members of Semnan University (Aynaz Talebi and M. Salehi), University of Coimbra (Rui Fausto and A. J. Lopes Jesus), Payame Noor University (Reza Golbedaghi), and Institute Sciences Adam Mickiewicz University (Maciej Kubicki). The authors consider it their duty to thank Rui Fausto (University of Coimbra) for his guidance during the revision of the manuscript.

Conflicts of Interest: The authors declare no conflicts of interest.

References

1. Schiff, H. Mittheilungen aus dem Universitätslaboratorium in Pisa: Eine neue Reihe organischer Basen. *Justus Liebigs Ann. Chem.* **1864**, *131*, 118–119. [CrossRef]
2. Fabbrizzi, L. Beauty in Chemistry: Making Artistic Molecules with Schiff Bases. *J. Org. Chem.* **2020**, *85*, 12212–12226. [CrossRef] [PubMed]
3. Raczuk, E.; Dmochowska, B.; Samaszko-Fiertek, J.; Madaj, J. Different Schiff Bases mdash—Structure, Importance and Classification. *Molecules* **2022**, *27*, 787. [CrossRef] [PubMed]
4. Vigato, P.A.; Tamburini, S. The challenge of cyclic and acyclic schiff bases and related derivatives. *Coord. Chem. Rev.* **2004**, *248*, 1717–2128.
5. Golbedaghi, R.; Tabanez, A.M.; Esmaeili, S.; Fausto, R. Biological Applications of Macrocyclic Schiff Base Ligands and Their Metal Complexes: A Survey of the Literature (2005–2019). *Appl. Organomet. Chem.* **2020**, *34*, e5884. [CrossRef]
6. Soroceanu, A.; Bargan, A. Advanced and Biomedical Applications of Schiff-Base Ligands and Their Metal Complexes: A Review. *Crystals* **2022**, *12*, 1436. [CrossRef]
7. Shaygan, S.; Pasdar, H.; Foroughifar, N.; Davallo, M.; Motiee, F. Cobalt (II) Complexes with Schiff Base Ligands Derived from Terephthalaldehyde and ortho-Substituted Anilines: Synthesis, Characterization and Antibacterial Activity. *Appl. Sci.* **2018**, *8*, 385. [CrossRef]
8. Sinicropi, M.S.; Ceramella, J.; Iacopetta, D.; Catalano, A.; Mariconda, A.; Rosano, C.; Saturnino, C.; El-Kashef, H.; Longo, P. Metal Complexes with Schiff Bases: Data Collection and Recent Studies on Biological Activities. *Int. J. Mol. Sci.* **2022**, *23*, 14840. [CrossRef]
9. Naeimi, H.; Sadat Nazifi, Z.; Matin Amininezhad, S.; Amouheidari, M. Synthesis, characterization and in vitro antimicrobial activity of some new Schiff bases and their complexes. *J. Antibiot.* **2013**, *66*, 687–689. [CrossRef]
10. Iacopetta, D.; Ceramella, J.; Catalano, A.; Mariconda, A.; Giuzio, F.; Saturnino, C.; Longo, P.; Sinicropi, M.S. Metal Complexes with Schiff Bases as Antimicrobials and Catalysts. *Inorganics* **2023**, *11*, 320. [CrossRef]
11. Amiri Rudbari, H.; Khorshidifard, M.; Askari, B.; Habibi, N.; Bruno, G. New asymmetric Schiff base ligand derived from allylamine and 2,3-dihydroxybenzaldehyde and its molybdenum(VI) complex: Synthesis, characterization, crystal structures, computational studies and antibacterial activity together with synergistic effect against Pseudomonas aeroginosa PTTC 1570. *Polyhedron* **2015**, *100*, 180–191.
12. Deilami, A.B.; Salehi, M.; Amiri, A.; Arab, A. New copper(II) and vanadium(IV) complexes based on allylamine-derived Schiff base ligand; synthesis, crystal structure, electrochemical properties and DFT calculations. *J. Mol. Struct.* **2019**, *1181*, 190–196. [CrossRef]
13. Deilami, A.B.; Salehi, M.; Arab, A.; Amiri, A. Synthesis, crystal structure, electrochemical properties and DFT calculations of three new Zn(II), Ni(II) and Co(III) complexes based on 5-bromo-2-((allylimino)methyl)phenol Schiff-based ligand. *Inorg. Chim. Acta* **2018**, *476*, 93–100. [CrossRef]
14. Kazemi, Z.; Rudbari, H.A.; Sahihi, M.; Mirkhani, V.; Moghadam, M.; Tangestaninejad, S.; Mohammadpoor-Baltork, I.; Gharaghani, S. Synthesis, characterization and biological application of four novel metal-Schiff base complexes derived from allylamine and their interactions with human serum albumin: Experimental, molecular docking and ONIOM computational study. *J. Photochem. Photobiol. B* **2016**, *162*, 448–462. [CrossRef] [PubMed]
15. Ramezanipoor, S.; Parvarinezhad, S.; Salehi, M.; Grześkiewicz, A.M.; Kubicki, M. Crystal structures, electrochemical properties and theoretical studies of three New Zn(II), Mn(III) and Co(III) Schiff base complexes derived from 2-hydroxy-1-allyliminomethyl-naphthalen. *J. Mol. Struct.* **2022**, *1257*, 132541. [CrossRef]
16. Miyasaka, H.; Saitoh, A.; Abe, S. Magnetic assemblies based on Mn(III) salen analogues. *Coord. Chem. Rev.* **2007**, *251*, 2622–2664. [CrossRef]
17. Chakrabarty, P.P.; Saha, S.; Schollmeyer, D.; Boudalis, A.K.; Jana, A.D.; Luneau, D. Azide-bridged manganese(III) one-dimensional chain: Synthesis, structure, and magnetic study. *J. Coord. Chem.* **2013**, *66*, 9–17. [CrossRef]

18. Ko, H.H.; Lim, J.H.; Kim, H.C.; Hong, C.S. Coexistence of Spin Canting and Metamagnetism in a One-Dimensional Mn(III) Complex Bridged by a Single End-to-End Azide. *Inorg. Chem.* **2006**, *45*, 8847–8849. [CrossRef]
19. Panja, A.; Shaikh, N.; Vojtíšek, P.; Gao, S.; Banerjee, P. Synthesis, crystal structures and magnetic properties of 1D polymeric $[Mn^{III}(salen)N_3]$ and $[Mn^{III}(salen)Ag(CN)_2]$ complexes. *New J. Chem.* **2002**, *26*, 1025–1028. [CrossRef]
20. Li, W.; Li, Z.; Li, L.; Liao, D.; Jiang, Z. A one-dimensional azido-bridged manganese(III) complex with bidentate Schiff base: Crystal structure and magnetic properties. *J. Solid State Chem.* **2007**, *180*, 2973–2977. [CrossRef]
21. Qian, K.; Xu, Y.; Wang, Z.; Yang, J. Synthesis, crystal structure, and magnetic properties of an azido-bridged Mn(II) complex $[C_3H_5NH_3][Mn(N_3)_3]$. *Z. Für Naturforschung B* **2017**, *72*, 409–413. [CrossRef]
22. Yoon, J.H.; Lee, W.R.; Ryu, D.W.; Lee, J.W.; Yoon, S.W.; Suh, B.J.; Kim, H.C.; Hong, C.S. End-to-End Azide-Bridged Manganese(III) Chain Compounds: Field-Induced Magnetic Phase Transitions and Variation of TC to 38 K depending on the Side Groups of the Schiff Bases. *Inorg. Chem.* **2011**, *50*, 10777–10785. [CrossRef]
23. Hong, C.S.; Koo, J.-e.; Son, S.-K.; Lee, Y.S.; Kim, Y.-S.; Do, Y. Unusual Ferromagnetic Couplings in Single End-to-End Azide-Bridged Cobalt(II) and Nickel(II) Chain Systems. *Chem. Eur. J.* **2001**, *7*, 4243–4252. [CrossRef] [PubMed]
24. Masoomi, M.Y.; Morsali, A. Morphological study and potential applications of nano metal–organic coordination polymers. *RSC Adv.* **2013**, *3*, 19191–19218. [CrossRef]
25. Islamoglu, T.; Goswami, S.; Li, Z.; Howarth, A.J.; Farha, O.K.; Hupp, J.T. Postsynthetic Tuning of Metal–Organic Frameworks for Targeted Applications. *Acc. Chem. Res.* **2017**, *50*, 805–813. [CrossRef] [PubMed]
26. Lu, K.; Aung, T.; Guo, N.; Weichselbaum, R.; Lin, W. Nanoscale Metal–Organic Frameworks for Therapeutic, Imaging, and Sensing Applications. *Adv. Mater.* **2018**, *30*, 1707634. [CrossRef] [PubMed]
27. Zhou, H.-C.J.; Kitagawa, S. Metal–Organic Frameworks (MOFs). *Chem. Soc. Rev.* **2014**, *43*, 5415–5418. [CrossRef]
28. Keypour, H.; Salehzadeh, S.; Pritchard, R.G.; Parish, R.V. Synthesis and Crystal Structure Determination of Some Asymmetrical and Symmetrical CR-Type Macrocyclic Schiff Base Complexes, with a Single Pendant Coordinating 2-Aminoethyl Arm. *Inorg. Chem.* **2000**, *39*, 5787–5790. [CrossRef]
29. Suydam, F.H. The C=N Stretching Frequency in Azomethines. *Anal. Chem.* **1963**, *35*, 193–195. [CrossRef]
30. Verma, A.L.; Venkateswarlu, P. Vibrational spectra and rotational isomerism of allyl amine. *J. Mol. Spectrosc.* **1971**, *39*, 227–241. [CrossRef]
31. İspir, E.; Kurtoğlu, M.; Purtaş, F.; Serin, S. Synthesis and Antimicrobial Activity of New Schiff Bases Having the –SiOR Group (R = CH_3 or CH_2CH_3), and their Transition Metal Complexes. *Transit. Met. Chem.* **2005**, *30*, 1042–1047. [CrossRef]
32. Darensbourg, D.J.; Frantz, E.B. Manganese(III) Schiff Base Complexes: Chemistry Relevant to the Copolymerization of Epoxides and Carbon Dioxide. *Inorg. Chem.* **2007**, *46*, 5967–5978. [CrossRef] [PubMed]
33. Sepehrfar, S.; Salehi, M.; Parvarinezhad, S.; Grześkiewicz, A.M.; Kubicki, M. New Cu(II), Mn(II) and Mn(III) Schiff base complexes cause noncovalent interactions: X-ray crystallography survey, Hirshfeld surface analysis and molecular simulation investigation against SARS-CoV-2. *J. Mol. Struct.* **2023**, *1278*, 134857. [CrossRef]
34. Yuan, M.; Zhao, F.; Zhang, W.; Wang, Z.-M.; Gao, S. Azide-Bridged One-Dimensional MnIII Polymers: Effects of Side Group of Schiff Base Ligands on Structure and Magnetism. *Inorg. Chem.* **2007**, *46*, 11235–11242. [CrossRef]
35. Freitag, R.; Conradie, J. Understanding the Jahn–Teller Effect in Octahedral Transition-Metal Complexes: A Molecular Orbital View of the $Mn(\beta\text{-diketonato})_3$ Complex. *J. Chem. Educ.* **2013**, *90*, 1692–1696. [CrossRef]
36. Bikas, R.; Shahmoradi, E.; Reinoso, S.; Emami, M.; Lezama, L.; Sanchiz, J.; Noshiranzadeh, N. The effect of the orientation of the Jahn–Teller distortion on the magnetic interactions of trinuclear mixed-valence Mn(II)/Mn(III) complexes. *Dalton Trans.* **2019**, *48*, 13799–13812. [CrossRef] [PubMed]
37. Yoon, J.H.; Lee, J.W.; Ryu, D.W.; Yoon, S.W.; Suh, B.J.; Kim, H.C.; Hong, C.S. One-Dimensional End-To-End Azide-Bridged MnIII Complexes Incorporating Alkali Metal Ions: Slow Magnetic Relaxations and Metamagnetism. *Chem. Eur. J.* **2011**, *17*, 3028–3034. [CrossRef] [PubMed]
38. Desiraju, G.; Steiner, T. *The Weak Hydrogen Bond in Structural Chemistry and Biology*; Oxford Univiversity Press: Oxford, UK, 2006.
39. Rigaku Oxford Diffraction. *CrysAlisPro, Version. 1.171.36.20*; Rigaku Oxford Diffraction: Yarnton, UK, 2015.
40. Altomare, A.; Cascarano, G.; Giacovazzo, C.; Guagliardi, A. Completion and refinement of crystal structures with SIR92. *J. Appl. Crystallogr.* **1993**, *26*, 343–350. [CrossRef]
41. Hirshfeld, F.L. Bonded-atom fragments for describing molecular charge densities. *Theor. Chim. Acta* **1977**, *44*, 129–138. [CrossRef]
42. Spackman, M.A.; Byrom, P.G. A novel definition of a molecule in a crystal. *Chem. Phys. Lett.* **1997**, *267*, 215–220. [CrossRef]
43. McKinnon, J.J.; Spackman, M.A.; Mitchell, A.S. Novel tools for visualizing and exploring intermolecular interactions in molecular crystals. *Acta Crystallogr. Sect. B Struct. Sci.* **2004**, *60*, 627–668. [CrossRef]
44. Spackman, M.A.; Jayatilaka, D. Hirshfeld surface analysis. *CrystEngComm* **2009**, *11*, 19–32. [CrossRef]
45. Spackman, P.R.; Turner, M.J.; McKinnon, J.J.; Wolff, S.K.; Grimwood, D.J.; Jayatilaka, D.; Spackman, M.A. CrystalExplorer: A program for Hirshfeld surface analysis, visualization and quantitative analysis of molecular crystals. *J. Appl. Crystallogr.* **2021**, *54*, 1006–1011. [CrossRef]
46. McKinnon, J.J.; Jayatilaka, D.; Spackman, M.A. Towards quantitative analysis of intermolecular interactions with Hirshfeld surfaces. *Chem. Commun.* **2007**, *37*, 3814–3816. [CrossRef] [PubMed]
47. Frisch, M.J.; Trucks, G.W.; Schlegel, H.B.; Scuseria, G.E.; Robb, M.A.; Cheeseman, J.R.; Scalmani, G.; Barone, V.; Petersson, G.A.; Nakatsuji, H.; et al. *Gaussian 16 Rev. A.03*; Gaussian, Inc.: Wallingford, CT, USA, 2016.

48. Lee, C.T.; Yang, W.T.; Parr, R.G. Development of the Colle-Salvetti Correlation-Energy Formula into a Functional of the Electron-Density. *Phys. Rev. B* **1988**, *37*, 785–789. [CrossRef]
49. Becke, A.D. Density-Functional Exchange-Energy Approximation with Correct Asymptotic-Behavior. *Phys. Rev. A* **1988**, *38*, 3098–3100. [CrossRef] [PubMed]
50. Vosko, S.H.; Wilk, L.; Nusair, M. Accurate Spin-Dependent Electron Liquid Correlation Energies for Local Spin Density Calculations: A Critical Analysis. *Can. J. Phys.* **1980**, *58*, 1200–1211. [CrossRef]
51. Krishnan, R.; Binkley, J.S.; Seeger, R.; Pople, J.A. Self-consistent molecular orbital methods. XX. A basis set for correlated wave functions. *J. Chem. Phys.* **1980**, *72*, 650–654. [CrossRef]
52. Francl, M.M.; Pietro, W.J.; Hehre, W.J.; Binkley, J.S.; Gordon, M.S.; DeFrees, D.J.; Pople, J.A. Self-consistent molecular orbital methods. XXIII. A polarization-type basis set for second-row elements. *J. Chem. Phys.* **1982**, *77*, 3654–3665. [CrossRef]
53. Schäfer, A.; Horn, H.; Ahlrichs, R. Fully optimized contracted Gaussian basis sets for atoms Li to Kr. *J. Chem. Phys.* **1992**, *97*, 2571–2577. [CrossRef]
54. Boucher, L.J.; Day, V.W. Manganese Schiff base complexes. 5. Synthesis and spectroscopy of some anion complexes of N,N′-ethylenebis(acetylacetone iminato)manganese(III). *Inorg. Chem.* **1977**, *16*, 1360–1367. [CrossRef]
55. Ferreira, G.A.; Nunes, C.M.; Lopes Jesus, A.J.; Fausto, R. The meta and para OH Substitution Effect on C-Phenyl-Nitrilimine Bond-Shift Isomers. *Eur. J. Org. Chem.* **2023**, *26*, e202300310. [CrossRef]
56. Lopes Jesus, A.J.; de Lucena Júnior, J.R.; Fausto, R.; Reva, I. Infrared Spectra and Phototransformations of meta-Fluorophenol Isolated in Argon and Nitrogen Matrices. *Molecules* **2022**, *27*, 8248. [CrossRef] [PubMed]
57. Kesharwani, M.K.; Brauer, B.; Martin, J.M.L. Frequency and Zero-Point Vibrational Energy Scale Factors for Double-Hybrid Density Functionals (and Other Selected Methods): Can Anharmonic Force Fields Be Avoided? *J. Phys. Chem. A* **2015**, *119*, 1701–1714. [CrossRef] [PubMed]
58. Filimonov, D.A.; Lagunin, A.A.; Gloriozova, T.A.; Rudik, A.V.; Druzhilovskii, D.S.; Pogodin, P.V.; Poroikov, V.V. Prediction of the Biological Activity Spectra of Organic Compounds Using the Pass Online Web Resource. *Chem. Heterocycl. Compd.* **2014**, *50*, 444–457. [CrossRef]
59. Parasuraman, S. Prediction of activity spectra for substances. *J. Pharmacol. Pharmacother.* **2011**, *2*, 52–53.

inorganics

MDPI

Article

Thionitrosyl Complexes of Rhenium and Technetium with PPh_3 and Chelating Ligands—Synthesis and Reactivity

Domenik Nowak, Adelheid Hagenbach, Till Erik Sawallisch and Ulrich Abram *

Institute of Chemistry and Biochemistry, Freie Universität Berlin, Fabeckstr. 34/36, D-14195 Berlin, Germany
* Correspondence: ulrich.abram@fu-berlin.de

Abstract: In contrast to corresponding nitrosyl compounds, thionitrosyl complexes of rhenium and technetium are rare. Synthetic access to the thionitrosyl core is possible by two main approaches: (i) the treatment of corresponding nitrido complexes with S_2Cl_2 and (ii) by reaction of halide complexes with trithiazyl chloride. The first synthetic route was applied for the synthesis of novel rhenium and technetium thionitrosyls with the metals in their oxidation states "+1" and "+2". $[M^{V}NCl_2(PPh_3)_2]$, $[M^{V}NCl(PPh_3)(L^{OMe})]$ and $[M^{VI}NCl_2(L^{OMe})]$ (M = Re, Tc; $\{L^{OMe}\}^-$ = (η^5-cyclopentadienyl)tris(dimethyl phosphito-*P*)cobaltate(III)) complexes have been used as starting materials for the synthesis of $[Re^{II}(NS)Cl_3(PPh_3)_2]$ (**1**), $[Re^{II}(NS)Cl_3(PPh_3)(OPPh_3)]$ (**2**), $[Re^{II}(NS)Cl(PPh_3)(L^{OMe})]^+$ (**4a**), $[Re^{II}(NS)Cl_2(L^{OMe})]$ (**5a**), $[Tc^{II}(NS)Cl(PPh_3)(L^{OMe})]^+$ (**4b**) and $[Tc^{II}(NS)Cl_2(L^{OMe})]$ (**5b**). The triphenylphosphine complex **1** is partially suitable as a precursor for ongoing ligand exchange reactions and has been used for the synthesis of $[Re^{I}(NS)(PPh_3)(Et_2btu)_2]$ (**3a**) (HEt_2btu = *N*,*N*-diethyl-*N′*-benzoyl thiourea) containing two chelating benzoyl thioureato ligands. The novel compounds have been isolated in crystalline form and studied by X-ray diffraction and spectroscopic methods including IR, NMR and EPR spectroscopy and (where possible) mass spectrometry. A comparison of structurally related rhenium and technetium complexes allows for conclusions about similarities and differences in stability, reaction kinetics and redox behavior between these *4d* and *5d* transition metals.

Keywords: rhenium; technetium; thionitrosyl complexes; synthesis; X-ray diffraction; EPR; NMR

1. Introduction

The two "group 7" elements technetium and rhenium possess radioactive isotopes (^{99m}Tc, $^{186,188}Re$), which make them more than interesting for nuclear medical applications [1–11]. The metastable nuclide ^{99m}Tc (pure γ-emitter, half-life 6 h, γ-energy: 411 keV) is sometimes denoted as the "workhorse" of diagnostic nuclear medicine, since more than 80 percent of clinical applications in this field in more than 10,000 hospitals worldwide are performed with ^{99m}Tc-containg pharmaceuticals [12]. This dominating position is mainly due to the favorable nuclear properties of this nuclide: (i) an almost optimal γ-energy without considerable amounts of accompanying particle radiation, (ii) a half-life which is long enough to examine metabolic processes but short enough to avoid a considerable radiation burden to the patient and (iii) the formation of ^{99m}Tc in a so-called generator system from ^{99}Mo [13], which ensures a permanent availability of this nuclide in the clinics. A similar generator system starting from ^{188}W as mother nuclide is also available for the supply of ^{188}Re ($E_{\beta,max}$: 2.12 MeV, $E_{\beta,av}$: 784 keV, half-life: 16.9 h), a beta-emitting isotope with potential in nuclear medical therapy [14]. For both isotopes, kit-like preparations of the pharmaceuticals have been developed, which allow for a rapid and cost-efficient production of the individual administrations [1,14].

^{99m}Tc and ^{188}Re are also under discussion as an almost perfect example of a "matched pair" of nuclides for theragnostic applications [15,16], which is a modern approach in personalized medicine, where an optimized dosage of a therapeutic drug becomes possible by the parallel monitoring of the therapeutic effects by a diagnostic counterpart.

Although the syntheses of ^{99m}Tc and ^{188}Re compounds proceed at an approximate nanomolar level due to the low (chemical) concentrations obtained from the corresponding nuclide generators, fundamental chemical and structural studies concerning the formed compounds are commonly conducted with natural rhenium and the long-lived technetium isotope ^{99}Tc, a weak beta emitter (E_{max} = 297 keV, half-life 2.11 $\times$ 10^5 years), which is also available in macroscopic amounts. During such studies, occasionally at the first glance, unusual solutions for the development of novel radiopharmaceuticals have been found, including several organometallic approaches [17–23] and drugs on the basis of nitrido [24–27] or hydrazido compounds [28–30]. Finally, the usability of a potential solution is decided by the successful transfer to the low concentration level of ^{99m}Tc and the stability of the product under aqueous conditions. Consequently, the potential of a chemical solution for a final medical application cannot easily be predicted by a simple, intuitive evaluation of a substance or a substance class. A good example is "^{99m}Tc-sestamibi", a cationic organotechnetium(I) complex with six isocyanide ligands. As a compound with six technetium-carbon bonds, it was most probably not in the first row for candidates for myocardial imaging. Nowadays, it is one of the most used radiopharmaceuticals worldwide and, up to now, more than 40 million patients have been examined with this drug [17]. Also, nitrosyl complexes of technetium were considered for such purposes and related ^{99m}Tc compounds have been tested for their biodistribution [31]. The related structural chemistry was performed with the long-lived isotope ^{99}Tc and starting from these early days in the 1980s, our knowledge about nitrosyltechnetium compounds has significantly increased.

In contrast to the relatively large number of nitrosyl complexes of the elements rhenium [32–40] and technetium [40–60], there are only a few reports about thionitrosyls of these two elements [61–80]. This should be mainly due to the lack of a readily available monomeric nitrogen sulfide. Although the thionitrosyl radical has been described spectroscopically and some NS^+ salts have been isolated [81–84], the synthesis of such compounds remains a challenge and they have only occasionally been used as precursors for the synthesis of thionitrosyl complexes [61,83,84]. More convenient are reactions starting from cyclic compounds such as tetrasulfur tetranitride, S_4N_4, or trithiazyltrichloride, $(NSCl)_3$ (Figure 1), which readily undergo thermal or photochemical decomposition under release of NS^+ or $NSCl^+$ building blocks. But it should be noted that N_4S_4 is a potentially dangerous compound, which tends to explode on heating or upon shock. Thus, it is not a favored agent at least for reactions with the radioactive technetium. Finally, the most promising synthetic approach to thionitrosyl complexes is the addition of a "sulfur atom" to a nitrido ligand. A variety of "sulfur sources" have been used including elemental sulfur, $SOCl_2$, dithionite or NCS^- ligands. Most successful, however, are reactions of nitrido complexes with S_2Cl_2, which give low-valent thionitrosyl species in good yields. About reactions following the latter approach, we report in the present paper starting from different rhenium and technetium complexes with the metals in the oxidation states "+5" or "+6".

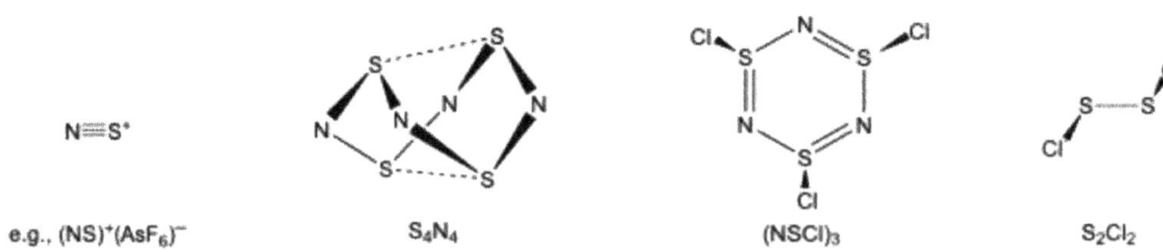

Figure 1. Common starting materials for the synthesis of thionitrosyl complexes.

2. Results and Discussion

2.1. Triphenylphosphine Complexes

Treatment of the rhenium(V) complex [ReNCl$_2$(PPh$_3$)$_2$] with an excess of disulfur dichloride in refluxing dichloromethane results in a subsequent dissolution of the sparingly soluble starting material and a dark-red, almost black solution is formed, from which the thionitrosyl complexes [Re(NS)Cl$_3$(PPh$_3$)$_2$] (**1**) and [Re(NS)Cl$_3$(PPh$_3$)(OPPh$_3$)] (**2**) can be isolated. The parallel formation of phosphine and phosphine oxide complexes is not completely surprising with respect to the oxidizing conditions in the reaction mixture and has been observed previously during similar reactions with [TcNCl$_2$(PMe$_2$Ph)$_3$] [74]. The two products can readily be separated during the crystallization procedure since compound **2** is highly soluble in acetone, while complex **1** is almost insoluble in this solvent. During the slow evaporation of CH$_2$Cl$_2$/acetone solutions of mixtures of **1** and **2**, [Re(NS)Cl$_3$(PPh$_3$)$_2$] precipitates first as dark-red crystals, while the orange-yellow crystals of **2** do not deposit until almost all acetone is evaporated. The best method for the isolation of pure [Re(NS)Cl$_3$(PPh$_3$)$_2$], however, is the addition of an excess of PPh$_3$ during the synthesis, which obviously avoids the formation of significant amounts of the phosphine oxide complex (Scheme 1).

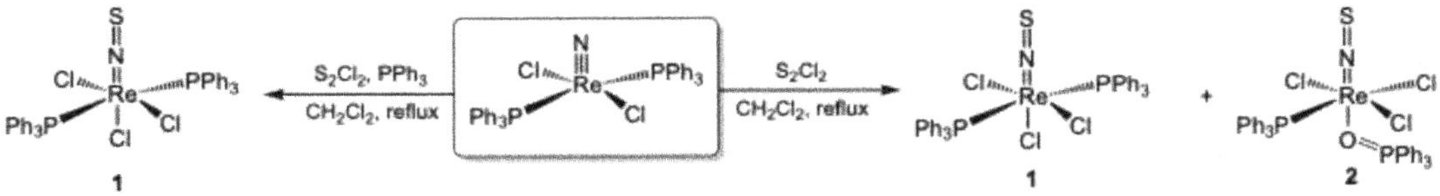

Scheme 1. Syntheses of [Re(NS)Cl$_3$(PPh$_3$)$_2$] (**1**) and [Re(NS)Cl$_3$(PPh$_3$)(OPPh$_3$)] (**2**).

The infrared spectra of the complexes show the characteristic $\nu_{(NS)}$ stretches at 1213 cm^{-1} (phosphine complex) and 1230 cm^{-1} (phosphine oxide complex). These values are in agreement with such recorded for other rhenium complexes and those calculated for similar chromium(I) species [62]. The difference of approximately 20 cm^{-1} reflects a stronger back-donation in compound **1** compared with **2**. Since the oxidation state of rhenium in both compounds is "+2", the observed difference indicates some changes in the coordination sphere of the transition metal. Particularly the electronic properties of the ligand in the *trans* position to the thionitrosyl ligand have a direct influence to the donor/acceptor behavior of NS$^+$.

Single-crystal X-ray diffraction confirms that the oxidation of one of the PPh$_3$ ligands is accompanied with a ligand rearrangement and the OPPh$_3$ ligand in **2** occupies the position *trans* to NS$^+$ instead of Cl$^-$ in **1**. The molecular structures of both complexes are depicted in Figure 2 and selected bond lengths and angles are summarized in Table 1. It becomes evident that the thionitrosyl units are arranged linearly in both compounds with Re–N–S angles between 167° and 178°. Such a bonding situation has been found for all hitherto structurally characterized thionitrosyl complexes. Also, the direction of a formed phosphine oxide ligand in the *trans* position to a multiply bonded ligand is not without precedence and has been observed before during the formation of [Tc(NS)Cl$_3$(PMe$_2$Ph)(OPMe$_2$Ph) from *cis,mer*-[TcNCl$_2$(PMe$_2$Ph)$_3$] [74]. More generally, the presence of one triphenylphosphine oxide ligand in the coordination sphere of a transition metal favors the *cis* coordination to PPh$_3$, since the steric and electronic effects, which direct two PPh$_3$ ligands commonly into *trans* positions to each other, are significantly lowered.

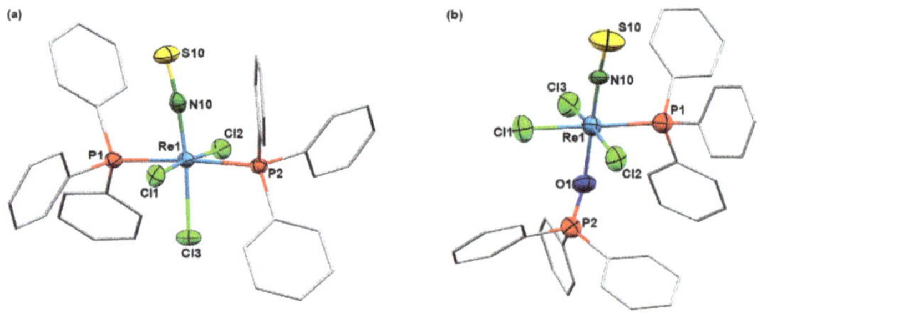

Figure 2. Molecular structures of (**a**) $[Re(NS)Cl_3(PPh_3)_2]$ (**1**) and (**b**) $[Re(NS)Cl_3(PPh_3)(OPPh_3)]$ (**2**). Thermal ellipsoids represent 50 percent probability.

Table 1. Selected bond lengths (Å) and angles (°) in $[Re(NS)Cl_3(PPh_3)_2]$ (**1**) and (b) $[Re(NS)Cl_3(PPh_3)(OPPh_3)]$ (**2**).

	Re1–N10	N10–S10	Re1–Cl1	Re1–Cl2	Re1–Cl3	Re1–P1	Re1–P2/O1	O1–P2	Re1–N10–S10	Re1–O1–P2
1	1.795(5)	1.518(5)	2.353(1)	2.340(1)	2.416(1)	2.540(2)	2.555(2)	-	174.4(3)	-
2 [(1)]	1.77(1) 1.77(1)	1.53(1) 1.52(1)	2.368(4) 2.357(4)	2.255(3) 2.332(3)	2.338(3) 2.366(3)	2.506(4) 2.584(4)	2.100(8) 2.099(8)	1.509(9) 1.497(9)	167.2(7) 178.4(6)	155.7(6) 164.5(5)

[(1)] Values for two crystallographically independent molecules.

$[Re(NS)Cl_3(PPh_3)_2]$ (**1**) as well as $[Re(NS)Cl_3(PPh_3)(OPPh_3)]$ (**2**) contain rhenium in the oxidation state "+2". Due to the corresponding $5d^5$ "low spin" configuration, they are paramagnetic with one unpaired electron. This allows for the detection of resolved EPR spectra in solution. Figure 3 shows the spectra of both compounds at various temperatures. The spectra in liquid solutions (Figure 3a) consist of six hyperfine lines due to interactions of the unpaired electron with $^{185,187}Re$, both having a nuclear spin of I = 5/2. Superhyperfine couplings due to interactions of the unpaired electron with the coordinating ^{31}P (I = 1/2), $^{35,37}Cl$ (I = 3/2) or ^{14}N (I = 1) nuclei could not be resolved. The frozen solution EPR spectra, which are shown in Figure 3b, confirm the presence of essentially axially symmetric, randomly oriented S = $\frac{1}{2}$ spin systems with resolved parallel and perpendicular sets of $^{185,187}Re$ hyperfine lines. The analysis of the spectral parameters indicates the presence of a small rhombic component in the perpendicular part of the spectrum of complex **1**, which has been taken into account during the simulation. The simulated spectra, which have been used to derive the spectral parameters, are depicted in the Supporting Material.

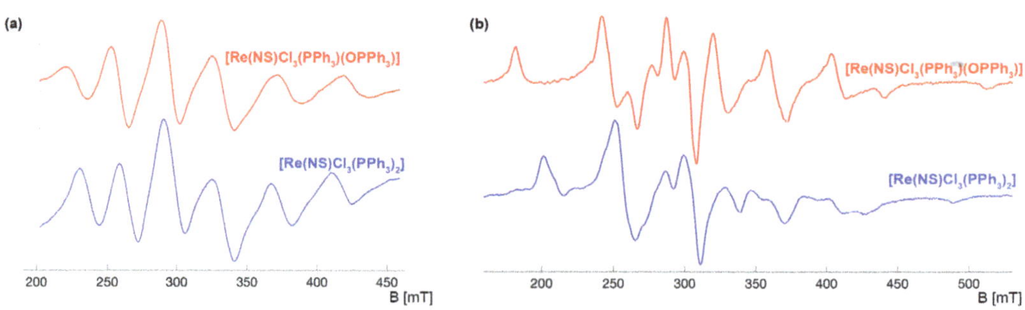

Figure 3. X-band EPR spectra of **1** and **2**: (**a**) at room temperature (parameters for **1** in CH_2Cl_2: g_0 = 2.025; a_0^{Re} = 342 × 10^{-4} cm^{-1}; parameters for **2** in acetone: g_0 = 2.017; a_0^{Re} = 372 × 10^{-4} cm^{-1}) and (**b**) in frozen solution at T = 77 K (parameters for **1** in CH_2Cl_2: g_x = 2.110, g_y = 1.991, g_z = 1.943; A_x^{Re} = 281 × 10^{-4} cm^{-1}, A_y^{Re} = 283 × 10^{-4} cm^{-1}, A_z^{Re} = 524 × 10^{-4} cm^{-1}; parameters for **2** in acetone: $g_{||}$ = 1.923, $g_{\perp}$ = 2.008; $A_{||}^{Re}$ = 594 × 10^{-4} cm^{-1}, $A_{\perp}^{Re}$ = 293 × 10^{-4} cm^{-1}).

As already mentioned, couplings with the ^{31}P nuclei of the coordinated phosphine ligands could not be resolved in the spectra of **1** and **2**. This finding is in contrast to the previously studied technetium(II) complexes [Tc(NS)Cl_3(PMe_2Ph)$_2$] and [Tc(NS)Cl_3(PMe_2Ph)($OPMe_2Ph$)], where the ^{99}Tc hyperfine signals of the parallel part show splittings into triplets (phosphine complex) or doublets (phosphine/phosphine oxide complex) and provide direct information for the occupation of their respective equatorial coordination spheres [74,77,85,86]. The larger line-widths of the EPR spectra obtained for the rhenium complexes prevent the resolutions of such couplings. Nevertheless, a remarkable trend in the EPR spectral parameters should be mentioned: the isotropic $^{185,187}Re$ coupling constants a_0^{Re} show a dependence on the number of chlorido ligands in the equatorial coordination sphere in a way that they increase from compound **1** (2 Cl^-: 342 × 10^{-4} cm^{-1}) via compound **2** (3 Cl^-: 372 × 10^{-4} cm^{-1}) to the [Re(NS)Cl_4]$^-$ anion (4 Cl^-: 457 × 10^{-4} cm^{-1} [66]). In the same sequence of complexes, the corresponding g_0 values decrease. Similar trends are also observed for the g values and coupling constants of the anisotropic spectra, which reflects considerable changes in the MO of the unpaired electron mainly due to the presence or absence of phosphine ligands.

2.2. Complexes with Chelating Ligands

Since the triphenylphosphine complexes **1** and **2** might have potential as precursors for the synthesis of more thionitrosyl complexes of rhenium, we tested some ligand exchange reactions with the chelating ligands shown in Figure 4. HEt_2btu is a potentially bidentate chelator. It readily deprotonates after the addition of a supporting base and stable complexes are known with almost all common transition metals. This also includes rhenium and technetium complexes with the metals in several oxidation states [87–94]. The tripodal, organometallic ligand {L^{OMe}}$^-$ belongs to a classical family of tripodal ligands, which has been designed by Wolfgang Kläui and tested for a plethora of various applications [95,96]. Also, a number of rhenium complexes with this ligand are known and it is the first example of a chelating ligand, which forms stable technetium complexes with the metal in seven different oxidation states [54,95–101].

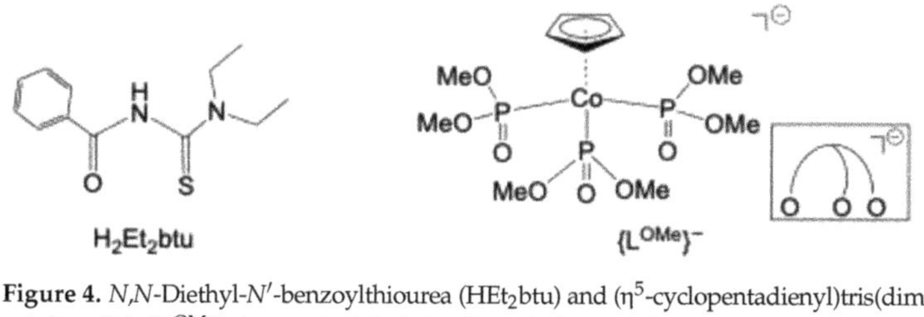

Figure 4. *N,N*-Diethyl-*N'*-benzoylthiourea (HEt_2btu) and (η^5-cyclopentadienyl)tris(dimethyl phosphito-P) cobaltate(III) ({L^{OMe}}$^-$) as typical chelating ligands for ligand exchange reactions.

Reactions of [Re(NS)Cl_3(PPh_3)$_2$] with HEt_2btu in a mixture of CH_2Cl_2 and EtOH proceeded at room temperature (Scheme 2). After the addition of a small amount of Et_3N as a supporting base, the red reaction mixture immediately turned dark brown and a red-brown solid could be isolated. Single crystals suitable for X-ray diffraction were obtained by the slow evaporation of a CH_2Cl_2 solution. The deprotonation of HEt_2btu and its coordination as an *S,O* chelate is strongly indicated by the IR spectrum of the product by the disappearance of the NH stretches of the ligands and a strong bathochromic shift of the $\nu_{C=O}$ band by almost 200 cm^{-1}. This is typical for chelate-bonded R_2btu^- ligands [87]. Reduction of the metal ion is frequently observed during reactions with sulfur-containing ligands and, thus, the product of the described reaction with HEt_2btu is the diamagnetic rhenium(I) complex [Re(NS)(PPh_3)(Et_2btu)$_2$] (**3a**). This allows for the measurement of NMR spectra. They support the composition of the product as is performed by the mass spectrum.

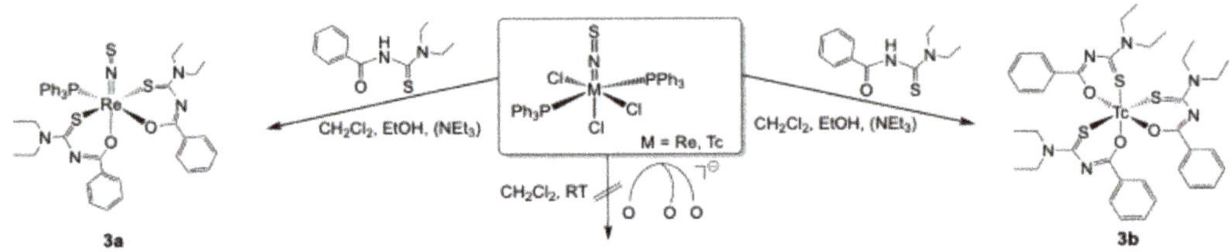

3a **3b**

Scheme 2. Ligand exchange reactions starting from $[Re(NS)Cl_3(PPh_3)_2]$ (**1**).

The results of the spectroscopic studies are confirmed by a single-crystal structure analysis of the product. Figure 5 depicts the molecular structure of compound **3a**, which shows the expected linear arrangement of the thionitrosyl ligand (Table 2). The equatorial coordination sphere of rhenium is occupied by an *S,O* chelate, a PPh_3 and the sulfur atom of the second Et_2btu^- ligand. Interestingly, the two Re–O single bonds are almost equal, which indicates similar structural *trans* influences induced by the NS^+ and the PPh_3 ligands. The N–S distance in the rhenium(I) complex is markedly longer than in the rhenium(II) complexes **1** and **2**, which might be a consequence of a stronger back-donation of the d^6 metal ion into anti-bonding orbitals of the thionitrosyl ligand. Unfortunately, the corresponding IR stretch, which is normally more indicative, cannot be assigned unambiguously for **3a** due to many overlapping bands in the respective range of the spectrum.

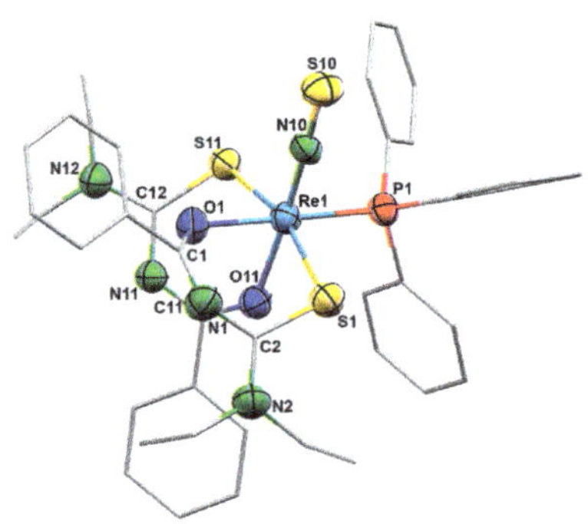

Figure 5. Molecular structure of $[Re(NS)(PPh_3)(Et_2btu)_2]$ (**3a**). Thermal ellipsoids represent 50 percent probability.

Table 2. Selected bond lengths (Å) and angles (°) $[Re(NS)(PPh_3)(Et_2btu)_2]$ (**3a**).

Re1–N10	N10–S10	Re1–P1	Re1–S1	Re1–O1	Re1–S11	Re1–O11	O1–C1	C1–N1	N1–C2
1.749(4)	1.571(4)	2.362(1)	2.401(1)	2.101(3)	2.439(1)	2.103(4)	1.296(6)	1.301(7)	1.353(7)
C2–N2	C2–S1	O11–C11	C11–N11	N11–C12	C12–N12	C12–S11		Re1–N10–S10	
1.319(7)	1.755(6)	1.266(6)	1.329(6)	1.338(7)	1.325(7)	1.758(6)		176.5(3)	

With respect to the clean reaction of $[Re(NS)Cl_3(PPh_3)_2]$ with HEt_2btu and the ready formation of the thionitrosylrhenium(I) complex **3a**, a surprising result was obtained for an analogous reaction of the corresponding technetium complex $[Tc(NS)Cl_3(PPh_3)_2]$. It proceeded under complete removal of the NS^+ ligand and oxidation of the metal ion. The resulting technetium(III) tris complex $[Tc(Et_2btu)_3]$ (Scheme 2) has been reported before (i) as the product of the direct reduction of pertechnetate with $SnCl_2$ in the presence of excess ligand [102], and (ii) by a subsequent reduction/ligand exchange procedure starting from $(NBu_4)[TcOCl_4]$ [87]. The observed lability of the thionitrosyl unit is a considerable drawback for a potential use of **1** as precursor in ligand exchange procedures and is accompanied by another disappointing experience during attempted reactions with Na{L^{OMe}}. Although reactions of an entire series of different technetium and rhenium complexes with the "Kläui ligand" give well-defined and stable products [95–101], attempted reactions with **1** at room temperature did not proceed. Heating of such reaction mixtures finally gave a small amount of the chelate [Re(NS)Cl(PPh_3)(L^{OMe})], but the yield was low and the product was accompanied by a number of side-products and impurities, which precluded the isolation of a pure compound in reasonable yields. Consequently, we preferred the second general approach to thionitrosyl compounds: the reaction of nitrido complexes with S_2Cl_2. Corresponding starting materials with {L^{OMe}}$^-$ ligands, [MNCl(PPh_3)(L^{OMe})] and [$MNCl_2$(L^{OMe})] complexes with M = Re or Tc, are readily available from simple procedures, and the reactions with S_2Cl_2 (Scheme 3) give the products [M(NS)Cl(PPh_3)(L^{OMe})]Cl (M = Re: **4a**; M = Tc: **4b**) and [M(NS)Cl_2(L^{OMe})] (M = Re: **5a**; M = Tc: **5b**) in reasonable yields and good purities.

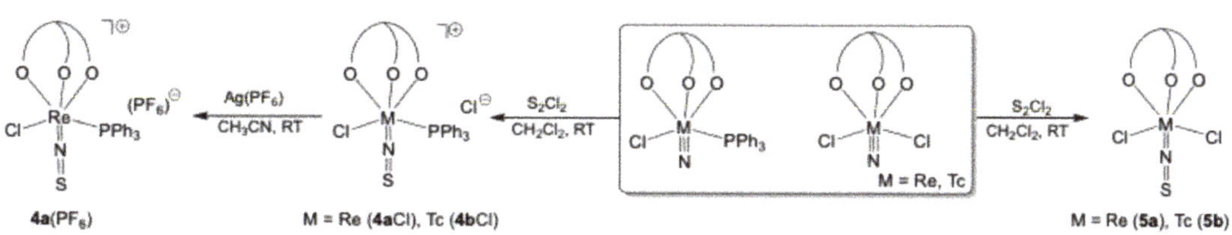

Scheme 3. Formation of thionitrosyl complexes containing the "Kläui ligand" {L^{OMe}}$^-$.

The reactions with all four starting complexes are restricted to the nitrido ligands and the remaining coordination spheres of the metals remain unchanged. Cationic thionitrosyls of rhenium(II) and technetium(II) are formed, when [M^VNCl(PPh_3)(L^{OMe})] complexes (M = Re, Tc) are used as precursors. The products can be precipitated from the reaction mixture as yellow (Re compound, **4a**Cl) and red (Tc complex, **4b**Cl) chloride salts by the addition of *n*-hexane. Microcrystalline samples were obtained after recrystallization of CH_2Cl_2/*n*-hexane mixtures. For the rhenium complex, the PF_6^- salt was prepared by metathesis with $AgPF_6$. Orange-yellow needles of [Re(NS)Cl(PPh_3)(L^{OMe})](PF_6), **4a**(PF_6), were suitable for X-ray diffraction. The structure of the complex cation is shown in Figure 6a and selected bond lengths and angles are summarized in Table 3. Expectedly, the {L^{OMe}}$^-$ ligands act as tripods in both complexes, which directs the remaining three ligands into *trans* positions of their oxygen atoms. The M–N–S angles are close to 180° in both compounds.

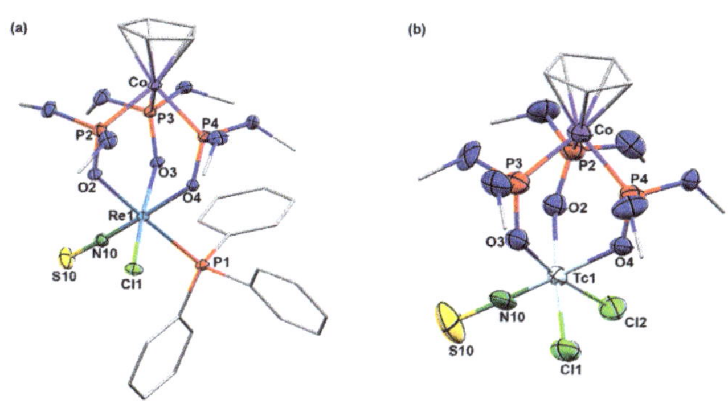

Figure 6. Molecular structures of (**a**) the complex cation of [Re(NS)Cl(PPh_3)(L^{OMe})](PF_6) (**4a**) and (**b**) [Tc(NS)Cl_2(L^{OMe})] (**5b**). Thermal ellipsoids represent 50 percent probability.

Table 3. Selected bond lengths (Å) and angles (°) in $[Re(NS)Cl(PPh_3)(L^{OMe})](PF_6)$ (**4a**(PF_6)) and (b) $[Tc(NS)Cl_2(L^{OMe})]$ (**5b**).

	M1–N10	N10–S10	M1–Cl1	M1–Cl2	M1–P1	M1–O2	M1–O3	M1–O4	M1–N10–S10
4a	1.752(5)	1.554(5)	2.337(1)	-	2.475(1)	2.071(3)	2.032(3)	2.083(3)	175.4(3)
5b [(1)]	1.732(7) 1.75(1)	1.540(7) 1.54(1)	2.339(3) 2.330(3)	2.343(2) 2.344(3)	-	2.060(6) 2.068(5)	2.047(6) 2.057(6)	2.095(5) 2.088(5)	171.7(6) 172.0(8)

[(1)] Values for two crystallographically independent molecules.

Similar reaction patterns are observed for $[ReNCl_2((L^{OMe})]$ and $[TcNCl_2((L^{OMe})]$. Both precursors contain the transition metals in their oxidation states "+6", while the products $[Re(NS)Cl_2(L^{OMe})]$ (**5a**) and $[Tc(NS)Cl_2(L^{OMe})]$ (**5b**) are Re(II) and Tc(II) complexes (Scheme 3). An X-ray crystal structure determination has been performed for the technetium complex and its molecular structure is shown in Figure 6b. The general structural features found for the cation **4a** also apply to technetium complex **5b** (Table 3).

As compounds **1** and **2**, the $\{L^{OMe}\}^-$ complexes of the present study, are paramagnetic with a d^5 "low spin" configuration, this results in $S = \frac{1}{2}$ spin systems and allows us to record resolved EPR spectra in liquid and frozen solutions. The spectra have essentially axial symmetry with well-resolved $^{185,187}Re$ and ^{99}Tc hyperfine couplings in their parallel and perpendicular parts. Figure 7 shows exemplarily the spectra obtained for $[Tc(NS)Cl(PPh_3)(L^{OMe})]Cl$ (**4b**Cl).

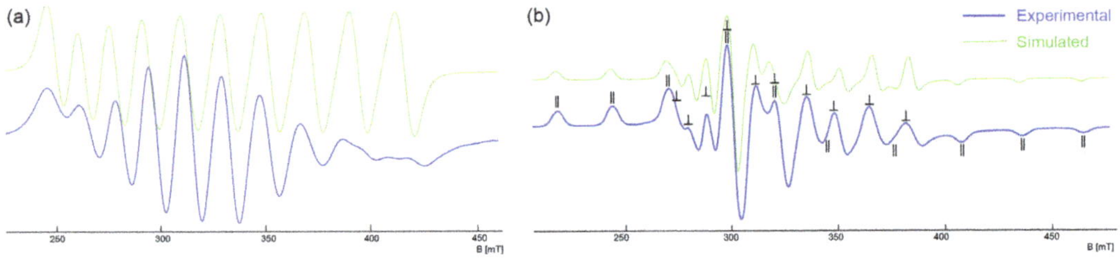

Figure 7. Liquid solution EPR spectra of $[Tc(NS)Cl(PPh_3)(L^{OMe})]$ in CH_2Cl_2 (**a**) at room temperature and (**b**) at T = 77 K (assignment of the ^{99}Tc lines to the parallel and perpendicular parts of the spectrum is indicated).

^{99}Tc has a nuclear spin of I = 9/2, which results in a ten-line pattern in the spectrum in liquid solution, while two sets of ten ^{99}Tc hyperfine lines (each one in the parallel and one in the perpendicular part of the spectrum) are characteristic for the anisotropic frozen-solution spectrum as can be described by the Spin Hamiltonian (1), where $g_{\|}$, $g_{\perp}$, $A_{\|}^{M.}$ and $A_{\perp}^{M}$ are the principal values of the ^{99}Tc and $^{185,187}Re$ hyperfine tensors A^M. The spectral parameters are summarized in Table 4.

$$\hat{H}_{sp} = \beta_e\left[g_{\|}B_z\hat{S}_z + g_{\perp}\left(B_x\hat{S}_x + B_y\hat{S}_y\right)\right] + A_{\|}^{M}\hat{S}_z\hat{I}_z + A_{\perp}^{M}\left(\hat{S}_x\hat{I}_x + \hat{S}_y\hat{I}_y\right) \quad (1)$$

Table 4. EPR parameters of the Re(II) and Tc(II) complexes with $\{L^{OMe}\}^-$ ligands, coupling constants are given in 10^{-4} cm^{-1}.

	g_0	a_0^M	$g_{\|}$	$g_{\perp}$	$A_{\|}^M$	$A_{\perp}^M$
$[Re(NS)Cl(PPh_3)(L^{OMe})]Cl$ (**4a**Cl)	1.995	420	1.842	1.986	644	330
$[Tc(NS)Cl(PPh_3)(L^{OMe})]Cl$ (**4b**Cl)	2.007	168	1.966	2.023	252	112
$[Re(NS)Cl_2(L^{OMe})]$ (**5a**)	1.990	455	1.945	1.760	787	398
$[Tc(NS)Cl_2(L^{OMe})]$ (**5b**)	1.996	177	1.943	2.021	285	129

Superhyperfine interactions due to couplings of the unpaired electron with the ^{31}P nuclei of the PPh_3 ligands in compounds **4** are not visible. This prevents us from obtaining direct information about the extent of spin delocalization into the ligand orbitals. Indirect information, however, can be deduced from the $^{185,187}Re$ coupling parameters. As described for compounds **1** and **2**, also for the $\{L^{OMe}\}^-$ thionitrosyl complexes, the isotropic and anisotropic A^{Tc} and A^{Re} coupling parameters decrease, when phosphine ligands are present in the equatorial coordination spheres of the metals instead of Cl^-. This can be understood by the transfer of electron density to the π-acceptor PPh_3. The experimental line-widths of the spectra, however, do not allow for the resolution of such couplings for the complexes of the present study.

3. Materials and Methods

$[ReNCl_2(PPh_3)_2]$ [103], $[TcNCl_2(PPh_3)_2]$ [104], $(NBu_4)[TcNCl_4]$ [105], $(NBu_4)[ReNCl_4]$ [106], $[ReNCl(PPh_3)(L^{OMe})]$ [98], $[TcNCl(PPh_3)(L^{OMe})]$ [58], $[ReNCl_2(L^{OMe})]$ [107], $[TcNCl_2(L^{OMe})]$ [58], $[Tc(NS)Cl_3(PPh_3)_2]$ [108] and HEt_2btu [109] were prepared according to literature procedures. All other chemicals were reagent-grade and purchased commercially. Reactions with air- or moisture-sensitive compounds were performed with the standard Schlenk technique. A laboratory approved for the handling of radioactive material was used for the synthesis of the technetium compounds. Normal glassware gives adequate protection against the weak beta radiation of ^{99}Tc as long as small amounts of this isotope as in the present paper are used.

3.1. Syntheses

*$[Re(NS)Cl_3(PPh_3)_2]$ (**1**)*: $[ReNCl_2(PPh_3)_2]$ (2 g, 2.6 mmol) was suspended in 40 mL CH_2Cl_2 and an excess of S_2Cl_2 (0.5 mL, 6.26 mmol) and PPh_3 (680 mg, 5.2 mmol) was added. The mixture was heated under reflux for approximately 30 min, which resulted in a slow dissolution of the starting material and the formation of a dark-red solution. The progress of the reaction can be controlled by TLC. After a complete consumption of $[ReNCl_2(PPh_3)_2]$, the solvent and the remaining S_2Cl_2 were removed in vacuum. The resulting solid was crystallized from a CH_2Cl_2/acetone (1:1) mixture giving dark-red crystals. Yield: 1.49 g (66%). Elemental analysis: Calcd. For $C_{36}H_{30}Cl_3NP_2ReS$: C, 50.1; H, 3.5; N, 1.6; S, 3.7%. Found: C 49.8; H, 3.5; N, 1.6; S, 4.0%. IR (ATR, cm^{-1}): 3057 (w), 1981 (w), 1670 (w), 1586 (w), 1574 (w), 1482 (s), 1434 (s), 1316 (m), 1213 (s), 1187 (s), 1159 (m), 1089 (s), 1074 (m), 1028 (m), 998 (m), 928 (w), 855 (w), 758 (m), 744 (s), 711 (m), 692 (s), 617 (w). EPR (77 K, CH_2Cl_2): $g_x = 2.110$, $g_y = 1.991$, $g_z = 1.943$; $A_x^{Re} = 281 \times 10^{-4}$ cm^{-1}, $A_y^{Re} = 283 \times 10^{-4}$ cm^{-1}, $A_z^{Re} = 524 \times 10^{-4}$ cm^{-1}. EPR (RT, CH_2Cl_2): $g_0 = 2.025$; $a_0^{Re} = 342 \times 10^{-4}$ cm^{-1}. ESI^+ MS (*m/z*): 885.002 $[M+Na]^+$ (calcd. 885.009), 900.975 $[M+K]^+$ (calcd. 900.983).

*$[Re(NS)Cl_3(PPh_3)(OPPh_3)]$ (**2**)*: The product was obtained as a side product of the synthesis of $[Re(NS)Cl_3(PPh_3)_2]$ when no additional PPh_3 was added. It could be isolated as an orange-red solid following compound **1** during the crystallization procedure. Yield: 0.6 g (26%). IR (ATR, cm^{-1}): 3056 (w), 1589 (w), 1482 (m), 1435 (s), 1231 (s), 1188 (w), 1140 (vs), 1119 (vs), 1026 (m), 998 (m), 924 (w), 852 (w), 747 (s), 723 (vs), 690 (vs). EPR (77 K, acetone): $g_{\parallel} = 1.923$, $g_{\perp} = 2.008$; $A_{\parallel}^{Re} = 594 \times 10^{-4}$ cm^{-1}, $A_{\perp}^{Re} = 293 \times 10^{-4}$ cm^{-1}. EPR (RT, acetone): $g_0 = 2.017$; $a_0^{Re} = 372 \times 10^{-4}$ cm^{-1}.

*$[Re(NS)(PPh_3)(Et_2btu)_2]$ (**3a**)*: $[Re(NS)Cl_3(PPh_3)_2]$ (100 mg, 0.1 mmol) was dissolved in 30 mL CH_2Cl_2 and a solution of HEt_2btu (60 mg, 0.25 mmol) in 20 mL Ethanol containing 3 drops of triethylamine was added. The mixture was stirred at room temperature for 30 min. During this time, the color changed to pale brown. The volatiles were removed in vacuum and the remaining solid was subsequently washed with cold methanol, diethyl ether and *n*-hexane. The raw product was dissolved in CH_2Cl_2 for crystallization. Slow evaporation of the solvent gave red-brown crystals. Yield: 31 mg (25%). Elemental analysis: Calcd. for $C_{42}H_{45}N_5O_2PReS_3$: C, 52.3; H, 4.7; N, 7.3; S, 10.0%. Found: C 52.0; H, 4.9; N, 6.8; S, 10.2%. IR (ATR, cm^{-1}): 3058 (w), 2972 (w), 2927 (w), 1599 (w), 1587 (w), 1528 (m), 1499 (s), 1446 (m), 1416 (s), 1394 (s), 1353 (s),1311 (m), 1294 (m),

1251 (s),1210 (w), 1140 (m), 1092 (m), 1071 (m), 1025 (m), 997 (m), 895 (m), 883 (m), 827 (m), 797 (m), 752 (w), 742 (m), 702 (s), 692 (s), 671 (m), 661 (m), 619 (w). ^{1}H NMR (400 MHz, $CDCl_3$, ppm): 8.10 (d, J = 7.6 Hz, 2H, Ph), 7.65 (d, J = 7.3 Hz, 6H, Ph), 7.37–7.13 (m, 15H), 7.01 (t, J = 7.9 Hz, 2H, Ph), 4.37 (dq, J = 14.2, 7.1 Hz, 1H, CH_2), 4.14–3.80 (m, 4H, CH_2), 3.73 (q, J = 7.3, 6.8 Hz, 1H, CH_2), 3.62 (q, J = 6.3 Hz, 1H, CH_2), 3.51 (dd, J = 12.9, 6.5 Hz, 1H, CH_2), 1.52 (s, 2H), 1.39 (dd, J = 9.6, 4.2 Hz, 3H, CH_3), 1.27 (q, J = 12.3, 9.7 Hz, 3H, CH_3), 1.03 (t, J = 6.5 Hz, 3H, CH_3), 0.90 (d, J = 6.3 Hz, 3H, CH_3). J = 8.0 Hz, 12H, CH_3). ^{31}P NMR (CDCl3, ppm): 21.67 (s). ESI$^+$ MS (*m/z*): = 965.237 [M]$^+$ (calcd. 965.203) 988.225 [M+Na]$^+$ (calcd. 988.192), 1004.202 [M+K]$^+$ (calcd. 1004.166).

[Re(NS)Cl(PPh_3)(L^{OMe})]Cl (**4a**Cl): [ReNCl(PPh_3)(L^{OMe})] (250 mg, 0.25 mmol) was suspended in 30 mL CH_2Cl_2, an excess of S_2Cl_2 (0.1 mL) was added and the mixture was heated under reflux for 30 min. The solvent was reduced to 15 mL and *n*-hexane (150 mL) was added. The resulting precipitate was washed with hexane and diethyl ether and dried. Recrystallization from a CH_2Cl_2/*n*-hexane (2:1) mixture gave a pale-yellow solid. Yield: 211 mg (83%). Elemental analysis: Calcd. for $C_{29}H_{38}Cl_2CoNO_9P_4ReS$: C, 34.3; H, 3.8; N, 1.4; S, 3.2%. Found: C 34.0; H, 4.0; N, 1.4; S, 3.1%. IR (ATR, cm^{-1}): 3056 (w), 2992 (w), 2948 (w), 1634 (w), 1483 (w), 1457 (w), 1436 (m), 1223 (m), 1177 (m), 1036 (vs), 1014 (vs), 943 (s), 912 (s), 853 (s), 793 (s), 746 (vs), 693 (s), 599 (vs). EPR (77 K, CH_2Cl_2): $g_{\|}$ = 1.842, $g_{\perp}$ = 1.986; $A_{\|}^{Re}$ = 644 × 10^{-4} cm^{-1}, $A_{\perp}^{Re}$ = 330 × 10^{-4} cm^{-1}. EPR (RT, $CHCl_3$): g_0 = 1.995; a_0^{Re} = 420 × 10^{-4} cm^{-1}. ESI$^+$ MS (*m/z*): 981.013 [M]$^+$ (calcd. 980.979).

*[Re(NS)Cl(PPh_3)(L^{OMe})](PF_6) (**4a**(PF_6))*: [Re(NS)Cl(PPh_3)(L^{OMe})]Cl (50 mg, 0.05 mmol) was dissolved in acetonitrile and an excess of $AgPF_6$ (19 mg, 0.075 mmol) was added. The solution was stirred at room temperature and the formed colorless precipitate was removed by filtration. Orange-yellow needles suitable for X-ray diffraction formed upon the slow evaporation of a solution of the complex in CH_2Cl_2/*n*-hexane. Yield: 52 mg (93%). The essential spectroscopic data are identical with those of **4a**Cl.

*[Re(NS)Cl_2(L^{OMe})] (**5a**)*: [ReNCl_2(L^{OMe})] (34 mg, 0.05 mmol) was dissolved in 5 mL CH_2Cl_2 and 3 drops of S_2Cl_2 were added. The mixture was stirred at room temperature. The progress of the reaction can readily be monitored by EPR spectroscopy and typically after 15 min the signals of the starting material have disappeared. The solvent and residual S_2Cl_2 were removed in vacuum and the solid residue was washed subsequently with *n*-hexane and diethyl ether. The product was recrystallized from CH_2Cl_2/diethyl ether giving a pale-brown, microcrystalline solid. Yield: 20 mg (53%). IR (ATR, cm^{-1}): 3111 (w), 2958 (m), 2874 (w), 1460 (w), 1425 (w), 1381 (w), 1260 (w), 1222 (s), 1175 (s), 1012 (vs), 945 (w), 860 (m), 796 (vs), 743 (s), 595 (m). EPR (77 K, CH_2Cl_2): $g_{\|}$ = 1.945, $g_{\perp}$ = 1.760; $A_{\|}^{Re}$ = 787 × 10^{-4} cm^{-1}, $A_{\perp}^{Re}$ = 398 × 10^{-4} cm^{-1}. EPR (RT, $CHCl_3$): g_0 = 1.990; a_0^{Re} = 455 × 10^{-4} cm^{-1}. ESI$^+$ MS (*m/z*): 776.846 [M+Na]$^+$ (ber.: 776.847); 792.821 [M+K]$^+$ (ber.: 792.821).

*[Tc(NS)Cl(PPh_3)(L^{OMe})]Cl (**4b**Cl)*: [TcNCl(PPh_3)(L^{OMe})] (72 mg, 0.08 mmol) was dissolved in 2 mL CH_2Cl_2 (2 mL) and S_2Cl_2 (11 mg, 0.08 mmol) was added under permanent stirring. The stirring was continued at room temperature for 1 h and the obtained dark-red mixture was filtered to remove some insoluble solid (most probably elemental sulfur). All volatiles were removed in vacuum and the remaining solid crystallized from a CH_2Cl_2/*n*-pentane (1:1) mixture. Storing such a mixture in a refrigerator for several days gave a red, microcrystalline powder. Yield: 11 mg (15%). IR (KBr, cm^{-1}): 2910 (vs), 2723 (s), 1726 (m), 1630 (w) 1479 (vs), 1381 (vs), 1227 (vs), 1171 (m), 1040 (w), 949 (s), 928 (s), 851 (w), 802 (w), 723 (m), 610 (w), 542 (m). EPR (77 K, CH_2Cl_2): $g_{\|}$ = 1.966, $g_{\perp}$ = 2.023; $A_{\|}^{Tc}$ = 252 × 10^{-4} cm^{-1}, $A_{\perp}^{Tc}$ = 112 × 10^{-4} cm^{-1}. EPR (RT, CH_2Cl_2): g_0 = 2.007; a_0^{Tc} = 168 × 10^{-4} cm^{-1}.

*[Tc(NS)Cl_2(L^{OMe})] (**5b**)*: [TcNCl_2(L^{OMe})] (41 mg, 0.06 mmol) was dissolved in CH_2Cl_2 (10 mL) and treated with an excess of S_2Cl_2 (0.3 mL). After stirring for 30 min at room temperature, all volatiles were removed in vacuum. The remaining solid was washed with *n*-hexane and diethyl ether and dissolved in 1 mL CH_2Cl_2. Overlayering with *n*-hexane and storing in a refrigerator for several days gave pale-red crystals. Yield: 32 mg (75%).

IR (KBr, cm^{-1}): 3107 (w), 3086 (w), 2990 (w), 2841 (w), 1462 (w), 1423 (w), 1310 (vs), 1236 (vs), 1182 (vs), 1128 (vs), 1098 (vs), 986 (vs), 847 (m), 795 (s), 747 (s), 638 (m), 608 (s), 530 (w), 476 (w), 426 (w). EPR (77 K, CH_2Cl_2): $g_{\|}$ = 1.943, $g_{\perp}$ = 2.021, $A_{\|}^{Tc}$ = 285 × 10^{-4} cm^{-1}, $A_{\perp}^{Tc}$ = 129 × 10^{-4} cm^{-1}. EPR (RT, CH_2Cl_2) g_0 = 1.996; a_0^{Tc} = 177 × 10^{-4} cm^{-1}.

3.2. Spectroscopic and Analytical Methods

Elemental analyses of carbon, hydrogen, nitrogen and sulfur were determined using a Heraeus vario EL elemental analyzer. IR spectra were measured as KBr pellets on a Shimadzu IR Affinity-1 (technetium compounds) or a Thermo Scientific Nicolet iS10 ATR spectrometer (rhenium complexes). The NMR spectra were recorded on JEOL ECS-400 or ECZ-400 400 MHz spectrometers. ESI TOF mass spectra were measured with an Agilent 6210 ESI TOF (Agilent Technologies, Santa Clara, CA, USA). X-Band EPR spectra were recorded in solution with a Magnettech Miniscope MS400 spectrometer at 300 and 77 K. Simulation and visualization of the EPR spectra were conducted with the EasySpin tool box in MatLab [110,111].

3.3. X-ray Crystallography

The intensities for the X-ray determinations were collected on STOE IPDS-2T or Bruker CCD instruments with Mo/Kα radiation. The various temperatures applied are due to the experimental setup of the different diffractometers. Semi-empirical or numerical absorption corrections were carried out by the SADABS or X-RED32 programs [112,113]. Structure solution and refinement were performed with the SHELX programs [114,115] included in the WinGX [116] program package or OLEX2 [117]. Hydrogen atoms were calculated for idealized positions and treated with the "riding model" option of SHELXL. The solvent mask option of OLEX2 was applied to treat diffuse electron density due to disordered CH_2Cl_2 in the crystals of [Re(NS)Cl_3(PPh_3)($OPPh_3$)] (**2**). Details are given in the Supplementary Materials. The representation of molecular structures was conducted using the program DIAMOND [118].

4. Conclusions

Reactions of rhenium and technetium nitrido compounds with disulfur dichloride give access to novel thionitrosyl complexes of these elements having monodentate and chelating ligands in their coordination spheres. The new compounds are chemically related to corresponding nitrosyl complexes but do not possess their robustness and stability. Their thermal instability and the lack of a suitable one-step synthesis starting from pertechnetate set narrow limits for potential applications in nuclear medical procedures, as is discussed in the introduction.

Supplementary Materials: The following supporting information can be downloaded at https://www.mdpi.com/article/10.3390/inorganics12080210/s1, Table S1.1: Crystallographic data and data collection parameters; Figure S1.1: Ellipsoid representation of [Re(NS)Cl_3(PPh_3)$_2$] (**1**). The thermal ellipsoids are set at a 50% probability level. Hydrogen atoms are omitted for clarity; Table S1.2: Selected bond lengths (Å) and angles (°) in the [Re(NS)Cl_3(PPh_3)$_2$] (**1**); Figure S1.2: Ellipsoid representation of two crystallographically independent molecules of [Re(NS)Cl_3(PPh_3)($OPPh_3$)] (**2**). The thermal ellipsoids are set at a 50% probability level. Hydrogen atoms are omitted for clarity; Table S1.3.: Selected bond lengths (Å) and angles (°) in [Re(NS)Cl_3(PPh_3)($OPPh_3$)] (**2**); Figure S1.5: Ellipsoid representation of [Re(NS)(PPh_3)(Et_2btu)$_2$] (**3a**). The thermal ellipsoids are set at a 35% probability level. Hydrogen atoms are omitted for clarity; Table S1.6: Selected bond lengths (Å) and angles (°) in [Re(NS)(PPh_3)(Et_2btu)$_2$] (**3a**); Figure S1.3: Ellipsoid representation of [Re(NS)Cl(PPh_3)(L^{OMe})] (PF_6) (**4a**(PF_6)) including the positional disorder between the Re–N–S and Re–Cl bonds. The thermal ellipsoids are set at a 50% probability level. Hydrogen atoms are omitted for clarity; Table S1.4: Selected bond lengths (Å) and angles (°) in the $[Re(NS)Cl(PPh_3)(L^{OMe})]^+$ (**4a**) cation; Figure S1.4: Ellipsoid representation of two crystallographically independent molecules of [Tc(NS)Cl_2(L^{OMe})] (**5b**) including the positional disorder for the O atoms of the $\{L^{OMe}\}^-$ ligands and between the Tc–N–S and the Tc–Cl bonds. The thermal ellipsoids are set at a 50% probability level; Table S1.5: Selected bond lengths (Å) and angles (°) in [$Tc(NS)Cl_2(L^{OMe})$] (**5b**); Figure S2.1: IR (ATR) spectrum of [$Re(NS)Cl_3(PPh_3)_2$] (**1**); Figure S2.2: Solution EPR spectra of [$Re(NS)Cl_3(PPh_3)_2$] (**1**) in CH_2Cl_2, (a) at room temperature and (b) at T = 77 K; Figure S2.3: ESI(+) mass spectrum of [$Re(NS)Cl_3(PPh_3)_2$] (**1**) in CH_3CN; Figure S2.4: IR spectrum (ATR) of [$Re(NS)Cl_3(PPh_3)(OPPh_3)$] (**2**); Figure S2.5: Solution EPR spectra of [$Re(NS)Cl_3(PPh_3)(OPPh_3)$] (**2**) in acetone; (a) at room temperature and (b) at T = 77 K; Figure S2.6: ESI(+) mass spectrum of [$Re(NS)Cl_3(PPh_3)(OPPh_3)$] (**2**) in CH_3CN; Figure S2.7: IR spectrum (ATR) of [$Re(NS)(PPh_3)(Et_2btu)_2$] (**3a**); Figure S2.8: ^{1}H NMR spectrum of [$Re(NS)(PPh_3)(Et_2btu)_2$] (**3a**) in $CDCl_3$; Figure S2.9: ^{31}P NMR spectrum of [$Re(NS)(PPh_3)(Et_2btu)_2$] (**3a**) in $CDCl_3$; Figure S2.10: ESI(+) mass spectrum [$Re(NS)(PPh_3)(Et_2btu)_2$] (**3a**) in CH_3CN; Figure S2.11: IR spectrum (ATR) of [$Re(NS)Cl(PPh_3)(L^{OMe})$]Cl (**4a**Cl); Figure S2.12: Solution EPR spectra of [$Re(NS)Cl(PPh_3)(L^{OMe})$]Cl (**4a**Cl) in CH_2Cl_2; (a) at room temperature and (b) at T = 77 K; Figure S2.13: ESI(+) mass spectrum of [$Re(NS)Cl(PPh_3)(L^{OMe})$]Cl (**4a**Cl) in CH_3CN; Figure S2.14: IR spectrum (ATR) of [$Re(NS)Cl(PPh_3)(L^{OMe})$](PF_6) (**4a**(PF_6)); Figure S2.15: ESI(+) mass spectrum of [$Re(NS)Cl(PPh_3)(L^{OMe})$](PF_6) (**4a**(PF_6)) in CH_3CN; Figure S2.16: IR spectrum (ATR) of [$Re(NS)Cl_2(L^{OMe})$] (**5a**); Figure S2.17: Solution EPR spectra of [$Re(NS)Cl_2(L^{OMe})$] (**5a**) in CH_2Cl_2; (a) at room temperature and (b) at T = 77 K; Figure S2.18: ESI(+) mass spectrum of [$Re(NS)Cl(PPh_3)(L^{OMe})$](PF_6) (**4a**(PF_6)) in CH_3CN; Figure S2.19: IR spectrum (KBr) of [$Tc(NS)Cl(PPh_3)(L^{OMe})$]Cl (**4b**Cl); Figure S2.20: Solution EPR spectra of [$Tc(NS)Cl(PPh_3)(L^{OMe})$]Cl (**4b**Cl) in CH_2Cl_2; (a) at room temperature and (b) at T = 77 K; Figure S2.21: IR spectrum (KBr) of [$Tc(NS)Cl_2(L^{OMe})$] (**5b**); Figure S2.22: Solution EPR spectra of [$Tc(NS)Cl_2(L^{OMe})$] (**5b**). in CH_2Cl_2; (a) at room temperature and (b) at T = 77 K.

Author Contributions: Conceptualization, U.A. and D.N.; methodology, D.N. and A.H.; validation, A.H., T.E.S. and D.N.; formal analysis, D.N. and T.E.S.; investigation, D.N.; resources, U.A.; data curation, D.N. and A.H.; writing—original draft preparation, U.A.; writing—review and editing, D.N.; A.H. and T.E.S.; visualization, D.N. and U.A.; supervision, U.A.; project administration, U.A.; funding acquisition, U.A. All authors have read and agreed to the published version of the manuscript.

Funding: This research was funded by Freie Universität Berlin and the Deutsche Forschungsgemeinschaft (Core Facility BIOSUPRAMOL).

Data Availability Statement: Additional data are contained within the article and Supplementary Materials.

Acknowledgments: We would like to acknowledge the assistance of the Core Facility BioSupraMol supported by the DFG.

Conflicts of Interest: The authors declare no conflicts of interest.

References

1. *Technetium-99m Radiopharmaceuticals: Manufacture of Kits*; IAEA Technical Reports Series No. 466; International Atomic Energy Agency: Vienna, Austria, 2008; pp. 126–129.
2. Abram, U.; Alberto, R. Technetium and rhenium—coordination chemistry and nuclear medical applications. *J. Braz. Chem. Soc.* **2006**, *17*, 1486–1500. [CrossRef]
3. Bartholomä, M.D.; Louie, A.S.; Valliant, J.F.; Zubieta, J. Technetium and gallium derived radiopharmaceuticals: Comparing and contrasting the chemistry of two important radiometals for the molecular imaging era. *Chem. Rev.* **2010**, *110*, 2903–2920. [CrossRef] [PubMed]
4. Bhattacharyya, S.; Dixit, M. Metallic radionuclides in the development of diagnostic and therapeutic radiopharmaceuticals. *Dalton Trans.* **2011**, *40*, 6112–6128. [CrossRef] [PubMed]
5. Liu, S.; Chakraborty, S. ^{99m}Tc-centered one-pot synthesis for preparation of ^{99m}Tc radiotracers. *Dalton Trans.* **2011**, *40*, 6077–6086. [CrossRef] [PubMed]
6. Dilworth, J.R.; Pascu, S.I. The Radiopharmaceutical Chemistry of Technetium and Rhenium. In *The Chemistry of Molecular Imaging*; Long, N., Wong, W.-T., Eds.; John Wiley & Sons: Chichester, UK, 2015; pp. 163–174.
7. Kuntic, V.; Brboric, J.; Vujic, Z.; Uskokovic-Markovic, S. Radioisotopes Used as Radiotracers in vitro and in vivo Diagnostics: A Review. *Asian J. Chem.* **2016**, *28*, 235–411. [CrossRef]
8. Papagiannopoulou, D. Technetium-99m radiochemistry for pharmaceutical applications. *J. Labelled Compd. Radiopharm.* **2017**, *60*, 502–520. [CrossRef] [PubMed]
9. Duatti, A. Review on ^{99m}Tc radiopharmaceuticals with emphasis on new advancements. *Nucl. Med. Biol.* **2021**, *92*, 202–216. [CrossRef] [PubMed]

10. Alberto, R.; Nadeem, Q. ^{99m}Technetium-Based Imaging Agents and Developments in ^{99}Tc Chemistry. In *Metal Ions in Bio-Imaging Techniques*; Sigel, A., Freisinger, E., Sigel, K.O., Eds.; De Gruyter: Berlin/Munich, Germany; Boston, MA, USA, 2021; pp. 196–238.
11. Alberto, R. Role of Pure Technetium Chemistry: Are There Still Links to Applications in Imaging? *Inorg. Chem.* **2023**, *62*, 20539–20548. [CrossRef] [PubMed]
12. World Nuclear Association. Radioisotopes in Medicine. Available online: https://world-nuclear.org/information-library/non-power-nuclear-applications/radioisotopes-research/radioisotopes-in-medicine (accessed on 1 July 2024).
13. Dash, A.; Knapp, F.F., Jr.; Pillai, M.R.A. ^{99}Mo/^{99m}Tc separation: An assessment of technology options. *Nucl. Med. Biol.* **2013**, *40*, 167–176. [CrossRef]
14. Lepareur, N.; Lacœuille, F.; Bouvry, C.; Hindré, F.; Garcion, E.; Chérel, M.; Noiret, N.; Garin, E.; Knapp, F.F. Rhenium-188 Labeled Radiopharmaceuticals: Current Clinical Applications in Oncology and Promising Perspectives. *Front. Med.* **2019**, *6*, 132. [CrossRef]
15. Cutler, C.S.; Hennkens, H.M.; Sisay, N.; Huclier-Markai, S.; Jurisson, S. Radiometals for Combined Imaging and Therapy. *Chem. Rev.* **2013**, *113*, 858–883. [CrossRef]
16. Price, E.W.; Orvig, C. Matching chelators to radiometals for radiopharmaceuticals. *Chem. Soc. Rev.* **2014**, *43*, 260–290. [CrossRef]
17. Kronauge, J.F.K.; Mindiola, D.J. The value of Stable Metal-Carbon Bonds in Nuclear Medicine and the Cardiolite Story. *Organometallics* **2016**, *35*, 3432–3435. [CrossRef]
18. Morais, G.R.; Paulo, A.; Santos, I. Organometallic Complexes for SPECT Imaging and/or Radionuclide Therapy. *Organometallics* **2012**, *31*, 5693–5714. [CrossRef]
19. Alberto, R. From oxo to carbonyl and arene complexes; A journey through technetium chemistry. *J. Organomet. Chem.* **2018**, *869*, 264–269. [CrossRef]
20. Benz, M.; Braband, H.; Schmutz, P.; Halter, J.; Alberto, R. From Tc^{VII} to Tc^{I}; facile syntheses of bis-arene complexes $[^{99(m)}Tc(arene)_2]^+$ from pertechnetate. *Chem. Sci.* **2015**, *6*, 165–169. [CrossRef]
21. Meola, G.; Braband, H.; Jordi, S.; Fox, T.; Blacque, O.; Spingler, B.; Alberto, R. Structure and reactivities of rhenium and technetium bis-arene sandwich complexes $[M(\eta^6\text{-arene})_2]^+$. *Dalton Trans.* **2017**, *46*, 14631–14637. [CrossRef]
22. Nadeem, Q.; Meola, G.; Braband, H.; Bollinger, R.; Blancque, O.; Hernandez-Valdes, D.; Alberto, R. To Sandwich Technetium: Highly Functionalized Bis-Arene Complexes $[^{99m}Tc(\eta^6\text{-arene})_2]^+$ Directly from Water and $[TcO_4^-]$. *Angew. Chem. Int. Ed. Engl.* **2020**, *59*, 1197–1200. [CrossRef] [PubMed]
23. Nadeem, Q.; Battistin, F.; Blancque, O.; Alberto, R. Naphtalene Exchange in $[Re(\eta^6\text{-napht})_2]^+$ with Pharmaceutical Leads to Highly Functionalized Sandwich Complexes $[M(\eta^6\text{-pharm})_2]^+$ (M = Re/^{99m}Tc). *Chemistry* **2022**, *28*, e202103566. [CrossRef]
24. Boschi, A.; Uccelli, L.; Marvelli, L.; Cittanti, C.; Giganti, M.; Martini, P. Technetium-99m Radiopharmaceuticals for Ideal Myocardial Perfusion Imaging: Lost and Found Opportunities. *Molecules* **2022**, *27*, 1188. [CrossRef]
25. Pasqualini, R.; Duatti, A.; Bellande, E.; Comazzi, V.; Brucato, V.; Hoffschir, D.; Fagret, D.; Comet, M. Bis (Dithiocarbamato) Nitrido Technetium-99m Radiopharmaceuticals: A Class of Neutral Myocardial Imaging Agents. *J. Nucl. Med.* **1994**, *35*, 334–341.
26. Boschi, A.; Uccelli, L.; Bolzati, C.; Duatti, A.; Sabba, N.; Moretti, E.; di Domenico, G.; Zavattini, G.; Refosco, F.; Giganti, M. Synthesis and Biologic Evaluation of Monocationic Asymmetric ^{99m}Tc-Nitride Heterocomplexes Showing High Heart Uptake and Improved Imaging Properties. *J. Nucl. Med.* **2003**, *44*, 806–814. [PubMed]
27. Salvarese, N.; Carta, D.; Marzano, C.; Gerardi, G.; Melendez-Alafort, L.; Bolzati, C. $[^{99m}Tc][Tc(N)(DASD)(PNPn)]^+$ (DASD = 1,4-Dioxa-8-Azaspiro[4,5]Decandithiocarbamate, PNP n = Bisphosphinoamine) for Myocardial Imaging: Synthesis, Pharmacological and Pharmacokinetic Studies. *J. Med. Chem.* **2018**, *61*, 11114–11126. [CrossRef] [PubMed]
28. Meszaros, L.K.; Dose, A.; Biagini, S.C.G.; Blower, P.J. Hydrazinonicotinic acid (HYNIC)—Coordination chemistry and applications in radiopharmaceutical chemistry. *Inorg. Chim. Acta* **2010**, *363*, 1059–1069. [CrossRef]
29. Abrams, M.J.; Juweid, M.; tenKate, C.I.; Schwartz, D.A.; Hauser, M.M.; Gaul, F.E.; Fuccello, A.J.; Rubin, R.H.; Strauss, H.W.; Fischman, A.J. Technetium-99m-human polyclonal IgG radiolabeled via the hydrazine nicotinamide derivative for imaging focal sites of infection in rats. *J. Nucl. Med.* **1990**, *31*, 2022–2028. [PubMed]
30. Jiang, Y.; Tian, Y.; Feng, B.; Zhao, T.; Yu, X.; Zhao, Q. A novel molecular imaging probe [^{99m}Tc]Tc-HYNIC-FAPI targeting cancer-associated fibroblasts. *Sci. Rep.* **2023**, *13*, 3700. [CrossRef] [PubMed]
31. Cheah, C.T.; Newman, J.L.; Nowotnik, D.P.; Thornback, J.R. Synthesis and biological studies of the [^{99m}Tc]tetrachloronitrosyltechnetium(II) anion—An alternative low valent technetium starting material. *Int. J. Rad. Appl. Instrum. Nucl. Med. Biol.* **1987**, *14*, 573–575. [CrossRef] [PubMed]
32. Machura, B. Structural and spectroscopic properties of rhenium nitrosyl complexes. *Coord. Chem. Rev.* **2005**, *249*, 2277–2307. [CrossRef]
33. Dilworth, J.R. Rhenium chemistry—Then and Now. *Coord. Chem. Rev.* **2021**, *436*, 213822. [CrossRef]
34. Mahmood, A.; Akgun, Z.; Peng, Y.; Mueller, P.; Jiang, Y.; Berke, H.; Jones, A.G.; Nicholson, T. The synthesis and characterization of rhenium nitrosyl complexes. The X-ray crystal structures of $[ReBr_2(NO)(NCMe)_3]$, $[Re(NO)(N_5)](BPh_4)_2]$ and $[ReBr_2(NO)(NCMe)\{py\text{-}CH2\text{-}NHCH_2CH_2\text{-}N(CH_2\text{-}py)_2\}]$. *Inorg. Chim. Acta* **2013**, *405*, 455–460. [CrossRef]
35. Abram, U.; Ortner, K.; Hübener, R.; Voigt, A.; Caballho, R.; Vazquez-Lopez, E. Darstellung, Strukturen und EPR-Spektren der Rhenium(II)-Nitrosylkomplexe $[Re(NO)Cl_2(PPh_3)(OPPh_3)(OReO_3)]$, $[Re(NO)Cl_2(OPPh_3)_2(OReO_3)]$ und $[Re(NO)Cl_2(PPh_3)_3][ReO_4]$. *Z. Anorg. Allg. Chem.* **1998**, *624*, 1662–1668. [CrossRef]

36. Agbossou, F.; O'Connor, E.J.; Garner, C.M.; Quiros Mendez, N.; Fernandez, J.M.; Patton, A.T.; Ramsden, J.A.; Gladysz, J.A. Cyclopentadienyl Rhenium Complexes. *Inorg. Synth.* **1992**, *29*, 211–225.
37. Seidel, S.N.; Prommesberger, M.; Eichenseher, S.; Meyer, O.; Hampel, F.; Gladysz, J.A. Syntheses and structural analyses of chiral rhenium containing amines of the formula (η^5-C_5H_5)Re(NO)(PPh_3)((CH_2)$_n$NRR′) (n = 0, 1). *Inorg. Chim. Acta* **2010**, *363*, 533–548. [CrossRef]
38. Bernasconi, C.F.; Bhattacharya, S.; Wenzel, P.J.; Olmstead, M.M. Kinetic and Thermodynamic Acidity of [Cp(NO)(PPh_3)Re(2,5-dimethyl-3-thienyl)carbene]$^+$. Transition State Imbalance and Intrinsic Barriers. *Organometallics* **2006**, *25*, 4322–4330. [CrossRef]
39. Dilsky, S.; Schenk, W.A. Diastereomeric Halfsandwich Rhenium Complexes Containing Hemilabile Phosphane Ligands. *Eur. J. Inorg. Chem.* **2004**, *2004*, 4859–4870. [CrossRef]
40. Nicholson, T.; Chun, E.; Mahmood, A.; Mueller, P.; Davison, A.; Jones, A.G. Synthesis, spectroscopy and structural analysis of Technetium and Rhenium nitrosyl complexes. *Commun. Inorg. Chem.* **2015**, *3*, 31–39.
41. Blanchard, S.S.; Nicholson, T.; Davison, A.; Davis, W.; Jones, A.G. The synthesis, characterization and substitution reactions of the mixed technetium(I) nitrosyl complex *trans-trans*-[(NO)($NCCH_3$)Cl_2(PPh_3)$_2$Tc]. *Inorg. Chim. Acta* **1996**, *244*, 121–130. [CrossRef]
42. Balasekaran, S.M.; Hagenbach, A.; Drees, M.; Abram, U. [Tc^{II}(NO)(trifluoroacetate)$_4$F]$^{2-}$—Synthesis and reactions. *Dalton Trans.* **2017**, *46*, 13544–13552. [CrossRef] [PubMed]
43. Linder, K.E.; Davison, A.; Dewan, J.C.; Costello, C.E.; Melaknia, S. Nitrosyl complexes of technetium: Synthesis and characterization of [Tc^I(NO)($CNCMe_3$)$_5$](PF_6)$_2$ and Tc(NO)Br_2($CNCMe_3$)$_3$ and the crystal structure of Tc(NO)Br_2($CNCMe_3$)$_3$. *Inorg. Chem.* **1986**, *25*, 2085–2089. [CrossRef]
44. Ackermann, J.; Noufele, C.N.; Hagenbach, A.; Abram, U. Nitrosyltechnetium(I) Complexes with 2-(Diphenylphosphanyl)aniline. *Z. Anorg. Allg. Chem.* **2019**, *645*, 8–13. [CrossRef]
45. Ackermann, J.; Hagenbach, A.; Abram, U. {Tc(NO)(Cp)(PPh_3)}$^+$—A novel technetium(I) core. *Chem. Commun.* **2016**, *52*, 10285–10288. [CrossRef]
46. Ackermann, J.; Abdulkader, A.; Scholtysik, C.; Jungfer, M.R.; Hagenbach, A.; Abram, U. [Tc^I(NO)X(Cp)(PPh_3)] Complexes (X^- = I^-, I_3^-, SCN^-, $CF_3SO_3^-$, or CF_3COO^-) and Their Reactions. *Organometallics* **2019**, *38*, 4471–4478. [CrossRef]
47. Abdulkader, A.; Hagenbach, A.; Abram, U. [Tc(NO)Cl(Cp)(PPh_3)]—A Technetium(I) Compound with an Unexpected Synthetic Potential. *Eur. J. Inorg. Chem.* **2021**, *2021*, 3812–3818.
48. Schibli, R.; Marti, N.; Maurer, P.; Spingler, B.; Lehaire, M.-L.; Gramlich, V.; Barnes, C.L. Syntheses and Characterization of Dicarbonyl–Nitrosyl Complexes of Technetium(I) and Rhenium(I) in Aqueous Media: Spectroscopic, Structural, and DFT Analyses. *Inorg. Chem.* **2005**, *44*, 683–690. [CrossRef]
49. Brown, D.S.; Newman, J.L.; Thornback, J.R.; Davison, A. Structure of the tetra-*n*-butylammoium salt of tetrachloro(methanol) nitrosyltechnetium(II) anion. *Acta Cryst.* **1987**, *C43*, 1692–1694.
50. Brown, D.S.; Newman, J.L.; Thornback, J.R.; Pearlstein, R.M.; Davison, A.; Lawson, A. The synthesis and characterisation of the trichloronitrosyl(acetylacetonato)technetium(II) anion, a novel technetium(II) complex. *Inorg. Chim. Acta* **1988**, *150*, 193–196. [CrossRef]
51. Nicholson, T.; Hirsch-Kuchma, M.; Freiberg, E.; Davison, A.; Jones, A.G. The reaction chemistry of a technetium(I) nitrosyl complex with potentially chelating organohydrazines: The X-ray crystal structure of [$TcCl_2$(NO)($HNNC_5H_4N$)(PPh_3)]. *Inorg. Chim. Acta* **1998**, *279*, 206–209.
52. De Vries, N.; Cook, J.; Davison, A.; Nicholson, T.; Jones, A.G. Synthesis and characterization of a technetium(III) nitrosyl compound: Tc(NO)(Cl)($SC_{10}H_{13}$)$_3$. *Inorg. Chem.* **1990**, *29*, 1062–1064. [CrossRef]
53. Nicholson, T.; Mahmood, A.; Limpa-Amara, N.; Salvarese, N.; Takase, M.K.; Müller, P.; Akgun, Z.; Jones, A.G. Reactions of the tridentate and tetradentate amine ligands di-(2-picolyl)(*N*-ethyl)amine and 2,5-bis-(2-pyridylmethyl)-2,5 diazohexane with technetium nitrosyl complexes. *Inorg. Chim. Acta* **2011**, *373*, 301–305. [CrossRef]
54. Roca Jungfer, M.; Ernst, M.J.; Hagenbach, A.; Abram, U. [{Tc^I(NO)(L^{OMe})(PPh_3)Cl}$_2$Ag](PF_6) and [Tc^{II}(NO)(L^{OMe})(PPh_3)Cl](PF_6): Two Unusual Technetium Complexes with a "Kläui-type" Ligand. *Z. Anorg. Allg. Chem.* **2022**, *648*, e202100316.
55. Nicholson, T.L.; Mahmood, A.; Muller, P.; Davison, A.; Storm-Blanchard, S.; Jones, A.G. The synthesis and structural characterization of the technetium nitrosyl complexes [TcCl(NO)(SC_5H_4N)(PPh_3)$_2$] and [Tc(NO)(SC_5H_4N)$_2$(PPh_3)]. *Inorg. Chim. Acta* **2011**, *365*, 484–486. [CrossRef] [PubMed]
56. Balasekaran, S.M.; Spandl, J.; Hagenbach, A.; Köhler, K.; Drees, M.; Abram, U. Fluoridonitrosyl Complexes of Technetium(I) and Technetium(II). Synthesis, Characterization, Reactions, and DFT Calculations. *Inorg. Chem.* **2014**, *53*, 5117–5128. [CrossRef]
57. Nicholson, T.; Hirsch-Kuchma, M.; Shellenbarger-Jones, A.; Davison, A.; Jones, A.G. The synthesis and characterization of a technetium nitrosyl complex with cis-{2-pyridyl,diphenylphosphine} coligands. The X-ray crystal structure of [$TcCl_2$(NO)($pyPPh_2$-*P,N*) ($pyPPh_2$-*P*)]. *Inorg. Chim. Acta* **1998**, *267*, 319–322. [CrossRef]
58. Grunwald, A.C.; Scholtysik, C.; Hagenbach, A.; Abram, U. One Ligand, One Metal, Seven Oxidation States: Stable Technetium Complexes with the "Kläui Ligand". *Inorg. Chem.* **2020**, *59*, 9396–9405. [CrossRef] [PubMed]
59. Nicholson, T.L.; Mahmood, A.; Refosco, F.; Tisato, F.; Müller, P.; Jones, A.G. The synthesis and X-ray structural characterization of *mer* and *fac* isomers of the technetium(I) nitrosyl complex [$TcCl_2$(NO)(PNPpr)]. *Inorg. Chim. Acta* **2009**, *362*, 3637–3640. [CrossRef] [PubMed]
60. Nicholson, T.; Müller, P.; Davison, A.; Jones, A.G. The synthesis and characterization of a cationic technetium nitrosyl complex: The X-ray crystal structure of [TcCl(NO)(DPPE)$_2$](PF_6) × CH_2Cl_2. *Inorg. Chim. Acta* **2006**, *359*, 1296–1298. [CrossRef]

61. Pandey, K.K. Coordination Chemistry of Thionitrosyl (NS), Thiazate (NSO^-), Disulfidothionitrate (S_3N^-), Sulfur Monoxide (SO), and Disulfur Monoxide (S_2O) Ligands. *Progr. Inorg. Chem.* **1992**, *40*, 445–502.
62. Døsing, A. The electronic structure and photochemistry of transition metal thionitrosyl complexes. *Coord. Chem. Rev.* **2016**, *306*, 544–557. [CrossRef]
63. Anhaus, J.; Siddiqi, Z.A.; Roesky, H.W. Reaction of Tetrasulfurtetranitride with Rhenium(VII)-chloronitride. The Crystal Structure of $[Ph_4As^+]_2[Cl_4Re(NS)(NSCl)^{2-}] \times CH_2Cl_2$. *Z. Naturforsch.* **1985**, *40b*, 740–744. [CrossRef]
64. Dirican, D.; Pfister, N.; Wozniak, M.; Braun, T. Reactivity of Binary and Ternary Sulfur Halides towards Transition-Metal Compounds. *Chem. Eur. J.* **2020**, *31*, 6945–6963. [CrossRef]
65. Dietrich, A.; Neumüller, B.; Dehnicke, K. $(PPh_4)_2[(SN)ReCl_3(\mu\text{-}N)(\mu\text{-}NSN)ReCl_3(THF)]$—Ein Nitrido-Thionitrosyl-Dinitridosulfato-Komplex des Rheniums. *Z. Anorg. Allg. Chem.* **2000**, *626*, 1268–1270. [CrossRef]
66. Reinel, M.; Höcher, T.; Abram, U.; Kirmse, R. Ein Beitrag zu Rhenium(II)-, Osmium(II)- und Technetium(II)-Thionitrosylkomplexe vom Typ $[M(NS)Cl_4py]$: Darstellung, Strukturen und EPR-Spektren. *Z. Anorg. Allg. Chem.* **2003**, *629*, 853–861. [CrossRef]
67. Voigt, A.; Abram, U.; Kirmse, R. Darstellung, Strukturen und EPR-Spektren der Rhenium(II)-Thionitrosylkomplexe *trans*-$[Re(NS)Cl_3(MePh_2P)_2]$ und *trans*-$[Re(NS)Br_3(Me_2PhP)_2]$. *Z. Anorg. Allg. Chem.* **1999**, *625*, 1658–1663. [CrossRef]
68. Hauck, H.-G.; Willing, W.; Müller, U.; Dehnicke, K. $[ReCl_2(NS)(NSCl)(Pyridin)_2]$, ein Thionitrosyl-chlorthionitrenkomplex des Rheniums. *Z. Anorg. Allg. Chem.* **1986**, *534*, 77–84. [CrossRef]
69. Hübener, R.; Abram, U.; Strähle, J. Isothiocyanato complexes of rhenium II. Synthesis, characterization and structures of $ReN(NCS)_2(Me_2PhP)_3$ and $Re(NS)(NCS)_2(Me_2PhP)_3$. *Inorg. Chim. Acta* **1994**, *216*, 223–228. [CrossRef]
70. Ritter, S.; Abram, U. Gemischtligand-Komplexe des Rheniums. VI. Darstellung und Strukturen der Rhenium Thionitrosyl-Komplexe *mer*-$[Re(NS)Cl_2(Me_2PhP)_3] \times CH_2Cl_2$ und *trans*-$[Re(NS)Cl_3(Me_2PhP)_2]$. *Z. Anorg. Allg. Chem.* **1994**, *620*, 1223–1228.
71. Ritter, S.; Abram, U. Gemischtligandkomplexe des Rheniums. IX. Reaktionen am Nitridoliganden von $[ReN(Me_2PhP)(Et_2dtc)_2]$. Synthese, Charakterisierung und Kristallstrukturen von $[Re(NBCl_3)(Me_2PhP)(Et_2dtc)_2]$, $[Re(NGaCl_3)(Me_2PhP)(Et_2dtc)_2]$ und $[Re(NS)Cl(Me_2PhP)_2(Et_2dtc)]$. *Z. Anorg. Allg. Chem.* **1995**, *622*, 965–973. [CrossRef]
72. Ruf, C.; Behrens, U.; Lork, E.; Mews, R. Reactions of halides with *trans*-$[Re(CO)_4(MeCN)(NS)][AsF_6]_2$: Syntheses and structure of *trans*-$[Re(CO)_4(Cl)(NS)][AsF_6]$ and $[(OC)_5ReNS\text{–}NS\text{–}N[Re(CO)_5]_S\{N\text{–}SNRe(CO)_5\}\text{–}CH_2\text{–}CH_2][AsF_6]_2$, an unusual trinuclear bis(thiazyl)rhenium complex. *Chem. Commun.* **1996**, 939–940. [CrossRef]
73. Baldas, J.; Bonnyman, J.; Mackay, M.F.; Williams, G.A. Structural studies of technetium complexes. V. The preparation and crystal structure of Dichlorobis(diethyldithiocarbamato)thionitrosyltechnetium(III). *Austr. J. Chem.* **1984**, *37*, 751–759. [CrossRef]
74. Kaden, L.; Lorenz, B.; Kirmse, R.; Stach, J.; Behm, H.; Beurskens, P.T.; Abram, U. Synthesis, characterization and x-ray molecular and crystal structure of $Tc(NS)Cl_3(Me_2PhP)(Me_2PhPO)$-a first example of mixed phosphine/phosphine oxide coordination. *Inorg. Chim. Acta* **1990**, *169*, 43–48. [CrossRef]
75. Baldas, J.; Colmanet, S.F.; Williams, G.A. Preparation and Structure of Dibromobis(N,N-diethyldithiocarbamato)-thionitrosyltechnetium(III). *Austr. J. Chem.* **1991**, *44*, 1125–1132. [CrossRef]
76. Lu, J.; Clarke, M.J. Modulation of Tc–NX (X = O or S) bonds by π-acceptor ligands. *J. Chem. Soc. Dalton Trans.* **1992**, 1243–1248. [CrossRef]
77. Abram, U.; Schulz Lang, E.; Abram, S.; Wegmann, J.; Dilworth, J.R.; Kirmse, R.; Woolins, J.D. Technetium(V) and rhenium(V) nitrido complexes with bis(diphenyl-thiophosphoryl)amide, $N(SPPh_2)^{2-}$. *J. Chem. Soc. Dalton Trans.* **1997**, 623–630. [CrossRef]
78. Hiller, W.; Hübener, R.; Lorenz, B.; Kaden, L.; Findeisen, M.; Stach, J.; Abram, U. Structural and spectroscopic studies on *mer*-dichlorotris(dimethylphenylphosphine)(thionitrosyl)technetium(I), *mer*-$[Tc(NS)Cl_2(Me_2PhP)_3]$. *Inorg. Chim. Acta* **1991**, *181*, 161–165. [CrossRef]
79. Lu, J.; Clarke, M.J. Sulfur atom transfer with reduction of a $[Tc^{VI}{\equiv}N]^{3+}$ core to a $[Tc^{I}\text{-}N{\equiv}S]^{2+}$ core. Crystal structure of mer-dichlorotris(4-picoline)(thionitrosyl)technetium. *Inorg. Chem.* **1990**, *29*, 4123–4125. [CrossRef]
80. Abram, U.; Hübener, R.; Wollert, R.; Kirmse, R.; Hiller, W. Synthesis, characterization and reactions of $[Tc(NS)X_4]^-$ complexes (X = Cl, Br, NCS). *Inorg. Chim. Acta* **1993**, *206*, 9–14. [CrossRef]
81. Dressler, K. Ultraviolett- und Schumannspektren der neutralen und ionisierten Moleküle PO, PS, NS, P_2. *Helv. Phys. Acta* **1955**, *28*, 563–590.
82. O'Hare, P.A.G. Dissociation Energies, Enthalpies of Formation, Ionization Potentials, and Dipole Moments of NS and NS^+. *J. Chem. Phys.* **1970**, *52*, 2992–2996. [CrossRef]
83. Mews, R. The Thionitrosyl Cation NS^+ as a Synthetic Reagent. *Angew. Chem. Int. Ed. Engl.* **1976**, *15*, 691–692. [CrossRef]
84. Clegg, W.; Glemser, O.; Harms, K.; Hartmann, G.; Mews, R.; Noltemeyer, M.; Sheldrick, G.M. Crystal structures of thionitrosyl hexafluoroantimonate(V) and thionitrosyl undecafluorodiantimonate(V) at 293 K and of thionitrosyl undecafluorodiantimonate(V) at 121.5 K: The effect of thermal motion on the apparent NS bond length. *Acta Cryst.* **1981**, *37b*, 548–552. [CrossRef]
85. Kaden, L.; Lorenz, B.; Kirmse, R.; Stach, J.; Abram, U. Darstellung und Charakterisierung von Thionitrosylkomplexen des Technetiums(I) und –(II). *Z. Chem.* **1985**, *25*, 29–30. [CrossRef]
86. Abram, U.; Kirmse, R.; Köhler, K.; Lorenz, B.; Kaden, L. $Tc(NX)Y_3(Me_2PhP)_2$ Complexes (X = O or S; Y = Cl or Br). Preparation, Characterization and EPR Studies. *Inorg. Chim. Acta* **1987**, *129*, 15–20. [CrossRef]
87. Nguyen, H.H.; Abram, U. Rhenium and Technetium Complexes with *N,N*-Dialkyl-*N′*-benzoylthioureas. *Inorg. Chem.* **2007**, *46*, 5310–5319. [PubMed]

88. Hayes, T.R.; Powell, A.S.; Benny, P.D. Synthesis and stability of 2 + 1 complexes of *N,N*-diethylbenzoylthiourea with $[M^{I}(CO)_3]^+$ (M = Re, ^{99m}Tc). *J. Coord. Chem.* **2015**, *68*, 3432–3448. [CrossRef]
89. Abram, U.; Abram, S.; Alberto, R.; Schibli, R. Ligand exchange reactions starting from $[Re(CO)_3Br_3]^{2-}$. Synthesis, characterization and structures of rhenium(I) tricarbonyl complexes with thiourea and thiourea derivatives. *Inorg. Chim. Acta* **1996**, *248*, 193–202. [CrossRef]
90. Borges, A.P.; Possato, B.; Hagenbach, A.; Machado, A.E.H.; Deflon, V.M.; Abram, U.; Maia, P.I.S. Re(V) complexes containing the phenylimido (NPh^{2-}) core and *SNS*-thiosemicarbazide ligands. *Inorg. Chim. Acta* **2021**, *516*, 120110. [CrossRef]
91. Mukiza, J.; Gerber, T.I.A.; Hosten, E.C.; Betz, R. Crystal structure of *fac*-$\kappa_{O,S}$-(Z)-1,1-diethyl-3-(hydroxido(phenyl)methylene) thiourea-($\kappa_{S'}$(Z)-1,1-diethyl-3-(hydroxido(phenyl)methylene)-thiourea)-tricarbonyl rhenium(I), $C_{27}H_{31}N_4O_5ReS_2$. *Z. Kristallogr. New Cryst. Struct.* **2015**, *230*, 50–52. [CrossRef]
92. Nguyen, H.H.; Abram, U. Rhenium and technetium complexes with tridentate S,N,O ligands derived from benzoylhydrazine. *Polyhedron* **2009**, *28*, 3945–3952. [CrossRef]
93. Salsi, F.; Portapilla, G.B.; Simon, S.; Roca Jungfer, M.; Hagenbach, A.; de Albuquerque, S.; Abram, U. Effect of Fluorination on the Structure and Anti-*Trypanosoma cruzy* Activity of Oxorhenium(V) Complexes with S,N,S-Tridentate Thiosemicarbazones and Benzoylthioureas. Synthesis and Structures of Technetium(V) Analogues. *Inorg. Chem.* **2019**, *58*, 10129–10138. [CrossRef]
94. Roca Jungfer, M.; Elsholz, L.; Abram, U. Technetium Hydrides Revisited: Syntheses, Structures, and Reactions of $[TcH_3(PPh_3)_4]$ and $[TcH(CO)_3(PPh_3)_2]$. *Organometallics* **2021**, *40*, 3095–3112. [CrossRef]
95. Kläui, W. The Coordination Chemistry and Organometallic Chemistry of Tridentate Oxygen Ligands with π-Donor Properties. *Angew. Chem. Int. Ed. Engl.* **1990**, *29*, 627–637. [CrossRef]
96. Leung, W.-H.; Zhang, Q.-F.; Yi, X.-Y. Recent developments in the coordination and organometallic chemistry of Kläui oxygen tripodal ligands. *Coord. Chem. Rev.* **2007**, *251*, 2266–2279. [CrossRef]
97. Kramer, D.J.; Davison, A.; Jones, A.G. Structural models for $[M(CO)_3(H_2O)_3]^+$ (M = Tc, Re): Fully aqueous synthesis of technetium and rhenium tricarbonyl complexes of tripodal oxygen donor ligands. *Inorg. Chim. Acta* **2001**, *312*, 215–220. [CrossRef]
98. Leung, W.-H.; Chan, E.Y.Y.; Lai, T.C.Y.; Wong, W.-T. Synthesis and reactivity of nitrido-rhenium and -osmium complexes with an oxygen tripod ligand. *J. Chem. Soc. Dalton Trans.* **2000**, 51–56. [CrossRef]
99. So, Y.-M.; Chiu, W.-H.; Cheung, W.-M.; Ng, H.-Y.; Lee, H.K.; Sung, H.H.-Y.; Williams, I.D.; Leung, W.H. Heterobimetallic rhenium nitrido complexes containing the Kläui tripodal ligand $[Co(\eta^5\text{-}C_5H_5)\{P(O)(OEt)_2\}_3]^-$. *Dalton Trans.* **2015**, *44*, 5479–5487. [CrossRef]
100. Banberry, H.J.; Hussain, W.; Evans, I.G.; Hamor, T.A.; Jones, C.J.; McCleverty, J.A.; Schulte, H.-J.; Engles, B.; Kläui, W. The syntheses of high oxidation state metal complexes containing the tripodal ligand $[(\eta^5\text{-}C_5H_5)Co\{P(OMe)_2(O)\}_3]^-$ and the X-ray crystal structure of $[(\eta^5\text{-}C_5H_5)Co\{P(OMe)_2(O)\}_3\ ReO_3]$. *Polyhedron* **1990**, *9*, 2549–2551. [CrossRef]
101. Dyckhoff, B.; Schulte, H.-J.; Englert, U.; Spaniol, T.P.; Kläui, W.; Schubiger, P.A. Rhenium-Komplexe von Sauerstoffchelatliganden: Ein Weg zu neuen Radiopharmaka? *Z. Anorg. Allg. Chem.* **1992**, *614*, 131–141. [CrossRef]
102. Abram, U.; Abram, S. Synthese und Charakterisierung neuartiger Technetiumkomplexe mit 1,1-disubstituierten Benzoylthioharnstoffen. *Z. Chem.* **1983**, *23*, 228. [CrossRef]
103. Sullivan, B.P.; Brewer, J.C.; Gray, H.B.; Linebarrier, D.; Mayer, J.M. Nitrido and Oxo Complexes of Rhenium(V). *Inorg. Synth.* **1992**, *29*, 146–150.
104. Kaden, L.; Lorenz, B.; Schmidt, K.; Sprinz, H.; Wahren, M. Nitridokomplexe des Technetium(V). *Isotopenpraxis* **1981**, *17*, 174–175.
105. Baldas, J.; Boas, J.F.; Bonnyman, J.; Williams, G.A. Studies of technetium complexes. Part 6. The preparation, characterisation, and electron spin resonance spectra of salts of tetrachloro- and tetrabromonitridotechnetate(VI): Crystal structure of tetraphenylarsonium tetrachloronitridotechnetate(VI). *J. Chem. Soc. Dalton Trans.* **1984**, *11*, 2395–2400. [CrossRef]
106. Abram, U.; Braun, M.; Abram, S.; Kirmse, R.; Voigt, A. $[NBu_4][ReNCl_4]$: Facile synthesis, structure, electron paramagnetic resonance spectroscopy and reactions. *J. Chem. Soc. Dalton Trans.* **1998**, *2*, 231–238. [CrossRef]
107. Grunwald, A.C. Metal Complexes with Tripodal and Chelating Thiourea Ligands towards Nuclear Medical Imaging. Doctoral Thesis, Freie Universität Berlin, Berlin, Germany, 2020. Available online: https://refubium.fu-berlin.de/handle/fub188/30387 (accessed on 1 July 2024).
108. Nowak, D. Thionitrosylkomplexe des Rheniums und Technetiums. Doctoral Thesis, Freie Universität Berlin, Berlin, Germany, 2022. Available online: https://refubium.fu-berlin.de/handle/fub188/36899 (accessed on 1 July 2024).
109. Kleinpeter, E.; Beyer, L. ^{1}H-NMR-Untersuchung der behinderten Rotation um die C-N-Bindung in 1,1′-Diäthyl-3-benzoylharnstoff-Derivaten. *J. Prakt. Chem.* **1975**, *317*, 938–942. [CrossRef]
110. Stoll, S.; Schweiger, A. EasySpin, a comprehensive software package for spectral simulation and analysis in EPR. *J. Magn. Res.* **2006**, *178*, 42–55. [CrossRef] [PubMed]
111. *MATLAB Version: 9.13.0 (R2022b)*; The MathWorks Inc.: Natick, MA, USA, 2022.
112. Sheldrick, G. *SADABS, vers. 2014/5*; University of Göttingen: Göttingen, Germany, 2014.
113. Coppens, P. The Evaluation of Absorption and Extinction in Single-Crystal Structure Analysis. In *Crystallographic Computing*; Muksgaard: Copenhagen, Denmark, 1979.
114. Sheldrick, G.M. A short history of SHELX. *Acta Crystallogr.* **2008**, *A64*, 112–122. [CrossRef] [PubMed]
115. Sheldrick, G.M. Crystal structure refinement with SHELXL. *Acta Crystallogr.* **2015**, *C71*, 3–8.
116. Farrugia, L.J. WinGX and ORTEP for Windows: An update. *J. Appl. Cryst.* **2012**, *45*, 849–854. [CrossRef]

117. Dolomanov, O.V.; Bourhis, L.J.; Gildea, R.J.; Howard, J.A.; Puschmann, H. OLEX2: A complete structure solution, refinement and analysis program. *J. Appl. Crystallogr.* **2009**, *42*, 339–341. [CrossRef]
118. Putz, H.; Brandenburg, K. *Diamond—Crystal and Molecular Structure Visualization*; Vers. 4.6.8.; Crystal Impact: Bonn, Germany, 2022.

inorganics

MDPI

Article

Conformational, Electrochemical, and Antioxidative Properties of Conjugates of Different Ferrocene Turn-Inducing Scaffolds with Hydrophobic Amino Acids

Monika Kovačević [1,*], Sunčica Roca [2,*], Dijana Jadreško [3], Jasna Mrvčić [4], Karla Hanousek Čiča [4], Mojca Čakić Semenčić [1] and Lidija Barišić [1]

1 Department of Chemistry and Biochemistry, Faculty of Food Technology and Biotechnology, University of Zagreb, Pierottijeva 6, 10000 Zagreb, Croatia; mojca.cakic@pbf.unizg.hr (M.Č.S.); lidija.barisic@pbf.unizg.hr (L.B.)
2 NMR Centre, Ruđer Bošković Institute, Bijenička 54, 10000 Zagreb, Croatia
3 Laboratory for Physical Chemistry of Traces, Division for Marine and Environmental Research, Ruđer Bošković Institute, Bijenička 54, 10000 Zagreb, Croatia; djadresko@irb.hr
4 Department of Food Engineering, Faculty of Food Technology and Biotechnology, University of Zagreb, Pierottijeva 6, 10000 Zagreb, Croatia; jasna.mrvcic@pbf.unizg.hr (J.M.); karla.hanousek.cica@pbf.unizg.hr (K.H.Č.)
* Correspondence: monika.kovacevic@pbf.unizg.hr (M.K.); sroca@irb.hr (S.R.); Tel.: +385-1-4605-275 (M.K.); +385-1-4571-224 (S.R.)

Abstract: The incorporation of different ferrocene scaffolds into the peptide sequences induces the formation of hydrogen-bond-based secondary structural elements that are frequently observed in natural peptides and proteins. There are three simple ferrocene scaffolds for conjugation with amino acids and peptides that serve as templates for ferrocene peptidomimetics, namely ferrocene-1,1′-dicarboxylic acid (Fcd, **I**), 1′-aminoferrocene-1-carboxylic acid (Fca, **III**), and ferrocene-1,1′-diamine (Fcda, **V**). Here, we have investigated their ability to induce the turn structure upon conjugation with Val, Leu, and Phe. Furthermore, we also wanted to determine whether the branched side chains of Val, Leu, and Phe interfere with intramolecular hydrogen bonding (IHB). For these purposes, we performed a detailed spectroscopic analysis by measuring the concentration, temperature, and solvent dependence of the IR, NMR, and CD spectra. The effect of the different ferrocene scaffolds on the antioxidant activity of the prepared peptides was tested using the DPPH and ABTS methods, and was further rationalized using electrochemical measurements. It was found that the ferrocene scaffold has the greatest influence on the hydrogen bonding pattern, while the influence of the side branches of the amino acids is less relevant.

Keywords: ferrocene; peptidomimetics; antioxidant activity; valine; leucine; phenylalanine; conformational analysis; hydrogen bonds; cyclic voltammetry

1. Introduction

Since the conformational stability, biological activities, and therapeutic properties of peptides and proteins are related to their secondary structure, the main goal in the design and preparation of synthetic peptides is to enable the formation of secondary structure elements [1–4]. Ferrocene peptides are known for their ability to adopt the turn- and β-sheet-like structures, similar to those found in natural peptides, as a result of intramolecular hydrogen bonding [5–8].

Kraatz and Metzler-Nolte have established three different types of 1,*n*′-disubstituted peptide systems based on the parent compound [9]. Since the choice of the ferrocene scaffold (**I**, **III**, and **V**) dictates the orientation of the peptide chain and affects the rigidity of the molecule depending on the pattern of hydrogen bonds established, it is possible to prepare three types of ferrocene peptides, i.e.: **II**, **IV**, and **VI** (Figure 1).

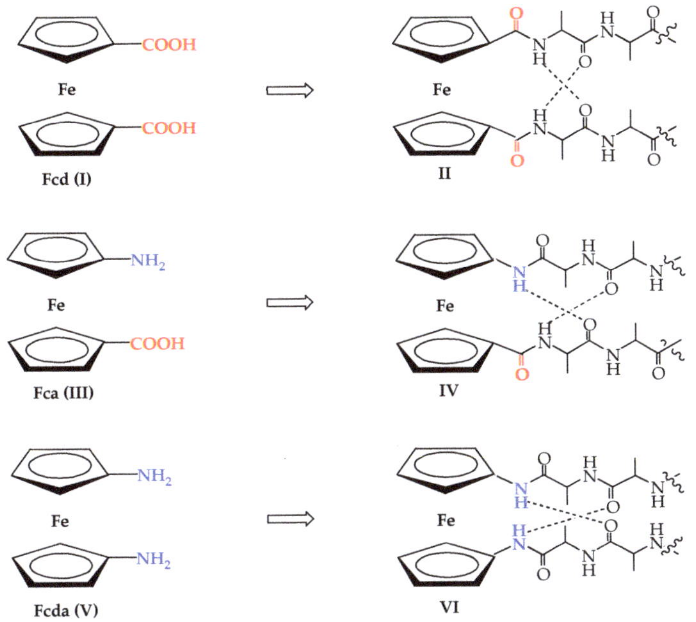

Figure 1. Ferrocene–alanine peptides **II**, **IV**, and **VI** derived from Fcd (**I**), Fca (**III**), and Fcda (**V**) (IHB denoted as dashed lines).

For example, amino acid or peptide conjugates **II** derived from the Fcd (**I**) most commonly form intramolecular interchain hydrogen bonds that form two ten-membered rings (two β-turns) in the resulting C_2-symmetric peptide structure. (Figure 1). This structural motif is known as the 'Herrick' conformation. Alternative motifs are less frequently observed, such as the structure with only one interchain IHB closing the seven-membered γ-turn ring, the so-called 'van Staveren's' conformation or 'Xu's' open conformation, in which no intramolecular hydrogen bonding interactions occur [10–13].

The essential feature of Fca (**III**) is that, in contrast to Fcd (**I**) and Fcda (**V**), it enables the formation of a 12-membered IHB ring between antiparallel podand peptide chains, as found in natural peptides [14–17].

Compared to peptides **II** and **IV**, conjugates derived from ferrocene-1,1′-diamine (**V**) have been less studied [5,18–20]. The formation of 14-membered IHB rings between podand peptide chains, similar to those in antiparallel β-sheets, makes it a desirable structural element.

It has been shown that isosteric modification of bioactive compounds with ferrocene enhances antimicrobial, antioxidant, and anticancer activity due to the exceptional properties of ferrocene, e.g., superaromaticity, air and heat stability, electrophilicity, solubility in organic solvents, non-toxicity, and lipophilicity [21–24]. Therefore, the biological potential of the new peptides **1–9** is also being investigated.

Peptide sequences containing a high number of amino acids with hydrophobic side chains, so-called β-branched amino acids (Leu, Val, Phe, or Ile), are known to form β-sheet or α-helical structures supported by strong inter- or intramolecular non-covalent interactions [25].

Mitochondria-penetrating peptides (MPPs), a subclass of cell-penetrating peptides (CPPs), are short amino acid sequences that are able to penetrate the plasma membrane, i.e., the mitochondrial lipid bilayer. This is facilitated by the hydrophobic residues of the hydrophobic amino acids, which provide a high degree of lipophilicity [26]. It is known that hydrophobic interactions cause the folding of proteins and stabilize the structures of single proteins, multiproteins, and protein–ligand systems [27].

The presence of hydrophobic amino acids in naturally occurring or synthetically produced antimicrobial peptides (AMPs) facilitates their distribution in microbial membranes [28]. Saint Jean et al. used the peptide C18G as a model system in which peptide variants were synthesized by replacing Leu residues with other hydrophobic amino acids and 2-aminoisobutyric acid. This led to the conclusion that hydrophobic groups in the AMP sequence play an important role in binding and the ability of the peptide to exhibit antibacterial activity. Therefore, controlling the hydrophobic groups in AMPs could be an approach to tune the activity to an optimal therapeutic window [29].

The study on the relationship between structure and antioxidant activity has shown that the antioxidant capacity of peptides is closely related to some structural features such as molecular mass, amino acid composition, sequences, and hydrophobicity [30]. A higher content of hydrophobic amino acids compared to hydrophilic amino acids was found in peptides with high antioxidant activity, which is considered to be a key factor for the radical scavenging ability of peptides [31,32]. The groups of Chanput and Li investigated the QSAR (Quantitative Structure–Activity Relationships) concept for antioxidant peptides, and more specifically the antioxidant activities of tripeptide libraries against the peroxidation of linoleic acid. Chanput et al. pointed out that hydrophobic amino acids such as Ile, Leu, Phe, Trp, Tyr, and Val are required for the antioxidant activity of peptide segments [33], while Li et al. showed that the *N*-terminal amino acid should be a highly hydrophobic amino acid (such as Ala, Gly, Val, and Leu), while the middle amino acid should have high hydrogen-bonding ability (such as basic Arg, Lys, and His) [34,35].

The work of Ohashi et al. discussed the correlation between six different antioxidant assays in the structure–activity relationship of antioxidant peptides based on the free radical scavenging activities of two series of tripeptide libraries [36]. The highest superoxide radical and reducing power activities were shown in the nanofiltrat fraction (1–4 kDa) from dark tuna muscle by-products containing antioxidant amino acids such as Tyr, Phe, Pro, Ala, His, and Leu, as reported by Saidi [37].

Ferrocene dipeptides derived from Fca (**III**) and hydrophobic amino acids (Phe, Val, and Leu) showed antimicrobial and antioxidant activity as well as high resistance to proteolytic cleavage, suggesting that the ferrocene scaffold plays an important role in stabilizing the dipeptides against enzymatic degradation [38]. In the same study, we have determined that the chirality of the amino acid groups, polarity of the protecting groups, and hydrophobicity influence the antimicrobial and antioxidant activity of the compounds due to the different patterns of hydrogen bonds formed. In recent research, derivatives of Fcda (**V**) and various hydrophobic amino acids (L-/D-Val, L-/D-Leu, and L-/D-Phe) were prepared to investigate whether Fcda induces a β–turn upon conjugation with an amino acid and whether the bulkiness of the Phe, Val, and Leu side-branches affects the pattern of hydrogen bonds [20].

In view of the previously mentioned results of the conformational analysis of ferrocene peptides, which point to the role of the ferrocene scaffold as a nucleator of turn-like structures, we decided to investigate the influence of the structurally different ferrocene scaffolds [Fcd (**I**), Fca (**III**), and Fcda (**V**)] (Figure 2) and the bulkiness of amino acid side chains on the conformational and electrochemical properties of bioconjugates **II** (**1–3**), **IV** (**4–6**), and **VI** (**7–9**) using IR, NMR, and CD spectroscopy, as well as cyclic voltammetry.

To determine how the aforementioned factors influence biological activity, the obtained compounds were analyzed for their antioxidative properties using the DPPH and ABTS methods. In addition, the obtained antioxidant activity of compounds **1–9** was further rationalized using electrochemical measurements. To the best of our knowledge, we have not found any examples in the literature dealing with the comparability of three types of ferrocene synthons and their influence on antioxidant activity.

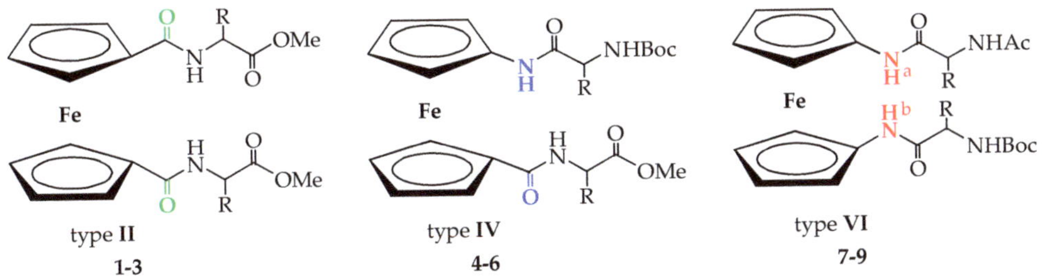

[**Ra** = –CH(CH$_3$)$_2$, **Rb** = –CH$_2$CH(CH$_3$)$_2$, **Rc** = –CH$_2$Ph]

Figure 2. Symmetrically disubstituted ferrocene peptides **II**, **IV**, and **VI**.

2. Results and Discussion

All spectroscopic measurements related to conformational analysis can be found in the Supplementary Materials. Although the synthesis of the target compounds has already been described in previous publications [6,16,20,39,40], we repeated all IR, NMR, and CD measurements to ensure the same conditions and reliable correlation.

2.1. Infrared Spectroscopy (IR Spectroscopy)

Infrared spectroscopy (IR) is an instrumental method that measures the wavelength and intensity of the absorption of infrared radiation and provides a reliable insight into the conformational space of peptides in a solution, based on NH and CO absorptions [41]. Stretching frequencies above 3400 cm^{-1} are attributed to free NH groups, while NH bands below 3400 cm^{-1} indicate their involvement in hydrogen bonding (Table 1, Figure 3). Absorption bands of carbonyl ester groups were detected below 1730 cm^{-1}, also indicating their involvement in hydrogen bonding [42].

In dichloromethane solutions of tripeptides **1–9** ($c = 5 \times 10^{-2}$ M), distinct signals were observed in the region of the NH stretching vibrations, with the signals of the associated NH dominating (<3400 cm^{-1}) (Table 1). The graph (Supplementary Materials Figure S1) showing the stretching vibrations in the region of the NH groups of the peptides prepared from Fcd (**I**) demonstrates that almost all NH groups are associated.

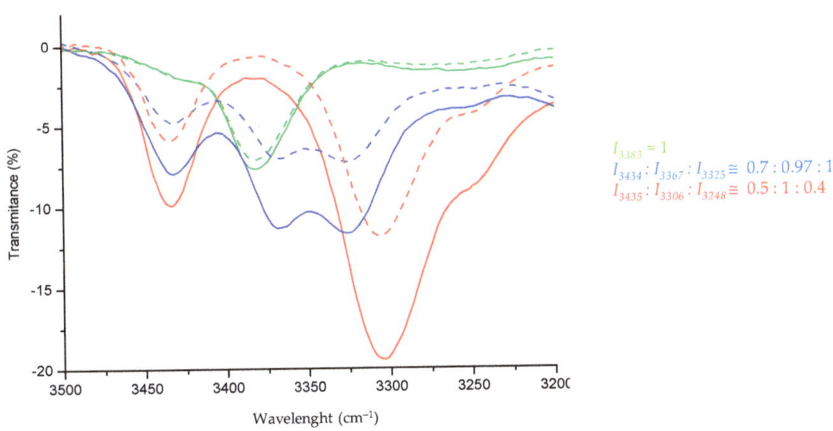

Figure 3. The NH stretching vibrations of **1** (—), **4** (—), and **7** (—) [R$_a$ = –CH(CH$_3$)$_2$], in IR spectra [CH$_2$Cl$_2$ [(—) $c = 5 \times 10^{-2}$ M; (···) c = 2.5×10^{-2} M].

Table 1. Stretching frequencies [ν (cm^{-1})] of NH and CO groups of **1–9** in CH_2Cl_2 ($c = 5 \times 10^{-2}$ M).

Compound	ν_{NH} (Free)	ν_{NH} (Associated)	ν_{CO} (Ester)
MeO−Val−CO−Fn *−CO−Val−OMe (**1**)	3430	3389	1729
MeO−Leu−CO−Fn−CO−Leu−OMe (**2**)	3429	3372	1729
MeO−Phe−CO−Fn−CO−Phe−OMe (**3**)	3434	3378	1730
MeO−Val−CO−Fn−NH−Val−Boc (**4**)	3433	3369, 3326	1720
MeO−Leu−CO−Fn−NH−Leu−Boc (**5**)	3434	3372, 3322	1724
MeO−Phe−CO−Fn−NH−Phe−Boc (**6**)	3430	3375, 3326	1725
Ac−Val−NH−Fn−NH−Val−Boc (**7**)	3422	3324	1710
Ac−Leu−NH−Fn−NH−Leu−Boc (**8**)	3425	3320	1708
Ac−Phe−NH−Fn−NH−Phe−Boc (**9**)	3416	3324	1709

* Fn = ferrocenylene.

The ratio of the bonded and non-bonded NH groups in the IR spectra of peptides **1**, **4**, and **7**, obtained by the conjugation of Val with ferrocene scaffolds **I**, **III**, and **V**, allows us to conclude which ferrocene scaffold favors the participation of NH groups in IHB (Figure 3). It can be seen that the formation of hydrogen bonds is most favored in the presence of Fcda (**V**) as the nucleating scaffold, since the proportion of the associated NH band is highest in peptide **7** (0.5:1). The lowest potential to induce IHBs is shown for the scaffold Fca (**III**), since the ratios of free and associated NH bands are 0.7:1.

Figure S1 shows that there is a pattern to how side-branching of the hydrophobic amino acids affects the intensity of the associated and free bands, as the ratio of free and associated NH bands shows that the bulky benzyl side chains strongly interfere with hydrogen bonding. In peptides **III** and **V**, the excessive ratio of free and associated NH bands (0.7:1 and 1:1) indicates a higher proportion of non-bonded states in conjugates **6** and **9**, with phenylalanine compared to conjugates with valine (**4** and **7**) and leucine (**5** and **8**) (0.5:1 and 0.6:1).

To further characterize the inter- or intramolecular nature of hydrogen bonding (association), it is necessary to test the concentration-dependence of the IR spectra. With a gradual dilution of the solution of the tested sample, the intermolecular hydrogen bonds tend to cleave, resulting in a decrease in the intensity of the associated NH bands compared to those of the free NH groups. In contrast, when intramolecular hydrogen bonds are formed, the ratio of free and associated bands remains unchanged. Here, the concentration-independent IR spectra of peptides **1–9** indicate the intramolecular character of the hydrogen bonds (Supplementary Materials S3, S8, S13, S18, S23, S28, S33, S38, and S43).

2.2. Nuclear Magnetic Resonance Spectroscopy

Nuclear magnetic resonance (NMR) is a method that can be used not only to determine the chemical structure of molecules, but also to gain a better insight into the patterns of hydrogen bonding. This method can be applied to distinguish whether individual NH groups are involved in hydrogen bonds. Thus, if the signal of the NH group is downfield shifted (above 7 ppm), this group is involved in hydrogen bonding [43,44].

A common feature of conjugates **1–9** is a high chemical shift of the amide protons of the ferrocene scaffold (NH_{Fn}), indicating their involvement in hydrogen bonding. Considering that the NH protons of the *N*-protecting groups (NH_{Ac} and NH_{Boc}) are upfield shifted ($\delta < 7.0$ ppm), they are not expected to be involved in IHBs (Table 2).

In peptides **1–3** (type **II**) with an amide bond between Val, Leu, or Phe, respectively, and a carbonyl group of the ferrocene scaffold (**I**), the NH signals are slightly upfield shifted (δ~7.5–7.8 ppm) compared to peptides **IV** (**4–6**) and **VI** (**7–9**) ($\delta > 9.0$ ppm), which originate from scaffolds **III** and **V** with an NH group bound to the ferrocene unit.

Table 2. Chemical shifts [δ/ppm] of the NH protons of **1–9** in $CDCl_3$-*d* ($c = 5 \times 10^{-2}$ M).

Compound	δ (ppm)				
	(NH_{CO})	(NH_{Fn} [a])	(NH_{Fn} [b])	(NH_{Ac})	(NH_{Boc})
MeO−Val−CO−Fn−CO−Val−OMe (**1**)	7.49				
MeO−Leu−CO−Fn−CO−Leu−OMe (**2**)	7.76				
MeO−Phe−CO−Fn−CO−Phe−OMe (**3**)	7.72				
MeO−Val−CO−Fn−NH−Val−Boc (**4**)	7.5	9.01			5.11
MeO−Leu−CO−Fn−NH−Leu−Boc (**5**)	7.62	9.33			5.01
MeO−Phe−CO−Fn−NH−Phe−Boc (**6**)	7.5	8.98			5.20
Ac−Val−NH−Fn−NH−Val−Boc (**7**)		9.21	9.04	6.75	5.21
Ac−Leu−NH−Fn−NH−Leu−Boc (**8**)		9.38	9.16	7.3	5.16
Ac−Phe−NH−Fn−NH−Phe−Boc (**9**)		9.16	9.1	6.9	5.34

[a,b] = denoted on Figure 2.

2.2.1. Concentration-Dependent NMR

Based on the concentration-independent IR specta, we assumed the intramolecular character of the hydrogen bonds in peptides **II** (**1–3**), **IV** (**4–6**), and **VI** (**7–9**), which has to be confirmed by examining the concentration-dependent ^{1}H NMR spectra. However, it should be taken into account that the dilution or cleavage of intermolecular hydrogen bonds causes an upfield shift of the amide protons. As can be seen from the plots in Figure 4, the gradual decrease in $CDCl_3$-*d* concentration (from high-50 mM to low-6.25 mM) did not lead to significant changes in the chemical shifts of the NH_{Fn} protons, which is further confirmation of the intramolecular nature of the hydrogen bonds, as suggested using IR spectroscopy. Even though the concentration-independent IR spectra of peptides **VI** (**7–9**) exclude the presence of intermolecular hydrogen bonds, the significantly changed chemical shifts of the acetamide NH_{Ac} protons upon dilution ($\Delta\delta$~0.6–1 ppm) still indicate the possible presence of intermolecular aggregates (Figure 4c).

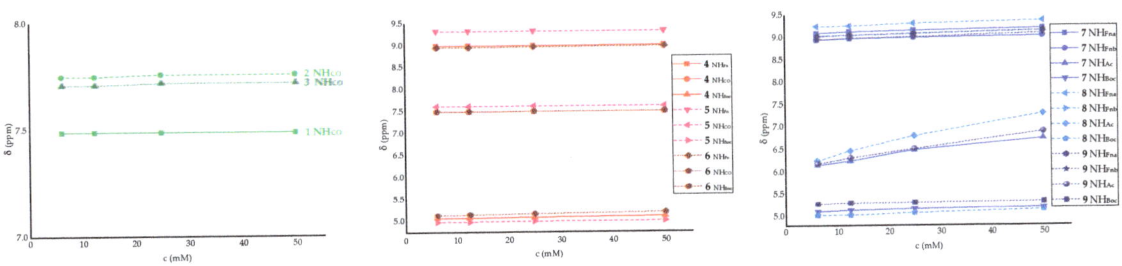

(**a**) (**b**) (**c**)

Figure 4. Concentration dependence of the amide proton chemical shifts for (**a**) **1–3**, (**b**) **4–6** and (**c**) **7–9**.

2.2.2. Temperature-Dependent NMR

To further confirm the intramolecular character of the hydrogen bonds in peptides **II**, **IV**, and **VI**, temperature-dependent imaging of a 12.5 mM peptide solution in $CDCl_3$-*d* in the range of 250–330 K was also performed (Figure 5). In our previous studies [17,18,20], we found that when peptide solutions are heated in the mentioned range, the chemical shifts of the free amide protons or those involved in the formation of IHB remain the same or change slightly, with the change in the proton shift range being up to 0.3 ppm. When amide protons are involved in intermolecular hydrogen bonding, the change in shift is more pronounced ($\Delta\delta$~1–4 ppm).

From the plots of the temperature dependence of peptide **II** (**1–3**), it is evident that the change in the chemical shift of NH is insignificant ($\Delta\delta < 0.4$ ppm), regardless of the constituent amino acid, thus confirming the intramolecular character of the hydrogen bonds in this type of peptide. The same difference in the displacement of amide protons is evident from the temperature dependence diagram of type **IV** (**4–6**) peptides. Considering that the signals of NH_{Fn} in peptides **VI** (**7–9**) have smaller upfield shifts ($\Delta\delta < 0.6$ ppm), it is confirmed that they are involved in strong IHBs. At the same time, the larger upfield shifts ($\Delta\delta > 1.2$ ppm) experienced by NH_{Ac} in dipeptides **8** and **9** are certainly due to their involvement in less stable structures.

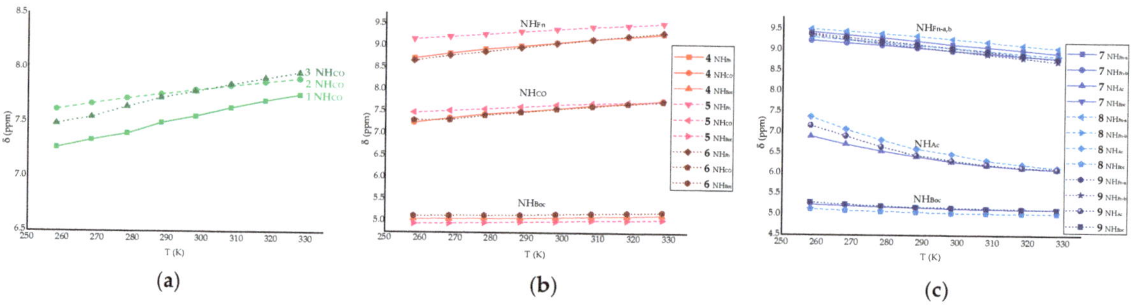

Figure 5. Temperature dependence of the amide proton chemical shifts for (**a**) **1–3**, (**b**) **4–6**, and (**c**) **7–9**.

Intermolecular hydrogen bonds, as well as hydrogen bonds with the solvent, are cleaved at an elevated temperature, which leads to an upfield shift in the NH group signal. In other words, the dependence on low temperatures in $CDCl_3$ is due to exposed and shielded NH groups, while the dependence on high temperatures always indicates NH groups that were shielded at the beginning of the measurement but which were exposed to the solvent upon heating due to dissociation of self-assembled aggregates or dissolution of the ordered conformation. The exposure of the NH groups of the peptides to the solvent is determined by the dependence of the chemical shifts on the temperature, the temperature coefficients ($\Delta\delta/\Delta T$). Larger values correspond to the initially shielded protons (involved in hydrogen bonding) exposed to the solvent during the unfolding of IHB-stabilized structures or the dissociation of aggregates at higher temperatures [45–48]. Therefore, the concentration-independent NH_{Fn} and NH_{CO} show larger temperature coefficients (-3.85 to -9.28 ppb K^{-1}), confirming their involvement in IHB-mediated folding. The large temperature coefficients (-11.57 to -17.71 ppb K^{-1}) of NH_{Ac} further confirms their involvement in self-assembly. Nevertheless, NH_{Boc} shows no significant dependence on temperature as it is not involved in HBs (Table 3).

Thus, the increased temperature dependencies of the chemical shifts of ferrocene peptides **II**, **IV**, and **VI**, which have been shown to be independent of concentration (ruling out the possibility of their intermolecular aggregations), reflect the originally shielded NH groups being exposed to the solvent by breaking the conformations ordered by intramolecular hydrogen bonds.

Table 3. Temperature coefficients ($\Delta\delta/\Delta T$) of the NH protons **1–9** in $CDCl_3$-*d* ($c = 5 \times 10^{-2}$ M).

Compound	$\Delta\delta/\Delta T$ (ppb K^{-1})				
	(NH_{CO})	(NH_{Fn} [a])	(NH_{Fn} [b])	(NH_{Ac})	(NH_{Boc})
MeO−Val−CO−Fn−CO−Val−OMe (**1**)	−7				
MeO−Leu−CO−Fn−CO−Leu−OMe (**2**)	−4.22				
MeO−Phe−CO−Fn−CO−Phe−OMe (**3**)	−6.11				
MeO−Val−CO−Fn−NH−Val−Boc (**4**)	−7.14	−8			−1.28
MeO−Leu−CO−Fn−NH−Leu−Boc (**5**)	−3.85	−5.28			−1.42
MeO−Phe−CO−Fn−NH−Phe−Boc (**6**)	−6.28	−9.28			−1.28
Ac−Val−NH−Fn−NH−Val−Boc (**7**)		−7.28	−6.14	−11.57	−1.57
Ac−Leu−NH−Fn−NH−Leu−Boc (**8**)		−6.57	−6.57	−17.71	−1.57
Ac−Phe−NH−Fn−NH−Phe−Boc (**9**)		−8.71	−9.14	−15.14	−2.42

[a,b] = denoted on Figure 2.

2.2.3. Solvent-Dependence of NMR Chemical Shifts

The strength of the IHBs resulting from the IR and NMR data presented above was tested using titration with DMSO-d_6. While a significant downfield shift of the NH protons exposed to polar DMSO-d_6 is expected, the NH protons involved in strong hydrogen bonds are shielded from DMSO, and their chemical shifts remain almost unchanged [49,50]. It was found that titration of 12.5 mM $CDCl_3$-*d* solutions of peptides **II**, **IV**, and **VI** with DMSO-d_6 significantly affected the chemical shifts of the NH protons of MeO−Phe−CO−Fn−CO−Phe−OMe (**3**) and the $NH_{Ac/Boc}$ protons in peptides **IV** (**4–6**) and **VI** (**7–9**) ($\Delta\delta > 0.6$ ppm), indicating their involvement in weak IHBs. In contrast, no change in chemical shift was observed for the downfield-shifted NH protons of the ferrocene scaffolds upon addition of DMSO-d_6, confirming their involvement in strong IHBs (Figure 6).

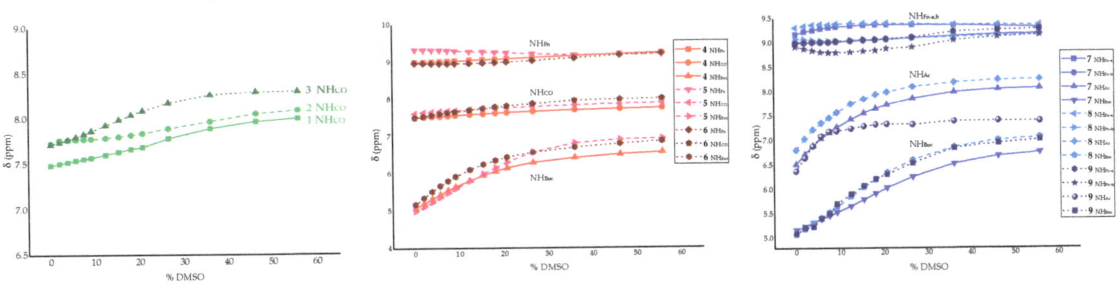

(a) (b) (c)

Figure 6. Solvent dependence of the amide proton chemical shifts of (**a**) **1–3**, (**b**) **4–6**, and (**c**) **7–9**.

2.3. Circular Dichroism (CD)

Circular dichroism (CD) is a spectroscopic method that enables the determination of secondary structure elements in chiral peptides. When a ferrocene scaffold is introduced into the chiral peptide environment to induce the formation of turn- and β-sheet-like structures, the free rotation of the ferrocene rings is disabled, resulting in the helical chirality of the ferrocene core, and the strong Cotton effect is observed in the ferrocene chromophore region (λ~480 nm) [51]. The sign of the Cotton effect is influenced by the protecting groups, the solvent and the type and sequence of the bound natural amino acids [52–54].

As shown in Table 4 and in Supplementary Materials S2, peptides **1–9**, which contain different ferrocene turn-inducing scaffolds, show positive Cotton effects.

Table 4. Cotton effects {[θ]/deg cm^2 dmol^{-1}} of **1–9** in CH_2Cl_2 ($c = 5 \times 10^{-2}$ M) and upon addition of 20% DMSO.

Compound	[θ]/deg cm^2 dmol^{-1}	
	CH_2Cl_2	CH_2Cl_2+DMSO
MeO−Val−CO−Fn−CO−Val−OMe (**1**)	4449	3779
MeO−Leu−CO−Fn−CO−Leu−OMe (**2**)	5595	5032
MeO−Phe−CO−Fn−CO−Phe−OMe (**3**)	2683	2577
MeO−Val−CO−Fn−NH−Val−Boc (**4**)	9923	5364
MeO−Leu−CO−Fn−NH−Leu−Boc (**5**)	10515	8530
MeO−Phe−CO−Fn−NH−Phe−Boc (**6**)	6663	3284
Ac−Val−NH−Fn−NH−Val−Boc (**7**)	25,958	21,125
Ac−Leu−NH−Fn−NH−Leu−Boc (**8**)	21,828	18,709
Ac−Phe−NH−Fn−NH−Phe−Boc (**9**)	15,001	10,994

Peptides **VI** (**7–9**) with ferrocene diamine scaffold (NH−Fn−NH), which were involved in stronger IHBs, also have significantly higher Cotton effects than peptides **IV** (**4–6**) with a CO−Fn−NH scaffold and **II** (**1–3**) with a CO−Fn−CO scaffold. Therefore, ferrocene diamine was found to induce the formation of the most stable chirally ordered structures among the three tested scaffolds.

In addition, a titration of 5 mM CH_2CL_2 solution with DMSO as a competing solvent was performed to test the stability of the folded structures. If the highly ordered structures are adopted, the addition of DMSO will not noticeably affect the intensity of the Cotton effect. As expected, the addition of 20% DMSO caused an attenuation of the Cotton effect by up to 50% for type **IV** peptides (**4–6**), and a less pronounced attenuation for type **II** (**1–3**) and type **VI** peptides (**7–9**) (Table 3, Figure 7).

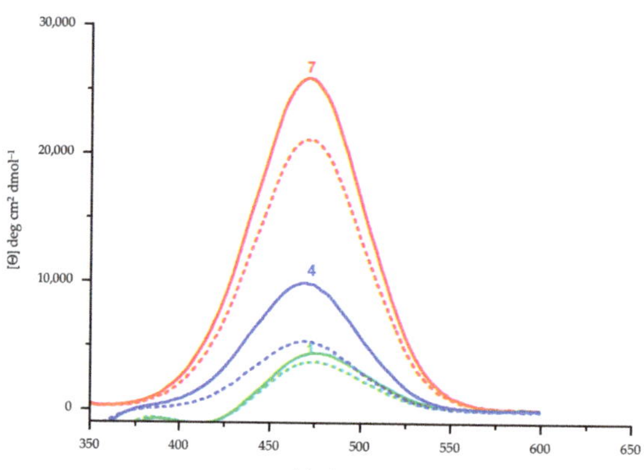

Figure 7. The Cotton effects in chirality-organized ferrocene peptides **1** (green), **4** (blue), and **7** (red) in solution {CH_2Cl_2 [(—) $c = 5 \times 10^{-3}$ M)], CH_2Cl_2 ($c = 5 \times 10^{-3}$ M) containing 20% of DMSO (···)}.

2.4. Antioxidant Activity

The antiradical activity of the tested ferrocene compounds (**1–9**) was estimated using assays to determine the compounds' ability to scavenge the following free radicals: 2′-diphenyl-1-picrylhydrazyl radical (DPPH), and 2,2′-azinobis(3-ethylbenzothiazoline-6-sulfonate) cationic radical (ABTS). The results of radical scavenging activity of the tested compounds at a concentration of 1 mM, expressed as mM Trolox equivalents, are shown in Table 5.

Table 5. Antioxidant activity of ferrocene peptides evaluated using a 1,1-diphenyl-2-picryl-hydrazyl radical scavenging assay (DPPH) and 2,2′-azinobis(3-ethylbenzothiazoline-6-sulfonate) cationic radical scavenging assay (ABTS) of **1–9** ($c = 1 \times 10^{-3}$ M).

Compound	mM Trolx	
	DPPH	ABTS
MeO−Val−CO−Fn−CO−Val−OMe (**1**)	0.047 ± 0.012	0.248 ± 0.093
MeO−Leu−CO−Fn−CO−Leu−OMe (**2**)	0.034 ± 0.019	0.137 ± 0.017
MeO−Phe−CO−Fn−CO−Phe−OMe (**3**)	0.050 ± 0.010	0.114 ± 0.017
MeO−Val−CO−Fn−NH−Val−Boc (**4**)	0.009 ± 0.001	1.651 ± 0.090
MeO−Leu−CO−Fn−NH−Leu−Boc (**5**)	0.008 ± 0.008	0.813 ± 0.036
MeO−Phe−CO−Fn−NH−Phe−Boc (**6**)	0.007 ± 0.000	1.147 ± 0.067
Ac−Val−NH−Fn−NH−Val−Boc (**7**)	0.026 ± 0.001	0.492 ± 0.024
Ac−Leu−NH−Fn−NH−Leu−Boc (**8**)	0.043 ± 0.021	0.878 ± 0.000
Ac−Phe−NH−Fn−NH−Phe−Boc (**9**)	0.028 ± 0.005	1.4617 ± 0.064

The ferrocene compounds tested showed better scavenging activity against ABTS radicals than against DPPH radicals at the same concentration. This is consistent with the literature data suggesting that ferrocene derivatives may be more active in reducing radicals (such as ABTS-+) than in donating electrons or hydrogen atoms to N-centered radicals (such as DPPH˙) [55], whose steric accessibility is critical for antioxidant compounds [56]. Of all the compounds tested, MeO−Phe−CO−Fn−CO−Phe−OMe (**3**) had the highest antioxidant activity, as determined using the DPPH method, having the same range as MeO−Val−CO−Fn−CO−Val−OMe (**1**) and Ac−Leu−NH−Fn−NH−Leu−Boc (**8**). The weakest antioxidant activity is exhibited by group **IV** peptides (**4–6**). Contrary, this peptide class, with −CO−Fn−NH− scaffold shows the best antioxidant activity as determined by the ABTS method. The next group by antioxidant activity is derived from the ferrocene scaffold −NH−Fn−NH− (**7–9**), while the peptides with the scaffold −CO−Fn−CO− (**1–3**) show the weakest antioxidant activity.

Our previous work has shown that peptides derived from Fca (CO−Fn−NH) and Fcda (NH−Fn−NH) have moderate antioxidant activity in the range of 0.1 mM Trolox equivalent as determined using the DPPH method. It is hypothesized that oxidative stress due to abnormal levels of reactive oxygen species (ROS) is involved in the process of carcinogenesis. Highly reactive intermediates of biological oxidation processes interact with intracellular structures such as proteins, nucleic acids, lipids, and membranes and cause their oxidative damage. For this reason, there has recently been an increasing interest in the synthesis and evaluation of various structures, including metal complexes, derived from different metals and organic ligands that serve as protection against oxidative stress. The use of ferrocene as an antioxidant described in alarming detail. Milaeva et al. focused on metal-based antioxidants (2,6-dialkylphenols, flavonoids, polyphenols, N-containing heterocycles, and macrocyclic compounds) and their *in vitro* and *in vivo* activities in cellular oxidation processes [57]. Obviously, the extent of activity of the tested compounds depends on their molecular structures. Liu presented his ongoing studies on antioxidants with ferrocene. He found that the ferrocene moiety can enhance the antioxidant effect even without the phenolic hydroxyl group [58]. The role of ferrocene in enhancing the antioxidant effect could be due to the fact that it scavenges radicals, reduces the oxidative potential, and increases the affinity to the DNA strand. Minić Jančić et al. determined the antioxidant activity of ferrocenes containing a six-membered cyclic urea ring using the following two methods: ABTS radical cation, and DPPH radical scavenging activity. They concluded that some of the synthesized ferrocenyl derivatives possess strong ABTS-+ scavenging activity [59]. Tabrizi et al. prepared the conjugates of aminoferrocene with caffeic acid (CA) and ferulic acid (FA). The evaluation of radical scavenging activity of these conjugates and their ligands towards DPPH-, superoxide anion ($O_2^{\cdot-}$-), NO-, and ABTS-+ using UV−VIS and electron spin resonance spectroscopy as well as DFT showed the higher antioxidant properties of aminoferrocene conjugates compared to their ligands [60]. These results indicate that the ferrocene moiety is crucial for the results of antioxidant activity and that its redox potential changes effectively depending on the Fn-CO or Fn-NH binding mode with amino acids and peptides, as expected based on the electron donor/acceptor ability of the N-H and C=O groups, the hydrogen bonding network, and the dipole moment orientation [61,62].

2.5. Electrochemical Study

Voltametric techniques have been successfully used to investigate redox reactions of many bioactive compounds, since they can mimic the redox mechanisms of organisms *in vitro* [63–66]. Therefore, taking into account the obtained biological/antiradical activity, compounds **1–9** were further evaluated for their electron transfer properties, using cyclic (CV) voltammetry at a glassy carbon (GC) electrode. Accordingly, it was expected that information on the electrochemical mechanisms of the investigated compounds, as well as on the influence of different substituents on redox potentials, would give better insight into their physiological mechanisms of action and provide new information on the structure–electrochemical–antioxidant activity relations.

The electrochemical properties of **1–9** were investigated using cyclic voltammetry in (0.1 M Et_4NClO_4) acetonitrile solution using GC as a working electrode. Figure 8 shows cyclic voltammograms for the oxidation of three groups of ferrocene derivatives (**1–3**, **4–6**, and **7–9**) recorded at a scan rate of 100 mV s^{-1}, within the potential range from −0.6 to 1.4 V. Cyclic voltammetry measurements showed that compounds **1–3** and **4–6** provide a similar voltametric response, i.e., it consists of one pair of redox peaks (A1/C1) at about $E_{1/2}$ = 0.4 V and $E_{1/2}$ = 0.2 V, respectively. These peaks correspond to the reversible charge transfer reaction of the ferrocene fragment on the GC electrode: $[Fe^{II}(C_5H_5)_2] \rightleftharpoons [Fe^{III}(C_5H_5)_2] + e^-$ [67–70]. The difference in peak potentials is due to the structural differences between the two groups of studied compounds, i.e., the presence of different substituents on the ferrocene core. More specifically, the presence of −CO−Fn−NH− scaffold in **4–6** causes a cathodic shift of the redox potential of peaks A1/C1, i.e., the electrochemical oxidation of ferrocene occurs at 250 mV more negative potential compared to compounds **1–3** as a result of the stronger electron-donating effect of -NH- group on Fc core, indicating lower energy required for primary electron transfer. On the other hand, a somewhat different electrochemical behavior was observed for compounds **7–9** under otherwise identical conditions (see Figure 8c). For these compounds, the redox potential of ferrocene shifts to more negative values by an additional 250 mV (relative to **4–6**) due to the presence of −NH−Fn−NH− scaffold, i.e., an additional electron-donating -NH- group on the ferrocene core. Moreover, besides the cathodic shift of the Fc redox potential, the voltammograms of **7–9** show two new oxidation peaks A2 and A3 at potentials around 0.9 V and 1.1 V, respectively. According to the literature [71], these anodic peaks probably correspond to the irreversible one-electron oxidation of the secondary amine, giving a radical cation, followed by a chemical reaction giving electro-inactive products. However, a deeper insight into the origin of these oxidation peaks and their identification are not the subject of this study.

Finally, these observations are somewhat consistent with the observed antioxidant properties of **1–9**, and support the fact that compounds with lower oxidation potential are more efficient radical scavengers. However, although it was expected that compounds **7–9** with the lowest oxidation potential would exhibit the strongest antioxidant activity, this was not the case. The most effective $ABTS^+$ radical quenchers were peptides with the -CO−Fn−NH- scaffold, i.e., those with moderately low oxidation potential. This phenomenon can be explained by the fact that the antioxidant action of investigated compounds towards $ABTS^+$ radicals may involve either electron (ET) or hydrogen atom transfer (HAT) (or both), and not exclusively the transfer of electrons. In other words, the oxidation potentials of investigated compounds were found to be more sensitive to the ferrocene scaffolds into the peptide sequences, which leads to a relative discrepancy between the scavenging activity and oxidation potential.

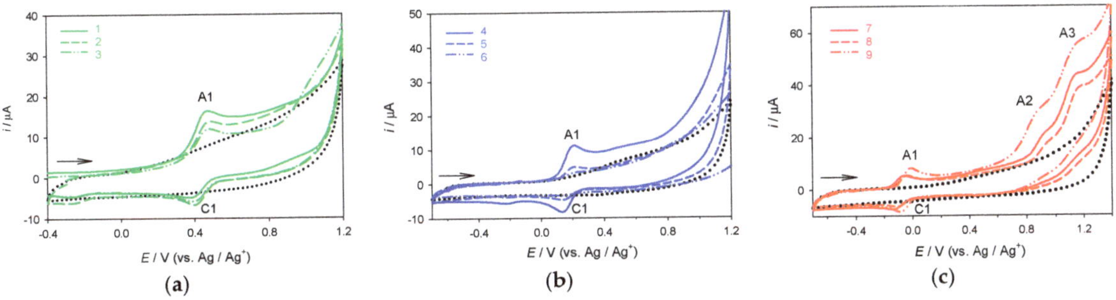

Figure 8. Cyclic voltammograms of 0.5 mmol/L solutions of (**a**) **1–3**, (**b**) **4–6** and (**c**) **7–9** on GC electrode, in CH_3CN (0.1 M Et_4NClO_4). The scan rate was 100 mV s^{-1}. Dotted (black) lines represent cyclic voltammograms of the supporting electrolyte; the arrows indicate the scan direction.

The electrochemical (CV) peak potentials for the investigated compounds taken at the scan rate of 100 mV s^{-1} are shown in Table 6.

Table 6. The values of oxidation (A1, A2, and A3) and reduction (C1) CV peak potentials of compounds **1–9** in CH_3CN (0.1 mol/L Et_4NClO_4) on GC electrode. The scan rate was 100 mV s^{-1}.

Compound	1	2	3	4	5	6	7	8	9
E_p, A1/V	0.45	0.47	0.47	0.20	0.20	0.21	−0.07	−0.04	−0.02
E_p, C1/V	0.39	0.40	0.40	0.14	0.14	0.15	−0.13	−0.10	−0.08
E_p, A2/V	-	-	-	-	-	-	0.93	0.95	0.86
E_p, A3/V	-	-	-	-	-	-	1.13	1.16	1.11

3. Materials and Methods

The IR spectra of the samples were recorded with a Perkin-Elmer Spectrum Two spectrophotometer as CH_2Cl_2 (Acros Organics, 99.5% for spectroscopy) solutions between NaCl windows. The concentration-dependent recording of the IR spectrum is also carried out using the stepwise binary dilution of a 50 mM ferrocene peptide solution in di-chloromethane.

^{1}H and ^{13}C NMR spectra were recorded in $CDCl_3$-*d* (Euristop, Paris, France, 99.80% D) or DMSO-d_6 (Euristop, Paris, France, 99.80% D), each with tetramethylsilane (TMS, δ = 0.0 ppm) as internal standard, using a Bruker AV600 spectrometer (Bruker BioSpin GmbH, Rheinstetten, Germany) at the Ruđer Bošković Institute A 5 mm broadband probe head with *z*-gradient coils operating at 600.130 MHz for ^{1}H and 150.903 MHz for ^{13}C was used. Concentration-dependent spectra are recorded by diluting the initial 50 mM peptide solution in chloroform, and the temperature-dependent spectra involve recording a 25 mM peptide solution in chloroform at 8 different temperatures (258–328 K).

NMR titration is performed by adding 10 μL DMSO-d_6 to a 25 mM peptide solution in $CDCl_3$-d. The spectra are recorded after each addition of DMSO-d_6 until the chemical shift of the observed NH groups no longer changes. When using $CDCl_3$-d/DMSO-d_6 mixture, calibration was performed using TMS as an internal standard.

CD spectra were recorded using a Jasco-810 spectropolarimeter in CH_2Cl_2 (Acros Organics, 99.5% for spectroscopy). Molar ellipticity coefficients (θ) are in degrees, concentration *c* is in molL^{-1}, and path length *l* is in cm, so the unit for (θ) is deg cm^2 dmol^{-1}. The CD activity of the peptides is investigated in the region of the ferrocene chromophore (λ~480 nm). For this purpose, a 5 mM solution of the peptide is taken up in dichloromethane, and 20% DMSO (Supelco, Darmstadt, Germany, >99.8% for spectroscopy Uvasol®) is added to the same sample. The strength of the Cotton effect before and after the addition of DMSO is examined, and the attenuation of the signal strength is used to draw conclusions about the chiral arrangement of the peptides produced in the solution.

Regarding the DPPH assay, the antiradical activity of ferrocene compounds (concentration of 1 mM) against DPPH (1,1-diphenyl-2-picrylhydrazyl) (Fluka, Buchs, Switzerland, >85%) radicals was measured according to the method of Brand-Williams et al. [72]. Samples (250 μL) were mixed with 0.1 mM DPPH working solution (2 mL). Absorbance at 517 nm was measured with a spectrophotometer after 30 min of incubation in the dark. Ethanol (Kemika, Zagreb, Croatia, p.a.) was used as a control. A calibration curve was constructed using Trolox as the reference antioxidant in the range of 0.01–0.05 mM (R2 = 0.9999). Results were expressed as millimoles Trolox equivalents (mM Trolox equivalents). Measurements were performed in triplicate.

ABTS radical cation scavenging activity. The ABTS radical scavenging activity of the tested compounds was estimated according to the method of Re et al. [73]. ABTS·+ was generated in the reaction of 7 mM stock solution of ABTS [2,2′-azinobis-(3-ethylbenzothiazoline-6-sulfonic acid) diammonium salt] (Sigma Aldrich, Steinheim, Germany, >98 %) with 140 mM potassium persulfate (Fluka, >99 %), and the mixture was left in the dark at room temperature for 16 h before use. Then, the ABTS-+ working solution was prepared by diluting the previously prepared mixture with 96% (*v*/*v*) ethanol (Kemika, Zagreb, Croatia, p.a.) until an initial absorbance value of 0.70 ± 0.02 at 734 nm was obtained. Briefly, 20 μL of sample was added to 2 mL of ABTS-+ working solution and the absorbance at 734 nm was measured after 6 min. Trolox (0.1–1.5 mM) was used as a reference antioxidant (R^2 = 0.9976). A control was prepared with the same volume of ethanol without test compounds or reference antioxidant. ABTS·+ scavenging activity of the samples was expressed in millimoles Trolox (mM Trolox). Measurements were performed in triplicate.

Voltametric measurements. Stock standard solutions of compounds **1–9** (c = 0.01 mol dm^{-3}) were prepared from dry pure substances in acetonitrile (CH_3CN, p.a. Kemika, Zagreb). For the supporting electrolyte, analytical-grade Et_4NClO_4 (Merck, Darmstadt, Germany) was used. Voltametric measurements were carried out using the computer-controlled electrochemical system "PGSTAT 101" (Eco Chemie, Utrecht, The Netherlands), controlled using the electrochemical software "NOVA 1.11". A three-electrode system (BioLogic, Claix, France) with a glassy carbon electrode (GCE) of 3 mm in diameter as a working electrode, Ag/Ag^+ (0.01 M $AgNO_3$ in CH_3CN) as a reference electrode and a platinum wire as a counter electrode was used. All potentials were expressed versus the Ag/Ag^+ (0.01 M $AgNO_3$) reference electrode. Before each run, the glassy carbon working electrode was polished with diamond suspension in spray (grain size 6 μm) and rinsed with ethanol and deionized water. The supporting electrolyte (0.1 M Et_4NClO_4 in CH_3CN) was placed in the electrochemical cell, and the required aliquot of the standard analyte solution was added. Before each experiment, the solution in the electrochemical cell was first degassed with high-purity nitrogen for 5 min, and the nitrogen blanket was maintained thereafter. All experiments were performed at room temperature. The presented results are reported as the mean value of three independent measurements. Cyclic voltammograms were taken in the potential range from −0.6 to 1.4 V.

4. Conclusions

We investigated the structure as well as the pattern of intramolecular hydrogen bonding in peptides containing three different ferrocene scaffolds, Fcd [−CO−Fn−CO− (**II**)], Fca [−CO−Fn−NH− (**IV**)], and Fcda [−NH−Fn−NH− (**VI**)], and the hydrophobic amino acids Val, Leu, and Phe by measuring concentration-dependent IR and NMR spectra, temperature-dependent NMR spectra, CD spectra, and the etitration of NMR and CD samples with DMSO.

The main conclusion from this study is that the hydrogen-bond-donating or accepting capacity of ferrocene scaffold influence the hydrogen bonding pattern more than bulkiness of the amino acid side chain. Conformational analysis revealed that the peptides derived from the hydrogen-bond-donating ferrocene-1,1′-diamine scaffold −NH−Fn−NH−(**7–9**) adopt the most ordered chiral arrangement based on the strongest intramolecular hydrogen bonds. Among the tested peptides, the most pronounced antioxidant activity determined using the ABTS method exhibited conjugates (**4–6**) derived from the −CO−Fn−NH−, while MeO−Phe−CO−Fn−CO−Phe−OMe (**3**) had the highest antioxidant activity, as determined using the DPPH method.

Voltametric analysis revealed that the oxidation of peptide derivatives of ferrocene was influenced by the ferrocene scaffold, showing that peptides **VI** (**7–9**) with a ferrocene diamine scaffold (-NH−Fn−NH-) require lower potential than peptides **IV** (**4–6**) and **II** (**1–3**) with -CO−Fn−NH- and -CO−Fn−CO- scaffold, respectively; which may also contribute to their antioxidant properties. However, the most effective $ABTS^+$ radical quenchers were peptides with a -CO−Fn−NH- scaffold (group **IV**, i.e., with moderately low oxidation potential, implying that the scavenging activity of compounds **1–9** is not mainly governed by their susceptibility to oxidation).

Supplementary Materials: The following supporting information can be downloaded at: https://www.mdpi.com/article/10.3390/inorganics12070195/s1. Figures S1–S49: Figure S1. The NH stretching vibrations in concentration-dependent IR spectra of **1–9** in DCM [CH_2Cl_2, (—)$c = 5 \times 10^{-2}$ M]; Figure S2. Cotton effects of **1–9** {CH_2Cl_2 [(—) $c = 5 \times 10^{-3}$ M), CH_2Cl_2 ($c = 5 \times 10^{-3}$ M) containing 20% of DMSO (···)}; Figure S3. The NH stretching vibrations in concentration-dependent IR spectra of **1** in DCM (CH_2Cl_2, [(—)$c = 5 \times 10^{-2}$ M, (—)$c = 2.5 \times 10^{-2}$ M, (—) $c = 1.25 \times 10^{-2}$ M, (—) $c = 6.13 \times 10^{-3}$ M, (—) $c = 3 \times 10^{-3}$ M]; Figure S4. Cotton effect of **1** {CH_2Cl_2 [(—) $c = 5 \times 10^{-3}$ M), CH_2Cl_2 ($c = 5 \times 10^{-3}$ M) containing 20% of DMSO (···)}; Figure S5. Concentration-dependent NH chemical shifts of compound **1** in $CDCl_3$; Figure S6. Temperature-dependent NH chemical shifts of compound **1** ($c = 2.5 \times 10^{-2}$ M); Figure S7. ^1H NMR spectra of compound **1** at varying concentrations of DMSO in $CDCl_3$ ($c = 2.5 \times 10^{-2}$ M); Figure S8. The NH stretching vibrations in concentration-dependent IR spectra of **2** in DCM (CH_2Cl_2, [(—)$c = 5 \times 10^{-2}$ M, (—)$c = 2.5 \times 10^{-2}$ M, (—) $c = 1.25 \times 10^{-2}$ M, (—) $c = 6.13 \times 10^{-3}$ M, (—) $c = 3 \times 10^{-3}$ M]; Figure S9. Cotton effect of **2** {CH_2Cl_2 [(—) $c = 5 \times 10^{-3}$ M), CH_2Cl_2 ($c = 5 \times 10^{-3}$ M) containing 20% of DMSO (···)}; Figure S10. Concentration-dependent NH chemical shifts of compound **2** in $CDCl_3$; Figure S11. Temperature-dependent NH chemical shifts of compound **2** ($c = 2.5 \times 10^{-2}$ M); Figure S12. ^1H NMR spectra of compound **2** at varying concentrations of DMSO in $CDCl_3$ ($c = 2.5 \times 10^{-2}$ M); Figure S13. The NH stretching vibrations in concentration-dependent IR spectra of **3** in DCM (CH_2Cl_2, [(—)$c = 5 \times 10^{-2}$ M, (—)$c = 2.5 \times 10^{-2}$ M, (—) $c = 1.25 \times 10^{-2}$ M, (—) $c = 6.13 \times 10^{-3}$ M, (—) $c = 3 \times 10^{-3}$ M]; Figure S14. Cotton effect of **3** {CH_2Cl_2 [(—) $c = 5 \times 10^{-3}$ M), CH_2Cl_2 ($c = 5 \times 10^{-3}$ M) containing 20% of DMSO (···)}; Figure S15. Concentration-dependent NH chemical shifts of compound **3** in $CDCl_3$; Figure S16. Temperature-dependent NH chemical shifts of compound **3** ($c = 2.5 \times 10^{-2}$ M); Figure S17. ^1H NMR spectra of compound **3** at varying concentrations of DMSO in $CDCl_3$ ($c = 2.5 \times 10^{-2}$ M); Figure S18. The NH stretching vibrations in concentration-dependent IR spectra of **4** in DCM (CH_2Cl_2, [(—)$c = 5 \times 10^{-2}$ M, (—)$c = 2.5 \times 10^{-2}$ M, (—) $c = 1.25 \times 10^{-2}$ M, (—) $c = 6.13 \times 10^{-3}$ M, (—) $c = 3 \times 10^{-3}$ M]; Figure S19. Cotton effect of **4** {CH_2Cl_2 [(—) $c = 5 \times 10^{-3}$ M), CH_2Cl_2 ($c = 5 \times 10^{-3}$ M) containing 20% of DMSO (···)}; Figure S20. Concentration-dependent NH chemical shifts of compound **4** in $CDCl_3$; Figure S21. Temperature-dependent NH chemical shifts of compound **4** ($c = 2.5 \times 10^{-2}$ M); Figure S22. ^1H NMR spectra of compound **4** at varying concentrations of DMSO in $CDCl_3$ ($c = 2.5 \times 10^{-2}$ M); Figure S23. The NH stretching vibrations in concentration-dependent IR spectra of **5** in DCM (CH_2Cl_2, [(—)$c = 5 \times 10^{-2}$ M, (—)$c = 2.5 \times 10^{-2}$ M, (—) $c = 1.25 \times 10^{-2}$ M, (—) $c = 6.13 \times 10^{-3}$ M, (—) $c = 3 \times 10^{-3}$ M]; Figure S24. Cotton effect of **5** {CH_2Cl_2 [(—) $c = 5 \times 10^{-3}$ M), CH_2Cl_2 ($c = 5 \times 10^{-3}$ M) containing 20% of DMSO (···)}; Figure S25. Concentration-dependent NH chemical shifts of compound **5** in $CDCl_3$; Figure S26. Temperature-dependent NH chemical shifts of compound **5** ($c = 2.5 \times 10^{-2}$ M); Figure S27. ^1H NMR spectra of compound **5** at varying concentrations of DMSO in $CDCl_3$ ($c = 2.5 \times 10^{-2}$ M); Figure S28. The NH stretching vibrations in concentration-dependent IR spectra of **6** in DCM (CH_2Cl_2, [(—)$c = 5 \times 10^{-2}$ M, (—)$c = 2.5 \times 10^{-2}$ M, (—) $c = 1.25 \times 10^{-2}$ M, (—) $c = 6.13 \times 10^{-3}$ M, (—) $c = 3 \times 10^{-3}$ M]; Figure S29. Cotton effect of **6** {CH_2Cl_2 [(—) $c = 5 \times 10^{-3}$ M), CH_2Cl_2 ($c = 5 \times 10^{-3}$ M) containing 20% of DMSO (···)}; Figure S30. Concentration-dependent NH chemical shifts of compound **6** in $CDCl_3$; Figure S31. Temperature-dependent NH chemical shifts of compound **6** ($c = 2.5 \times 10^{-2}$ M); Figure S32. ^1H NMR spectra of compound **6** at varying concentrations of DMSO in $CDCl_3$ ($c = 2.5 \times 10^{-2}$ M); Figure S33. The NH stretching vibrations in concentration-dependent IR spectra of **7** in DCM (CH_2Cl_2, [(—)$c = 5 \times 10^{-2}$ M, (—)$c = 2.5 \times 10^{-2}$ M, (—) $c = 1.25 \times 10^{-2}$ M, (—) $c = 6.13 \times 10^{-3}$ M, (—) $c = 3 \times 10^{-3}$ M];

Figure S34. Cotton effect of **7** {CH_2Cl_2 [(—) $c = 5 \times 10^{-3}$ M), CH_2Cl_2 ($c = 5 \times 10^{-3}$ M) containing 20% of DMSO (···)}; Figure S35. Concentration-dependent NH chemical shifts of compound **7** in $CDCl_3$; Figure S36. Temperature-dependent NH chemical shifts of compound **7** ($c = 2.5 \times 10^{-2}$ M); Figure S37. ^{1}H NMR spectra of compound **7** at varying concentrations of DMSO in $CDCl_3$ ($c = 2.5 \times 10^{-2}$ M); Figure S38. The NH stretching vibrations in concentration-dependent IR spectra of **8** in DCM (CH_2Cl_2, [(—)$c = 5 \times 10^{-2}$ M, (—)$c = 2.5 \times 10^{-2}$ M, (—) $c = 1.25 \times 10^{-2}$ M, (—) $c = 6.13 \times 10^{-3}$ M, (—) $c = 3 \times 10^{-3}$ M]; Figure S39. Cotton effect of **8** {CH_2Cl_2 [(—) $c = 5 \times 10^{-3}$ M), CH_2Cl_2 ($c = 5 \times 10^{-3}$ M) containing 20% of DMSO (···)}; Figure S40. Concentration-dependent NH chemical shifts of compound **8** in $CDCl_3$; Figure S41. Temperature-dependent NH chemical shifts of compound **8** ($c = 2.5 \times 10^{-2}$ M); Figure S42. ^{1}H NMR spectra of compound **8** at varying concentrations of DMSO in $CDCl_3$ ($c = 2.5 \times 10^{-2}$ M); Figure S43. The NH stretching vibrations in concentration-dependent IR spectra of **9** in DCM (CH_2Cl_2, [(—)$c = 5 \times 10^{-2}$ M, (—)$c = 2.5 \times 10^{-2}$ M, (—) $c = 1.25 \times 10^{-2}$ M, (—) $c = 6.13 \times 10^{-3}$ M, (—) $c = 3 \times 10^{-3}$ M]; Figure S44. Cotton effect of **9** {CH_2Cl_2 [(—) $c = 5 \times 10^{-3}$ M), CH_2Cl_2 ($c = 5 \times 10^{-3}$ M) containing 20% of DMSO (···)}; Figure S45. Concentration-dependent NH chemical shifts of compound **9** in $CDCl_3$; Figure S46. Temperature-dependent NH chemical shifts of compound **9** ($c = 2.5 \times 10^{-2}$ M); Figure S47. ^{1}H NMR spectra of compound **9** at varying concentrations of DMSO in $CDCl_3$ ($c = 2.5 \times 10^{-2}$ M).

Author Contributions: Conceptualization. M.K., S.R., D.J. and J.M.; methodology. L.B., M.K. and J.M.; validation. M.Č.S., K.H.Č. and S.R.; formal analysis. M.K., S.R., K.H.Č. and D.J.; investigation. M.K., S.R., K.H.Č., D.J. and M.Č.S.; resources. L.B.; writing—original draft preparation. M.K. and J.M.; writing—review and editing. L.B., S.R., M.Č.S. and D.J.; visualization. M.K.; supervision. L.B.; project administration. L.B.; funding acquisition. L.B. All authors have read and agreed to the published version of the manuscript.

Funding: This work has been fully supported by the Croatian Science Foundation under the project IP-2020-02-9162.

Data Availability Statement: The spectroscopic and biological evaluation data are provided as figures and tables and are included in this paper.

Conflicts of Interest: The authors declare no conflict of interest.

References

1. Dias, C.L.; Karttunen, M.; Chan, H.S. Hydrophobic interactions in the formation of secondary structures in small peptides. *Phys. Rev. E* **2011**, *84*, 041931. [CrossRef]
2. Xiong, H.; Buckwalter, B.L.; Shieh, H.-M.; Hecht, M.H. Periodicity of polar and nonpolar amino acids is the major determinant of secondary structure in self-assembling oligomeric peptides. *Proc. Natl. Acad. Sci. USA* **1995**, *92*, 6349–6353. [CrossRef] [PubMed]
3. Dill, K.A. Dominant forces in protein folding. *Biochemistry* **1990**, *29*, 7133–7155. [CrossRef]
4. Nowick, J.S. Exploring β-Sheet Structure and Interactions with Chemical Model Systems. *Acc. Chem. Res.* **2008**, *41*, 1319–1330. [CrossRef] [PubMed]
5. Kovačević, M.; Kodrin, I.; Cetina, M.; Kmetič, I.; Murati, T.; Čakić-Semenčić, M.; Roca, S.; Barišić, L. The conjugate of β-turn-nucleating ferrocene-1,1′-diamine and amino acids. A novel synthetic approach and conformational analysis. *Dalton Trans.* **2015**, *44*, 16405–16420. [CrossRef] [PubMed]
6. Van Staveren, D.R.; Weyhermüller, T.; Metzler-Nolte, N. Organometallic β-turn mimetics. A structural and spectroscopic study of inter-strand hydrogen bonding in ferrocene and cobaltocenium conjugates of amino acids and dipeptides. *Dalton Trans.* **2003**, *2*, 210–220. [CrossRef]
7. Čakić Semenčić, M.; Kodrin, I.; Molčanov, K.; Kovačević, M.; Rapić, V. Novel ferrocene imide derivatives: Synthesis, conformational analysis and X-ray structure. *Heliyon* **2022**, *8*, e09470. [CrossRef]
8. Moriuchi, T.; Ohmura, S.D.; Moriuchi-Kawakami, T. Chirality Induction in Bioorganometallic Conjugates. *Inorganics* **2018**, *6*, 111. [CrossRef]
9. Kirin, S.I.; Kraatz, H.B.; Metzler-Nolte, N. Systematizing structural motifs and nomenclature in 1,n′-disubstituted ferrocene peptides. *Chem. Soc. Rew.* **2006**, *35*, 348–354. [CrossRef]
10. Herrick, R.S.; Jarret, R.M.; Curran, T.P.; Dragoli, D.R.; Flaherty, M.B.; Lindyberg, S.E.; Slate, R.A.; Thornton, L.C. Ordered conformations in bis(amino acid) derivatives of 1,1′-ferrocenedicarboxylic acid. *Tetrahedron Lett.* **1996**, *37*, 5289–5292. [CrossRef]
11. Xu, Y.; Saweczko, P.; Kraatz, H.-B. 1,1′-Ferrocenoyl–oligoprolines. A synthetic, structural and electrochemical study. *J. Organomet. Chem.* **2001**, *637*, 335–342. [CrossRef]
12. Moriuchi, T.; Hirao, T. Dipeptide-induced chirality organization. *J. Incl. Phenom. Macrocycl. Chem.* **2012**, *74*, 23–40. [CrossRef]
13. Moriuchi, T.; Hirao, T. Design of ferrocene-dipeptide bioorganometallic conjugates to induce chirality-organized structures. *Acc. Chem. Res.* **2010**, *43*, 1040–1051. [CrossRef] [PubMed]

14. Barišić, L.; Čakić, M.; Mahmoud, K.A.; Liu, Y.-N.; Kraatz, H.-B.; Pritzkow, H.; Kirin, S.I.; Metzler-Nolte, N.; Rapić, V. Helically Chiral Ferrocene Peptides Containing 1′-Aminoferrocene-1-Carboxylic Acid Subunits as Turn Inducers. *Chem. Eur. J.* **2006**, *12*, 4965–4980. [CrossRef] [PubMed]
15. Čakić Semenčić, M.; Siebler, D.; Heinze, K.; Rapić, V. Bis- and Trisamides Derived From 1′-Aminoferrocene-1-carboxylic Acid and α-Amino Acids: Synthesis and Conformational Analysis. *Organometallics* **2009**, *28*, 2028–2037. [CrossRef]
16. Kovačević, M.; Molčanov, K.; Radošević, K.; Srček Gaurina, V.; Roca, S.; Čače, A.; Barišić, L. Conjugates of 1′-aminoferrocene-1-carboxylic acid and proline: Synthesis, conformational analysis and biological evaluation. *Molecules* **2014**, *21*, 12852–12880. [CrossRef] [PubMed]
17. Kovačević, M.; Čakić Semenčić, M.; Radošević, K.; Molčanov, K.; Roca, S.; Šimunović, L.; Kodrin, I.; Barišić, L. Conformational Preferences and Antiproliferative Activity of Peptidomimetics Containing Methyl 1′-Aminoferrocene-1-carboxylate and Turn-Forming Homo- and Heterochiral Pro-Ala Motifs. *Int. J. Mol. Sci.* **2021**, *22*, 13532. [CrossRef] [PubMed]
18. Kovačević, M.; Kodrin, I.; Roca, S.; Molčanov, K.; Shen, Y.; Adhikari, B.; Kraatz, H.B.; Barišić, L. Helically Chiral Peptides That Contain Ferrocene-1,1′-diamine Scaffolds as a Turn Inducer. *Chem. Eur. J.* **2017**, *23*, 10372–10395. [CrossRef] [PubMed]
19. Chowdhury, S.; Mahmoud, K.A.; Schatte, G.; Kraatz, H.-B. Amino acid conjugates of 1,1′-diaminoferrocene. Synthesis and chiral organization. *Org. Biomol. Chem.* **2005**, *3*, 3018–3023. [CrossRef]
20. Kovačević, M.; Markulin, D.; Zelenika, M.; Marjanović, M.; Lovrić, M.; Polančec, D.; Ivančić, M.; Mrvčić, J.; Molčanov, K.; Milašinović, V.; et al. Hydrogen Bonding Drives Helical Chirality *via* 10-Membered Rings in Dipeptide Conjugates of Ferrocene-1,1′-Diamine. *Int. J. Mol. Sci.* **2022**, *23*, 12233. [CrossRef]
21. Astruc, D. Why is Ferrocene so Exceptional? *Eur. J. Inorg. Chem.* **2017**, *2017*, 6–29. [CrossRef]
22. Patra, M.; Gasser, G. The medicinal chemistry of ferrocene and its derivatives. *Nat. Rev. Chem.* **2017**, *1*, 0066. [CrossRef]
23. Ludwig, B.S.; Correia, J.D.G.; Kühn, F.E. Ferrocene derivatives as anti-infective agents. *Coord. Chem. Rev.* **2019**, *396*, 22–48. [CrossRef]
24. Sharma, B.; Kumar, V. Has Ferrocene Really Delivered Its Role in Accentuating the Bioactivity of Organic Scaffolds? *J. Med. Chem.* **2021**, *64*, 16865–16921. [CrossRef]
25. Mueller, L.K.; Baumruck, A.C.; Zhdanova, H.; Tietze, A.A. Challenges and Perspectives in Chemical Synthesis of Highly Hydrophobic Peptides. *Front. Bioeng. Biotechnol.* **2020**, *8*, 162. [CrossRef]
26. Borrelli, A.; Tornesello, A.L.; Tornesello, M.L.; Buonaguro, F.M. Cell Penetrating Peptides as Molecular Carriers for Anti-Cancer Agents. *Molecules* **2018**, *23*, 295. [CrossRef]
27. Blaber, M.; Zhang, X.J.; Matthews, B.W. Structural Basis of Amino Acid α Helix Propensity. *Science* **1993**, *260*, 1637–1640. [CrossRef]
28. Saravanan, R.; Li, X.; Lim, K.; Mohanram, H.; Peng, L.; Mishra, B.; Basu, A.; Lee, J.M.; Bhattacharjya, S.; Leong, S.S. Design of short membrane selective antimicrobial peptides containing tryptophan and arginine residues for improved activity, salt-resistance, and biocompatibility. *Biotechnol. Bioeng.* **2014**, *111*, 37–49. [CrossRef] [PubMed]
29. Saint Jean, K.D.; Henderson, K.D.; Chrom, C.L.; Abiuso, L.E.; Renn, L.M.; Caputo, G.A. Effects of Hydrophobic Amino Acid Substitutions on Antimicrobial Peptide Behavior. *Probiotics Antimicrob. Prot.* **2018**, *10*, 408–419. [CrossRef]
30. Zou, T.-B.; He, T.-P.; Li, H.-B.; Tang, H.-W.; Xia, E.-Q. The Structure-Activity Relationship of the Antioxidant Peptides from Natural Proteins. *Molecules* **2016**, *21*, 72. [CrossRef]
31. Wang, B.; Gong, Y.D.; Li, Z.R.; Yu, D.; Chi, C.F.; Ma, J.Y. Isolation and characterisation of five novel antioxidant peptides from ethanol-soluble proteins hydrolysate of spotless smoothhound (*Mustelus griseus*) muscle. *J. Funct. Foods* **2014**, *6*, 176–185. [CrossRef]
32. Ren, Y.; Wu, H.; Li, X.; Lai, F.; Xiao, X. Purification and characterization of high antioxidant peptides from duck egg white protein hydrolysates. *Biochem. Biophys. Res. Commun.* **2014**, *452*, 888–894. [CrossRef] [PubMed]
33. Chanput, W.; Nakai, S.; Theerakulkait, C. Introduction of new computer softwares for classification and prediction purposes of bioactive peptides: Case study in antioxidative tripeptides. *Int. J. Food Prop.* **2010**, *13*, 947–959. [CrossRef]
34. Li, Y.-W.; Li, B.; He, J.; Qian, P. Structure-activity relationship study of antioxidative peptides by QSAR modeling: The amino acid next to C-terminus affects the activity. *J. Pept. Sci.* **2011**, *17*, 454–462. [CrossRef] [PubMed]
35. Li, Y.-W.; Li, B.; He, J.; Qian, P. Quantitative structureactivity relationship study of antioxidative peptide by using different sets of amino acids descriptors. *J. Mol. Struct.* **2011**, *998*, 53–61. [CrossRef]
36. Ohashi, Y.; Onuma, R.; Naganuma, T.; Ogawa, T.; Naude, R.; Nokihara, K.; Muramoto, K. Antioxidant properties of tripeptides revealed by a comparison of six different assays. *Food Sci. Technol. Res.* **2015**, *21*, 695–704. [CrossRef]
37. Saidi, S.; Deratani, A.; Belleville, M.P.; Amar, R.B. Antioxidant properties of peptide fractions from tuna dark muscle protein by-product hydrolysate produced by membrane fractionation process. *Food Res. Int.* **2014**, *65*, 329–336. [CrossRef]
38. Kovačević, M.; Čakić Semenčić, M.; Kodrin, I.; Roca, S.; Perica, J.; Mrvčić, J.; Stanzer, D.; Molčanov, K.; Milašinović, V.; Brkljačić, L. Biological Evaluation and Conformational Preferences of Ferrocene Dipeptides with Hydrophobic Amino Acids. *Inorganics* **2023**, *11*, 29. [CrossRef]
39. Galka, M.M.; Kraatz, H.-B. Electron Transfer Studies on Self-Assembled Monolayers of Helical Ferrocenoyl-Oligoproline-Cystamine Bound to Gold. *Chem. Phys. Chem.* **2002**, *3*, 356–359. [CrossRef]
40. Huang, H.; Mu, L.; He, J.; Cheng, J.-P. Ferrocenyl-Bearing Cyclopseudopeptides as Redox-Switchable Cation Receptors. *J. Org. Chem.* **2003**, *68*, 7605–7611. [CrossRef]

41. Kong, J.; Yu, S. Fourier transform infrared spectroscopic analysis of protein secondary structures. *Acta Biochim. Biophys. Sin.* **2007**, *39*, 549–559. [CrossRef] [PubMed]
42. Ganesh, S.; Jayakumar, R. Role of *N-t*-Boc group in helix initiation in a novel tetrapeptide. *J. Peptide Res.* **2002**, *59*, 249–256. [CrossRef] [PubMed]
43. Pardi, A.; Wagner, G.; Wuthrich, K. Protein conformation and proton nuclear-magnetic-resonance chemical shifts. *Eur. J. Biochem.* **1983**, *137*, 445–454. [CrossRef] [PubMed]
44. Gellman, S.H.; Dado, G.P.; Liang, G.-B.; Adams, B.R. Conformation-directing effects of a single intramolecular amide-amide hydrogen bond: Variable-temperature NMR and IR studies on a homologous diamide series. *J. Am. Chem. Soc.* **1991**, *113*, 1164–1173. [CrossRef]
45. Baxter, N.J.; Williamson, M.P. Temperature dependence of ^{1}H chemical shifts in proteins. *J. Biomol. NMR* **1997**, *9*, 359–369. [CrossRef] [PubMed]
46. Liu, J.-Y.; Sun, X.-Y.; Tang, Q.; Song, J.-J.; Li, X.-Q.; Gong, B.; Liu, R.; Lu, Z.-L. An unnatural tripeptide structure containing intramolecular double H-bond mimics a turn-hairpin conformation. *Org. Biomol. Chem.* **2021**, *19*, 4359–4363. [CrossRef] [PubMed]
47. Berger, N.; Wollny, L.J.B.; Sokkar, P.; Mittal, S.; Mieres-Perez, J.; Stoll, R.; Sander, W.; Sanchez-Garcia, E. Solvent-Enhanced Conformational Flexibility of Cyclic Tetrapeptides. *ChemPhysChem* **2019**, *20*, 1664–1670. [CrossRef] [PubMed]
48. Stevens, E.S.; Sugawara, N.; Bonora, G.M.; Toniolo, C. Conformational analysis of linear peptides. 3. Temperature dependence of NH chemical shifts in chloroform. *J. Am. Chem. Soc.* **1980**, *102*, 7048–7050. [CrossRef]
49. Sladojevich, F.; Guarna, A.; Trabocchi, A. Evaluation of stereochemically dense morpholine-based scaffolds as proline surrogates in β-turn peptides. *Org. Biomol. Chem.* **2010**, *8*, 916–924. [CrossRef]
50. Curran, T.P.; Marques, K.A.; Silva, M.V. Bis(amino acid) derivatives of 1,4-diamino2-butyne that adopt a C_2-symmetric turn conformation. *Org. Biomol. Chem.* **2005**, *3*, 4134–4138. [CrossRef]
51. Kovač, V.; Čakić Semenčić, M.; Kodrin, I.; Roca, S.; Rapić, V. Ferrocene-dipeptide conjugates derived from aminoferrocene and 1-acetyl-1′-aminoferrocene: Synthesis and conformational studies. *Tetrahedron* **2013**, *69*, 10497–10506. [CrossRef]
52. Pignataro, M.F.; Herrera, M.G.; Dodero, V.I. Evaluation of Peptide/Protein Self-Assembly and Aggregation by Spectroscopic Methods. *Molecules* **2020**, *25*, 4854. [CrossRef] [PubMed]
53. Rogers, D.M.; Jasim, S.B.; Dyer, N.T.; Auvray, F.; Réfrégiers, M.; Hirst, J.D. Electronic Circular Dichroism Spectroscopy of Proteins. *Chem* **2019**, *5*, 2751–2774. [CrossRef]
54. Moriuchi, T. Helical Chirality of Ferrocene Moieties in Cyclic Ferrocene-Peptide Conjugates. *Chem. Eur. J.* **2022**, *2022*, e202100902. [CrossRef]
55. Ternansky, R.J.; Draheim, S.E.; Pike, A.J.; Counter, F.T.; Eudaly, J.A.; Kasher, J.S. Structure-activity relationship within a series of pyrazolidinone antibacterial agents. 2. Effect of side-chain modification on in vitro activity and pharmacokinetic parameters. *J. Med. Chem.* **1993**, *36*, 3224–3229. [CrossRef] [PubMed]
56. Bugarinović, J.; Pešić, M.; Minić, A.; Katanić Stanković, J.; Ilic Komatina, D.; Pejovic, A.; Mihailovic, V.; Stevanović, D.; Nastasijevic, B.; Damljanović, I. Ferrocene-containing tetrahydropyrazolopyrazolones: Antioxidant and antimicrobial activity. *J. Inorg. Biochem.* **2018**, *189*, 134–142. [CrossRef] [PubMed]
57. Milaeva, E.R.; Filimonova, S.I.; Meleshonkova, N.N.; Dubova, L.G.; Shevtsova, E.F.; Bachurin, S.O.; Zefirov, N.S. Antioxidative Activity of Ferrocenes Bearing 2,6-Di-Tert-Butylphenol Moieties. *Bioinorg. Chem. Appl.* **2010**, *2010*, 165482. [CrossRef] [PubMed]
58. Liu, Z.Q. Enhancing Antioxidant Effect against Peroxyl Radical-induced Oxidation of DNA: Linking with Ferrocene Moiety! *Chem. Rec.* **2019**, *19*, 2385–2396. [CrossRef]
59. Minić Jančić, A.; Katanić Stanković, J.S.; Srećković, N.; Mihailović, V.; Ilić Komatina, D.; Stevanović, D. Ferrocene-containing tetrahydropyrimidin-2(1H)-ones: Antioxidant and antimicrobial activity. *J. Organomet. Chem.* **2022**, *967*, 122335. [CrossRef]
60. Tabrizi, L.; Nguyen, T.L.A.; Tran, H.D.T.; Pham, M.K.; Dao, D.Q. Antioxidant and Anticancer Properties of Functionalized Ferrocene with Hydroxycinnamate Derivatives-An Integrated Experimental and Theoretical Study. *J. Chem. Inf. Model.* **2020**, *60*, 6185–6203. [CrossRef]
61. Santi, S.; Biondi, B.; Cardena, R.; Bisello, A.; Schiesari, R.; Tomelleri, S.; Crisma, M.; Formaggio, F. Helical versus Flat Bis-Ferrocenyl End-Capped Peptides: The Influence of the Molecular Skeleton on Redox Properties. *Molecules* **2022**, *27*, 6128. [CrossRef]
62. Biondi, B.; Bisello, A.; Cardena, R.; Schiesari, R.; Facci, M.; Cerveson, L.; Rancan, M.; Formaggio, F.; Santi, S. Conformational Analysis and Through-Chain Charge Propagation in Ferrocenyl-Conjugated Homopeptides of 2,3-Diaminopropionic acid (Dap). *Eur. J. Inorg. Chem.* **2022**, *2022*, e202100. [CrossRef]
63. Rodrigues Pontiha, A.D.; Alves Jorge, S.M.; Chiorcea-Paquim, A.M.; Diculescu, V.C.; Oliveira-Brett, A.M. In situ evaluation of anticancer drug methotrexate-DNA interaction using a DNA-electrochemical biosensor and AFM characterization. *Phys. Chem. Chem. Phys.* **2011**, *13*, 5227–5234. [CrossRef]
64. Chiorcea-Paquim, A.M.; Rodrigues Pontiha, A.D.; Oliveira-Brett, A.M. Quadruplex-targeting anticancer drug BRACO-19 voltammetric and AFM characterization. *Electrochim. Acta* **2015**, *174*, 155–163. [CrossRef]
65. Diculescu, V.C.; Oliveira-Brett, A.M. In situ electrochemical evaluation of dsDNA interaction with the anticancer drug danusertib nitrenium radical product using the DNA-electrochemical biosensor. *Bioelectrochemistry* **2016**, *107*, 50–57. [CrossRef]
66. Novak Jovanović, I.; Komorsky-Lovrić, Š.; Lucić Vrdoljak, A.; Popović, A.R.; Neuberg, M. Voltammetric characterisation of anticancer drug irinotecan. *Electroanalysis* **2018**, *30*, 336–344. [CrossRef]

67. Bond, A.M.; McLennan, E.A.; Stojanovic, R.S.; Thomas, F.G. Assessment of conditions under which the oxidation of ferrocene can be used as a standard voltammetric reference process in aqueous media. *Anal. Chem.* **1987**, *59*, 2853–2860. [CrossRef]
68. Neghmouche, N.S.; Khelef, A.; Lanez, T. Electrochemistry characterization of ferrocene/ferricenium redox couple at glassy carbon electrode. *J. Fundam. Appl. Sci.* **2009**, *1*, 23–30. [CrossRef]
69. Batterjee, S.M.; Marzouk, M.I.; Aazab, M.E.; El-Hashash, M.A. The Electrochemistry of Some Ferrocene Derivatives: Redox Potential and Substituent Effects. *Appl. Organomet. Chem.* **2003**, *17*, 291–297. [CrossRef]
70. Rep, V.; Piškor, M.; Šimek, H.; Mišetić, P.; Grbčić, P.; Padovan, J.; Gabelica Marković, V.; Jadreško, D.; Pavelić, K.; Kraljević Pavelić, S.; et al. Purine and Purine Isostere Derivatives of Ferrocene: An Evaluation of ADME, Antitumor and Electrochemical Properties. *Molecules* **2020**, *25*, 1570. [CrossRef]
71. Adenier, A.; Chehimi, M.M.; Gallardo, I.; Pinson, J.; Vila, N. Electrochemical Oxidation of Aliphatic Amines and Their Attachment to Carbon and Metal Surfaces. *Langmuir* **2004**, *20*, 8243–8253. [CrossRef] [PubMed]
72. Brand-Williams, W.; Cuvelier, M.E.; Berset, C. Use of a Free Radical Method to Evaluate Antioxidant Activity. *LWT* **1995**, *28*, 25–30. [CrossRef]
73. Re, R.; Pellegrini, N.; Proteggente, A.; Pannala, A.; Yang, M.; Rice-Evans, C. Antioxidant activity applying an improved ABTS radical cation decolorization assay. *Free Radic. Biol. Med.* **1999**, *26*, 1231–1237. [CrossRef] [PubMed]

inorganics

MDPI

Article

Electronic and Steric Effects on Oxygen Reactivities of NiFeSe Complexes Related to O_2-Damaged [NiFeSe]-Hydrogenases' Active Site

Yuchen Qiao [1,†], Enting Xu [2,†], Yameng Hao [2], Xuemei Yang [2,*] and Ming Ni [1,*]

1 BGI Research, Shenzhen 518083, China; qiaoyuchen@genomics.cn
2 School of Science, Harbin Institute of Technology (Shenzhen), Shenzhen 518055, China; 23S058068@stu.hit.edu.cn (E.X.); yameng.hao@alumni.epfl.ch (Y.H.)
* Correspondence: yangxuemei@hit.edu.cn (X.Y.); niming@mgi-tech.com (M.N.)
† These authors contributed equally to this work.

Abstract: Hydrogen has the potential to serve as a new energy resource, reducing greenhouse gas emissions that contribute to climate change. Natural hydrogenases exhibit impressive catalytic abilities for hydrogen production, but they often lack oxygen tolerance. Oxygen-tolerant hydrogenases can work under oxygen by reacting with oxygen to form inactive states, which can be reactivated to catalytic states by oxygen atom removal. Herein, we synthesized three NiFeSe complexes: (NiSe(CH_3)FeCp, NiSe(CH_3)FeCp* and NiSe(Ph^{NMe2})FeCp) with features of active sites of [NiFeSe]-H_2ases, which are the oxygen-tolerant hydrogenases, and we investigated the influence of electronic and steric factors on the oxygen reaction of these "biomimetic" complexes. In our research, we found that they react with oxygen, forming 1-oxygen species, which is related to the O_2-damaged [NiFeSe] active site. Through a comparative analysis of oxygen reactions, we have discovered that electronic factors and steric hindrance on Se play a significant role in determining the oxygen reactivity of NiFe complexes related to hydrogenases' active sites.

Keywords: hydrogenases biomimetics; oxygen tolerance; electronic effect; steric effect

1. Introduction

With the world's increasing reliance on fossil fuels, the rise of greenhouse gases and looming threat of climate change, there has been a drive to develop alternative fuels and environmentally friendly energy solutions. While solar, wind and water powers are clean, pollution-free and inexhaustible, unfortunately, they are unstable, variable and intermittent, which makes them less convenient, stable or efficient for widespread use. One of the solutions to address this problem is to store or convert these unstable natural energies into stable chemical bonds [1]. Among various energy carriers, hydrogen stands out due to its high energy per unit mass, cleanliness and zero carbon emission [2]. By utilizing hydrogen as an energy carrier, we can store the solar power and/or wind power into the dihydrogen bonds and release the energies when needed, such as in hydrogen vehicles.

So far, the best catalyst reported for hydrogen production has been platinum, which is both expensive and relatively scarce in terms of resources [3]. In the quest to design economic, efficient and sustainable catalysts for hydrogen production, natural catalysts, hydrogenases, have captured the interest of researchers. Hydrogenases, which are metalloenzymes, reversibly catalyze hydrogen gas from protons and electrons [4]. There are three main classes of hydrogenases, which include the [Fe]-, [FeFe]- and [NiFe] hydrogenases [5–10]. The active sites of these metalloenzymes are composed of inexpensive transition metals (iron and nickel), sulfur and simple diatomic ligands (CO and CN) that are effective for π-delocalization and H-bonding. While these natural catalysts demonstrate impressive catalytic abilities, a great challenge in utilizing hydrogenases or their models for industrial purposes is their low oxygen tolerance [11–13].

Oxygen tolerance has been used to describe the ability of hydrogenase to catalyze hydrogen oxidation or evolution under aerobic conditions, and more recently, it has been suggested to be limited for the catalysis of hydrogen oxidation only due to the catalysis possibility of oxygen-sensitive hydrogenases for hydrogen evolution [14]. However, a previous broader definition is still accepted and used in publications [15–17]. Some hydrogenases are reported to be oxygen-tolerant [18–20], and one possible mechanism is that oxygen-tolerant hydrogenases can react with oxygen to form an inactive state that can be subsequently recovered to rejoin the catalytic cycle [21,22]. Among the three classes of hydrogenases, it is reported that the [NiFe] hydrogenases have the highest oxygen tolerance [23]. Within the class of [NiFe] hydrogenases, there are two subclasses: [NiFeS] and the [NiFeSe] hydrogenases (Figure 1a, adapted from a previous study [24]). While sharing similar structures, one of the terminal cysteines in [NiFeS] hydrogenases is replaced by the selenocysteine in [NiFeSe] hydrogenases. However, this small modification results in the difference that the [NiFeSe] hydrogenases exhibit better recovery from oxygenation, suggesting a possible oxygen tolerance [15,16,25,26].

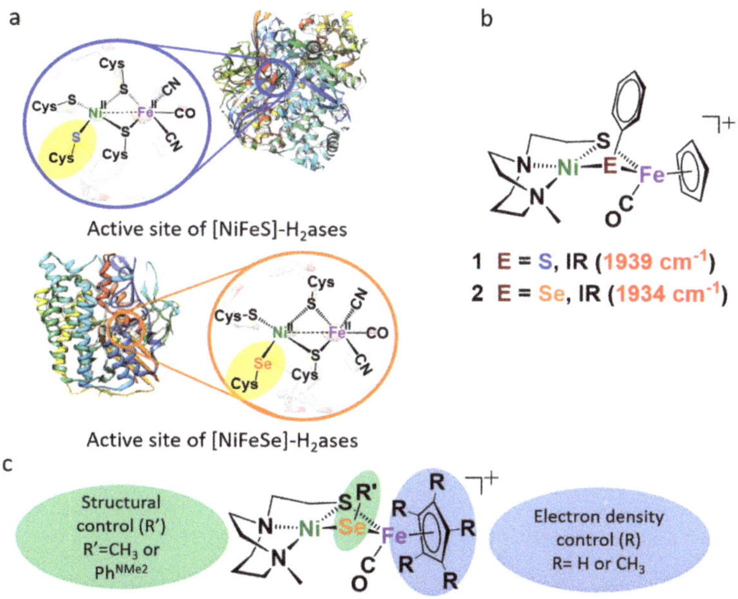

Figure 1. (**a**) The structures and active sites of [NiFeS] and [NiFeSe] hydrogenases. (**b**) A formatted structure of NiFe complexes related to [NiFe] hydrogenases' active sites (previous study [24]). (**c**) A sketch of modifications to the NiFeSe complex.

Although hydrogenases serve as natural catalysts for hydrogen production, they are inconvenient for direct industrial use due to the need for extraction and purification. Therefore, based on the active sites of hydrogenases, chemists and biochemists have designed and synthesized biomimetics [4,9,17,27,28] and bioinspired complexes [29–31], which show promise as catalysts. In addition, several biomimetics have been developed for the study of oxygenated [NiFeS] and [NiFeSe] hydrogenases, which helps to understand oxygen tolerance [24,32–36]. Specifically, in our previous study, the NiFe complex with selenium (Figure 1b) exhibited lower oxygen reactivity and higher oxygen recovery than its all-sulfur analog [24,35,36]. In the presence of oxygen, our sulfur complex was found to preferentially uptake two oxygen atoms that could not be recovered, while the selenium complex exhibited a preference for one oxygen atom that could be removed while adding a reducing agent.

To gain a deeper understanding of the O atom uptake, this study explores both steric hindrance on selenium and electronic modifications at the iron site. As depicted in Figure 1c, R′ was the phenyl group. However, R′ can also be substituted with other groups to change the chemical environment of the center atom, selenium. For example, once the R′ group was substituted by methyl, due to its smaller size, a less steric hindrance effect on the selenium atom should make it much easier to be attacked by oxygen. This would make the NiFeSe complex more oxygen-sensitive in order to generate more oxidated species once exposed under oxygen-containing environments (surprisingly, the results did not match with our expectations, *vide infra*). Additionally, the Cp ring that coordinated on the Fe side could be replaced by η^5-C_5Me_5 (Cp* ring). Because the Cp* group is the more electron-donating group compared to the Cp group, the electron density on Fe would increase and should be able to transfer to the chalcogenide, resulting in a higher electron density on the selenium side. Such an increase of the electron density on Se was expected to facilitate its 1-oxygen uptake, which was also confirmed by computational calculations. However, it should be pointed out that the great CO ligand was maintained, because it is not only a part for mimicking the hydrogenases' active site but also a great monitor for any structure change during the reaction, since it has an obvious band in the IR spectrum.

In summary, we anticipate that these modifications will alter the oxygen reactivity of the NiSeFe complex and lead to the formation of more 1-oxygen–selenium products. By comparing the oxygen reactions of these complexes, our goal is to analyze the effects of the electronic and steric factors on the O_2 reactivities of the NiFe complexes.

2. Results

2.1. Synthesis and Characterizations

Nickel dithiolate complexes are well-known for their reactivity as metalloligands, forming M-(μ_2-SR)$_2$-M′ bridges [37–39]. In 2010, Darensbourg's group reported several cleavage reactions of nickel bimetallic dimer $[NiN_2S]_2^{2+}$ using ligands such as imidazole, pyridine, etc. [40]. In our previous studies, those dimers were also able to be sliced by phenyl-based reductants, such as thiophenolate (SPh), benzeneselenolate (SePh) and their derivatives [24,35]. Based on these studies, the heterobimetallic complex was synthesized in our study. The $[NiNS_2]_2^{2+}$ dimer was divided by SeMe or $SePh^{NMe2}$, and the latter group was much bigger than the former, which could introduce a comparison of the steric hindrance in this system. The attained NiN_2S•SeMe was further reacted with $FeCp(CO)(CH_3CN)^+$ or $FeCp^*(CO)(CH_3CN)^+$ to obtain $NiSe(CH_3)FeCp$ or $NiSe(CH_3)FeCp^*$, respectively. This set, $NiSe(CH_3)FeCp$/$NiSe(CH_3)FeCp^*$, was chosen to investigate the electronic effect, since the Cp* was much more electron rich than Cp. The isolated NiN_2S•$SePh^{NMe2}$ was reacted with $FeCp(CO)(CH_3CN)^+$ to obtain $NiSe(Ph^{NMe2})FeCp$. This complex, in combination with $NiSe(CH_3)FeCp$, was used to analyze the steric effect on the O_2 reaction.

The synthesis routes and IR spectra of these heterobimetallics are shown in Figure 2. We hypothesized that, as the electron-donating ability of the substitution group on chalcogenide increases, more electron density should be able to transfer to iron through chemical bonds, which would result in more π-back-bonding formation between the iron atom and the coordinated carbon monoxide molecule and weaker carbon–oxygen triple bonds. This can be reflected by the decrease of the wavelength of carbon–oxygen bonding (v(CO)). In order to prove our hypothesis, we used an IR spectrophotometer to detect the v(CO) of each synthetic complex. The v(CO) of $NiSe(CH_3)FeCp^*$ was found at 1905 cm^{-1}, while $NiSe(CH_3)FeCp$ exhibited a value of v(CO) = 1929 cm^{-1}, which is in agreement with the fact that Cp* contributed more electrons due to its stronger electron-donating effect of methyl groups to the Fe atom rather than Cp. However, the electron-donating abilities of $SeCH_3$ (the top branch species in Figure 2) and $SePh^{NMe2}$ (the methyl group was substituted by a para-dimethyl amine benzyl group, the bottom branch species in Figure 2) were determined to be equivalent according to their identical v(CO) bands at 1929 cm^{-1}. Thus, when we compared these two complexes, we could exclude the electronic effect and consider the steric effect as the main factor.

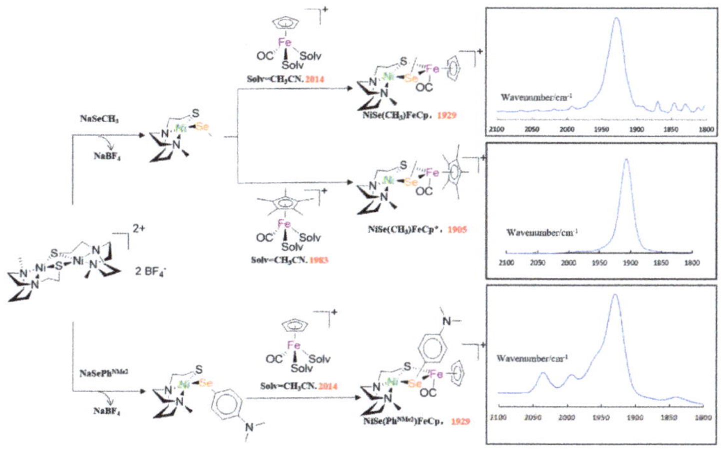

Figure 2. Synthetic routes of NiSeFe derivatives, their corresponding v(CO) values and IR spectra (the red numbers indicated the wavelength of the carbon–oxygen bond, v(CO)).

2.2. Oxygen Tolerance Test

At 22 °C, the heterobimetallic complexes were added to the O_2-saturated DCM solutions, and the reaction progresses were monitored by v(CO) using FTIR spectroscopy. We observed that the initially clear solution of complexes gradually transformed into a slightly cloudy solution. Following filtration, solvent removal and redissolving in DCM, the resulting solution was analyzed by high-resolution mass spectrometry. Based on previous studies, nickel–sulfur complexes were found to preferentially uptake two oxygen atoms, while nickel–selenium species exhibited a preference for one, suggesting the formation of a 1-oxygen uptake species in the three NiSeFe complexes [24,35]. Our previous work also showed that a 2-oxygen uptake NiSeFe complex, if generated, underwent comproportionating with the precursor to form two equivalents of the 1-oxygen complex. In this study, the NiSeFe complexes also followed the 1-oxygen uptake rule and matched with our expectations. It should be pointed out that, in our study, the O atom was bonded to Ni; however, in other, previous NiFe hydrogenase biomimetics studies, the O atom was also found to bind to the Fe moiety. For example, Ogo's group have published several significant studies on the O_2 reaction with biomimetic NiFe complexes in which the O_2 was activated to bond to Fe and perform O_2 reduction [41,42]. In our study, the O atom adopted a different insertion mode, although it was less reactive in O_2 reduction but it more gentle and potentially easier to be reversed to the original structure. Besides, while there was an oxidation state change in most reported studies, in our case, both the Fe and Ni maintained the two oxidation states.

Furthermore, it also suggested that the oxygen reactivity would be influenced by the electron-rich nature and steric hindrance of selenium, leading to the shorter reaction time and higher yield of oxygenates. The results of each O_2 reaction were characterized and analyzed *vide infra*.

2.2.1. The O_2 Reaction of $NiSe(CH_3)FeCp$

In the O_2 uptake reaction between O_2 and NiSe(R′)FeCp (R′ = Me or H), the time-dependent IR spectra for its product, the 1-oxygen uptake species, is presented in Figure 3. The reaction took approximately 35 min to reach completion. IR spectroscopy revealed the disappearance of the CO band at 1929 cm^{-1}, which represented the starting material, while a new band at 1947 cm^{-1} emerged along with the progression of the reaction. Upon the reaction lasting over 35 min, no further peak shifting and peak height changes could be observed. After isolation, the yield of oxygenated products, calculated by weight, was approximately 30%. Furthermore, in the mass spectra, the parent peak and isotope bundle for this complex, which was 490.95 *m/z*, matched with the predicted value (Figure S1).

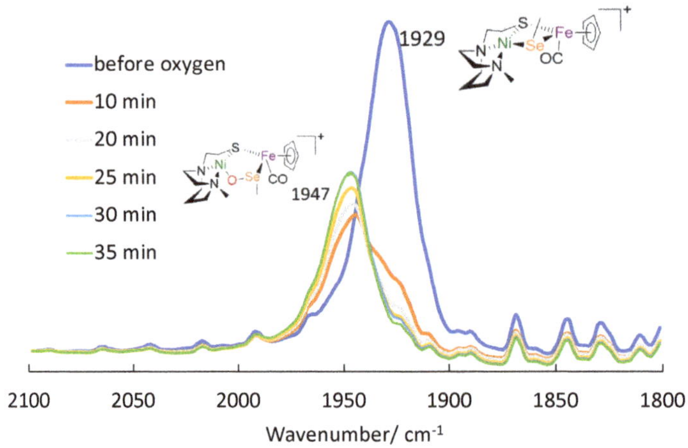

Figure 3. The IR spectra of the initial complex and final product of the O_2 uptake reaction between O_2 and NiSe(CH_3)FeCp.

2.2.2. The O_2 Reaction of NiSe(CH_3)FeCp*

Compared to the v(CO) band of NiSe(CH_3)FeCp mentioned above, the NiSe(CH_3)FeCp*, where the Cp group was switched to Cp* (C_5Me_5), exhibited the same optimized reaction time (35 min), whereas, based on the IR analysis as shown in Figure 4, the v(CO) band shifted from 1905 cm^{-1} to 1934 cm^{-1} after the reaction. It should be pointed out that this reaction was not finished until around 35 min, as shown in the inset of Figure 4b. The shoulder of 1905 (the band of the reactant) was still there at the reaction time of 25 min. Also, one thing to be pointed out in departure from all the other cases, is that the band of 1934 cm^{-1} was broader than those observed in the other oxygenated compounds, indicating that there might be more than one product; more details are analyzed below. The yield of the overall oxygenated products was approximately 58%.

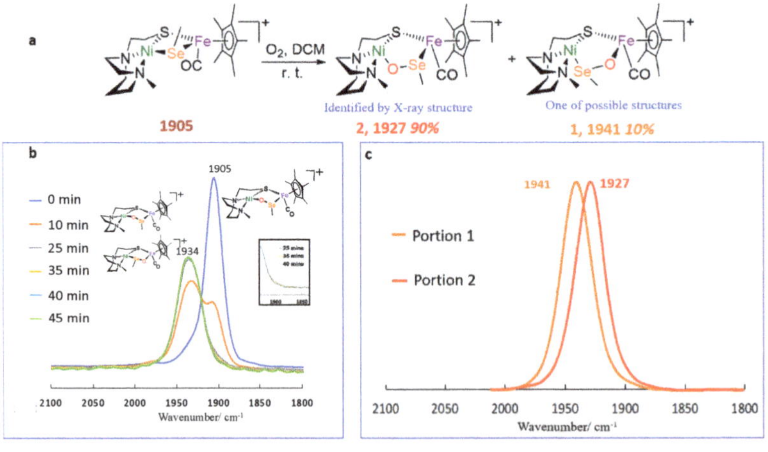

Figure 4. (**a**) Synthetic route of the O_2 uptake reaction between O_2 and NiSe(CH_3)FeCp*. (**b**) IR spectra of the initial complex and final product within the CO band range, the inset indicating the differences between 25, 35 and 40 min traces. (**c**) IR spectra of the two separated portions in the reaction mixture.

To separate the two components in the reaction mixture, we employed the column purification method by using a column filled by aluminum oxide (Al_2O_3). After the O_2 uptake reactions, dark-colored product mixture solutions were filtered and concentrated by vacuum vaporization. Subsequently, the volume-reduced filtrates were separately loaded onto different Al_2O_3 columns for further clean up.

We initially used pure DCM as the eluent to wash down portion 1 (v(CO) = 1941 cm^{-1}), which exhibited an orangish color. Then, a mixture of 10% methanol and 90% dichloromethane was used to elute down the following reddish-colored portion 2 (v(CO) = 1927 cm^{-1}). The yield ratio of portion 1 to portion 2 was approximately one to nine. It is important to note that, in order to visualize the difference better, the intensity of the v(CO) band at 1941 cm^{-1} was multiplied by nine in Figure 4c. We also attempted to influence the yield ratio using other O atom sources besides O_2, including hydrogen peroxide, trimethylamine N-oxide, tert-butyl OOH, etc., but those trials were all unsuccessful.

The parent peak and isotope bundles for this complex mixture, as shown in Figure S2a, matched with the predicted value 561.03 *m/z*. Isolated portion samples were also sent for analysis by mass spectrometry. As shown in Figure S2b,c, the same parent peak at 561.03 *m/z* was found in the $^{+}$ESI-MS spectra. The results strongly suggested that the two portions were likely isomers.

As shown in Figure 5a, the portion 2 species, NiOSeFe, still maintained the hetero-bimetallic structure, and the O atom was inserted into the Ni-Se bond, with the formation of new bond of Ni-O and O-Se due to the fact that the Ni-Se bond was broken. The distance between Ni and Fe was 3.604 Å, indicating that there was no metal–metal bond. From the side view of the structure, as shown in Figure 5b, it can be seen that the five-membered ring of Ni-O-Se-Fe-S was not in a plane but as a "crown". Such a structure might be due to the repelling of electrons or the steric hindrance. According to the structure of portion 2, we assume that the O atom was inserted between Fe and Se in portion 1 due to the higher nucleophile and oxygen reactivity of Se than S [26]. However, the structure of portion 1 was not confirmed, as it was not dominant in the mixture, and further studies are needed to confirm the structure.

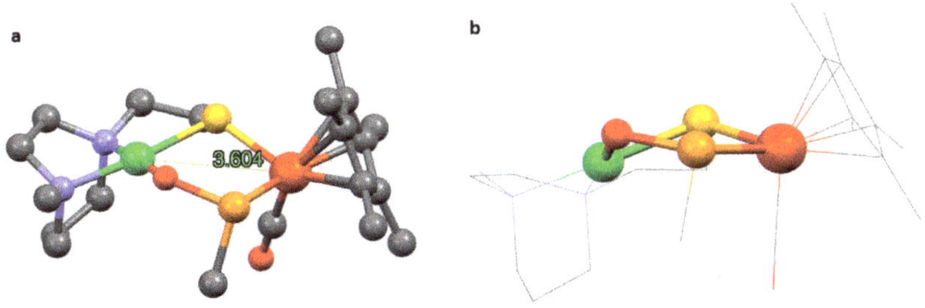

Figure 5. The crystal structure of portion 2, NiOSeFe, in a top-right view (**a**) and side view (**b**); hydrogens are omitted for clarity.

2.2.3. The O_2 Reaction of NiSe(Ph^{NMe2})FeCp

Same as the other biomimetics, the NiSe(Ph^{NMe2})FeCp was dissolved in the O_2-saturated DCM, and the oxygenation reaction was monitored by IR spectroscopy. According to the v(CO) in the IR spectra (Figure 6), within 40 min, the shift was transformed from 1929 cm^{-1} to 1950 cm^{-1}, with a final yield of 61% oxygenated product. Such product was further confirmed as the 1 oxygen uptake complex by mass spectrometry, as the *m/z* signals of 596.0067 matched perfectly with its calculated value (Figure S3).

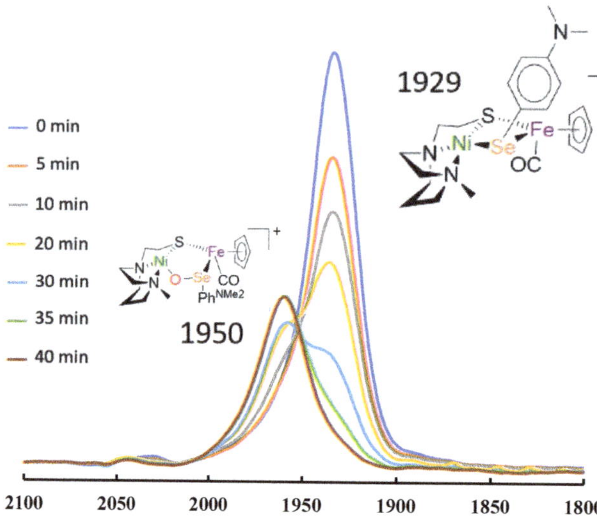

Figure 6. IR spectra of the O_2 reaction of NiSe(Ph^{NMe2})FeCp within the CO band range.

3. Discussion

Taking the three O_2 reactions mentioned above as the conclusion, their reaction parameters are summarized in Table 1. From the reaction time, there was no obvious difference between the three reactions. It should be pointed out that, different from the O_2 bubbling method that some literature has reported (our previous studies [24,35] (from M. Darensbourg's team) included), in this study, we used the O_2-saturated method, dissolving oxygen in DCM solution before the O_2 reaction. Such a modification excluded that the speed of O_2 diffusion in the solution affected the O_2 reaction. The O_2 reactions in this study all finished in 35–40 min, indicating that the differences in electronic or steric hindrance did not affect the reaction time in oxygenation.

Table 1. The overall oxygen reactions of the three NiSeFe complexes.

R′-R	Reaction Time (min)	ν_1(CO) (cm^{-1})	ν_2(CO) (cm^{-1})	Yield (%)
CH_3-Cp	35	1929	1947	30 ± 4
CH_3-Cp*	35	1905	1941, 1927	56 ± 3
Ph^{NMe2}-Cp	40	1929	1950	61 ± 4

3.1. Electronic Effect on the Fe Moiety

With the experimental data of NiSe(CH_3)FeCp and NiSe(CH_3)FeCp* in hand, we could analyze how different substitution groups on Fe caused their distinct O_2 uptake reactivity. Consistent with previous studies, mass spectrometry analysis proved that 1-O uptake was the only oxygenating type of the nickel–selenium complexes [24,35]. However, contrary to our expectations, the oxygen uptake reaction time period of electrons enriched in NiSe(CH_3)FeCp* did not vary at all. There are two assumptions that can be used to explain this experimental phenomenon. First, the electron density generated by the substituted selenium center was sufficient enough to provide the optimized conditions for oxygen to be added, regardless of the donor group on Fe. This step was not the time-limited step for this oxygenating reaction. Second, due to the relatively far distance between the Cp ring and reaction center, selenium, the change in electron density of the iron atom might not have influenced the oxygenating reaction rates at a significant level.

Although there was no substantial variance on the reaction rates of the Cp and Cp* complexes, the yields of their oxygenized compounds still diverged from each other (56%

vs. 30%). Based on our previous DFT (density functional theory) studies, the O_2 uptake intermediates were proposed to be Ni(III)-OO, with bond breaking of Ni-S or Ni-Se at the same time [35]. Despite no oxygen atom being directly added onto the Fe site, it played the role as a Se-R stabilizer and reaction accelerator, which promoted the oxygen reduction progress. For Cp* substitution, the existing five methyl groups on the five-member ring enlarged the electron density of Fe, which might better facilitate an O_2 reduction and steady its reaction intermediate.

To further compare the oxidation reactions of NiSe(CH_3)FeCp and NiSe(CH_3)FeCp*, DFT computations were also performed in this study to address the thermodynamic driving forces for O atom uptake and the different types of oxidized products. As summarized in Figure 7, the free energies, $\Delta G°$, for O insertion into the Ni-Se and Fe-Se bonds in NiSe(CH_3)FeCp were −24.1 and −6.2 kcal/mol, while, in NiSe(CH_3)FeCp*, the numbers were −29.9 and −9.1 kcal/mol. These numbers indicated that the oxidation reaction of NiSe(CH_3)FeCp* was easier than NiSe(CH_3)FeCp, with more negative free energies (ca. −3 to 5 kcal/mol), which was what we expected, since Cp* causes the Fe moiety to obtain a higher electron density than Cp, and it matched with our experimental results that the oxidation product yield of NiSe(CH_3)FeCp* (56%) was higher than the products of NiSe(CH_3)FeCp (30%). Compared to the 2-O insertion type, in both cases, the O atom thermodynamically preferred to insert between Ni and Se rather than Fe and Se.

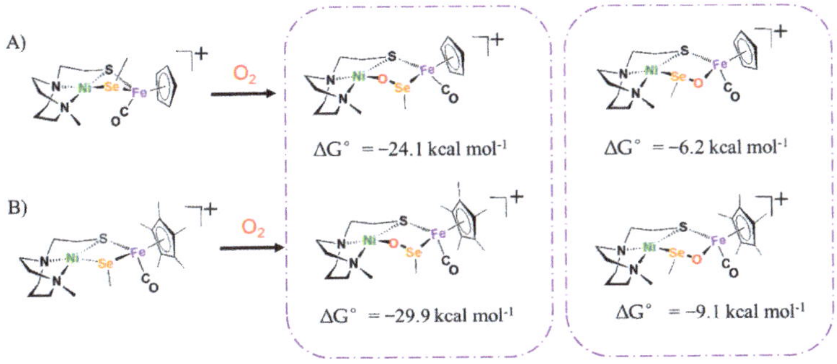

Figure 7. DFT calculated the free energy values, $\Delta G°$, for comparison of the oxygen uptake reactions of (**A**) NiSe(CH_3)FeCp and (**B**) the NiSe(CH_3)FeCp*.

3.2. Intermediate Delocalization Effect on Se

Although the reaction time (35–40 min) remained similar, the yield of NiOSe(Ph^{NMe2})FeCp (61%) was much greater than NiOSe(CH_3)FeCp (30%). Considering the fact that both analogs showed equal C≡O IR vibration values, those two groups' electron-providing capabilities were comparable. Under this circumstance, to some degree, we could compare the steric hindrance between them. However, it should be pointed out that the PhNMe$_2$ group was not suitable to investigate the purely steric effects, and the synthesis of new complexes with more hindered alkyl chains should be explored to better investigate this aspect.

Surprisingly, the experiments aforementioned contradicted our initial thought that the higher yield should result from less steric hindrance, which would make the oxygen molecule proximate Se easier. Due to the possibility of an electronic effect as well, in order to explain why CH_3 and Ph^{NMe2} revealed similar reaction rates but distinct yields, we looked at if any structural features would facilitate the formation of NiOSe(Ph^{NMe2})FeCp. As mentioned in the previous paragraph, the intermediate of such an oxygen uptake reaction was likely to be a peroxide species, which is a kind of radical species. Compared to Se-Ph^{NMe2}, where the radical had a chance to delocalize on the benzene ring and the *para*-nitrogen atom, which led to the increasing of the intermediate stability, the methyl group, due to its relatively small size and the absence of an electron delocalization ability, could not offer any protection for the intermediate radical when reacted with other radical quenchers in the reaction mixture. It is very likely that the radical was quickly consumed by the cations and anions that existed everywhere in the solvent to form variable byproducts, such as Se-Se and di-selenide, after the peroxide intermediates had formed. The demonstrated two-isomer formation additionally supported our assumption. In contrast, in the amino phenyl group case, the •Se- Ph^{NMe2} radical was able to delocalize in the π system in order to prolong its lifetime, which would definitely drive the reaction in the single-oxygen uptake direction. Further experiments will be necessary to prove the aforementioned assumptions.

4. Materials and Methods

4.1. General Materials and Techniques

All reagents and solvents were purchased from Sigma-Aldrich (St. Louis, MO, USA), VWR (Radnor, PA, USA) or Thermo Fisher Scientific (Waltham, MA, USA). All purified organic solvents were further dehydrated by a manual solvent purification system (M. Braun Inc., Stratham, NH, USA) packed with Alcoa F200-activated alumina desiccant. All reactions (except the O_2 oxidation reactions) and operations were performed in a glovebox or on a double-manifold Schlenk vacuum line under an inert gas supply.

4.2. IR Spectrometry Analysis and Reaction Progress Monitoring

Solution infrared spectra were evaluated on a Tensor 37 Fourier-transform IR (FTIR) spectrometer (Bruker, Billerica, MA, USA) using a CaF_2 cell with a 0.2 mm path length. Oxygenation completion was monitored by the change in the IR absorption peaks that corresponded to the C≡O stretching area. The time period the reaction lasted was recorded to represent the optimized reaction rate when there was no change in the intensity for both the reactant and product's C≡O stretching bands.

4.3. X-ray Structure Analysis and Data Processing

Data collections for X-ray structure determination were carried out using APEX2[4] or Venture (Bruker, Billerica, MA, USA) with a graphite mono-chromated radiation source (λ = 0.71073 Å). All crystals were coated in paraffin oil and mounted on a nylon loop and placed under streaming nitrogen gas (110/150K). The structures were solved by direct methods (SHELXS-97) and refined by standard Fourier techniques against F square with a full-matrix least squares algorithm using SHELXL-97 and the WinGX (1.80.05) software package. All non-hydrogen atoms were refined anisotropically. Hydrogen atoms were placed in calculated positions and refined with a riding model. Graphical representations were prepared with ORTEP-III. Crystallographic data (including structure factors) were deposited with the Cambridge Crystallographic Data Centre (no. 2035045).

4.4. DFT Calculation Methods

All of the spin-polarized calculations were based on density functional theory (DFT) and performed by the DMol3 package. The GGA (generalized gradient approximation) in the Perdew–Burke–Ernzerhof form, DSPP (semi-core pseudopotential method) with the double numerical basis sets and the polarization functional (DNP) were chosen for the calculations. The dispersion interaction was accounted by a DFT-D correction with a Grimme scheme. The SCF convergence for every electronic energy was set as 1.0×10^{-7} Ha. The geometry optimization convergence criteria were set up as follows: 1.0×10^{-7} Ha for the energy, 0.001 Ha Å^{-1} for the force and 0.001 Å for displacement, respectively. The free energy (G) for the elemental reaction step was analyzed by the formula

$$G = E + EZPE - TS$$

where E is the total energy of the structures, EZPE is the zero-point energy, T is the temperature and S the change in entropy. Finally, the reaction energies (G) of different intermediates were defined by the formula

$$\Delta G = Gi - Greactant$$

where Gi is the energy of the intermediate, and Greactant is the total energy of the reactant.

5. Conclusions

In our work, we synthesized and analyzed three different NiSeFe heterobimetallics in order to further investigate the buried mystery of what factors make nickel–selenium hydrogenase exhibit a certain oxygen tolerance. The O_2 uptake of the three "biomimetics" are conducted to compare the electronic and steric effects, as well as structural comparisons. Although similar reaction times were observed for all complexes to achieve the optimal yields, the actual yields differed. A higher yield of 1-oxygen species was observed when the Se-Ph^{NMe2} and Cp* groups were present in the biomimetic organometallic complexes. Therefore, we drew the following conclusions:

(a) The three complexes can react with oxygen to form a 1-oxygen species, which is related to the O_2-damaged [NiFeSe] hydrogenases' active sites.
(b) Since NiSe(Ph^{NMe2})FeCp and NiSe(CH_3)FeCp shared identical v(CO) values, Ph^{NMe2} had an electron-donating ability equivalent to that of the CH_3 group. Cp* demonstrated better electron-donating properties than Cp, resulting in increased π-backbonding from Fe to CO and a higher oxygen reactivity.
(c) The -SeMe group, due to its greater lability and reduced steric bulk, produced a pair of isomers that were detected at different IR wavenumbers.
(d) The -SeMe variant, owing to its small size and limited electron delocalization ability of the methyl group, exhibited lower oxygen reactivity—in other words, more decomposed byproducts and a lower yield of 1-oxygen uptake species.
(e) Besides all the effects aforementioned, other factors might influence the complex oxidation reactivity.

In conclusion, first, we found all the complexes could be oxygenated to 1-oxygen species, which had the potential to be recovered. Then, through a comparative analysis of oxygen reactions, we discovered that the higher electron donation and steric hindrance on selenium could increase the oxygen reactivity of our hydrogenase biomimetics. We hope our study can become a modest spur to stimulate researchers to dig deeper into the hydrogenase oxygen tolerance area to discover more appreciated charities to promote the study and production of hydrogen energy.

Supplementary Materials: The following supporting information can be downloaded at https://www.mdpi.com/article/10.3390/inorganics12060163/s1: Figure S1: High-resolution $^{+}$ESI-MS of the 1-O uptake of NiSe(CH_3)FeCp. Figure S2: High-resolution $^{+}$ESI-MS of the 1-O uptake of NiSe(CH_3)FeCp*. Figure S3: High-resolution $^{+}$ESI-MS of the 1-O uptake of NiSe(Ph^{NMe2})FeCp.

Author Contributions: Conceptualization, X.Y. and M.N.; methodology, Y.Q., E.X. and X.Y.; software, E.X. and Y.H.; validation, Y.Q., E.X. and X.Y.; formal analysis, Y.Q. and Y.H.; investigation, Y.Q. and Y.H.; resources, Y.Q. and Y.H.; data curation, Y.Q. and E.X.; writing—original draft preparation, X.Y.; writing—review and editing, Y.Q., Y.H. and M.N.; visualization, Y.Q.; supervision, X.Y. and M.N.; project administration, X.Y. and M.N.; funding acquisition, Y.Q., X.Y. and M.N. All authors have read and agreed to the published version of the manuscript.

Funding: This research was funded by the Shenzhen Science and Technology Innovation Commission, grant number RCBS20221008093334080, the Development and Reform Commission of Shenzhen, grant number XMHT20220103004, and the Shenzhen Science and Technology Program, grant number JCYJ20230807153500001.

Data Availability Statement: The original contributions presented in the study are included in the article/Supplementary Materials; further inquiries can be directed to the corresponding author/s.

Conflicts of Interest: Yuchen Qiao and Ming Ni are employed by the company BGI Research. All the authors declare no conflicts of interest.

References

1. Su, L.; Ajo-Franklin, C.M. Reaching full potential: Bioelectrochemical systems for storing renewable energy in chemical bonds. *Curr. Opin. Biotechnol.* **2019**, *57*, 66–72. [CrossRef] [PubMed]
2. Sarmah, M.K.; Singh, T.P.; Kalita, P.; Dewan, A. Sustainable hydrogen generation and storage—A review. *RSC adv.* **2023**, *13*, 25253–25275. [CrossRef] [PubMed]
3. Gao, X.; Dai, S.; Teng, Y.; Wang, Q.; Zhang, Z.; Yang, Z.; Park, M.; Wang, H.; Jia, Z.; Wang, Y.; et al. Ultra-Efficient and Cost-Effective Platinum Nanomembrane Electrocatalyst for Sustainable Hydrogen Production. *NanoMicro Lett.* **2024**, *16*, 108. [CrossRef] [PubMed]
4. Lubitz, W.; Ogata, H.; Rüdiger, O.; Reijerse, E. Hydrogenases. *Chem. Rev.* **2014**, *114*, 4081–4148. [CrossRef]
5. Heinekey, D.M. Hydrogenase enzymes: Recent structural studies and active site models. *J. Organomet. Chem.* **2009**, *694*, 2671–2680. [CrossRef]
6. Ogata, H.; Hirota, S.; Nakahara, A.; Komori, H.; Shibata, N.; Kato, T.; Kano, K.; Higuchi, Y. Activation process of [NiFe] hydrogenase elucidated by high-resolution X-ray analyses: Conversion of the ready to the unready state. *Structure* **2005**, *13*, 1635–1642. [CrossRef]
7. Shima, S.; Thauer, R.K. A third type of hydrogenase catalyzing H2 activation. *Chem. Rec.* **2007**, *7*, 37–46. [CrossRef]
8. Volbeda, A.; Amara, P.; Darnault, C.; Mouesca, J.M.; Parkin, A.; Roessler, M.M.; Armstrong, F.A.; Fontecilla-Camps, J.C. X-ray crystallographic and computational studies of the O2-tolerant [NiFe]-hydrogenase 1 from Escherichia coli. *Proc. Natl. Acad. Sci. USA* **2012**, *109*, 5305–5310. [CrossRef]
9. Wombwell, C.; Reisner, E. Synthetic Active Site Model of the [NiFeSe] Hydrogenase. *Chemistry* **2015**, *21*, 8096–8104. [CrossRef]
10. Wombwell, C.; Reisner, E. Synthesis, structure and reactivity of Ni site models of [NiFeSe] hydrogenases. *Dalton Trans.* **2014**, *43*, 4483–4493. [CrossRef]
11. Buhrke, T.; Lenz, O.; Krauss, N.; Friedrich, B. Oxygen tolerance of the H2-sensing [NiFe] hydrogenase from Ralstonia eutropha H16 is based on limited access of oxygen to the active site. *J. Biol. Chem.* **2005**, *280*, 23791–23796. [CrossRef] [PubMed]
12. Dey, S.; Rana, A.; Crouthers, D.; Mondal, B.; Das, P.K.; Darensbourg, M.Y.; Dey, A. Electrocatalytic O2 Reduction by [Fe-Fe]-Hydrogenase Active Site Models. *J. Am. Chem. Soc.* **2014**, *136*, 8847–8850. [CrossRef] [PubMed]
13. Frielingsdorf, S.; Fritsch, J.; Schmidt, A.; Hammer, M.; Löwenstein, J.; Siebert, E.; Pelmenschikov, V.; Jaenicke, T.; Kalms, J.; Rippers, Y.; et al. Reversible [4Fe-3S] cluster morphing in an O(2)-tolerant [NiFe] hydrogenase. *Nat. Chem. Biol.* **2014**, *10*, 378–385. [CrossRef] [PubMed]
14. Shafaat, H.S.; Rüdiger, O.; Ogata, H.; Lubitz, W. [NiFe] hydrogenases: A common active site for hydrogen metabolism under diverse conditions. *Biochim. Biophys. Acta* **2013**, *1827*, 986–1002. [CrossRef] [PubMed]
15. Marques, M.C.; Tapia, C.; Gutiérrez-Sanz, O.; Ramos, A.R.; Keller, K.L.; Wall, J.D.; De Lacey, A.L.; Matias, P.M.; Pereira, I.A.C. The direct role of selenocysteine in [NiFeSe] hydrogenase maturation and catalysis. *Nat. Chem. Biol.* **2017**, *13*, 544–550. [CrossRef] [PubMed]
16. Barbosa, T.M.; Baltazar, C.S.A.; Cruz, D.R.; Lousa, D.; Soares, C.M. Studying O(2) pathways in [NiFe]- and [NiFeSe]-hydrogenases. *Sci. Rep.* **2020**, *10*, 10540. [CrossRef]
17. Ahmed, M.E.; Dey, S.; Darensbourg, M.Y.; Dey, A. Oxygen-Tolerant H2 Production by [FeFe]-H2ase Active Site Mimics Aided by Second Sphere Proton Shuttle. *J. Am. Chem. Soc.* **2018**, *140*, 12457–12468. [CrossRef]
18. Pandelia, M.E.; Fourmond, V.; Tron-Infossi, P.; Lojou, E.; Bertrand, P.; Léger, C.; Giudici-Orticoni, M.T.; Lubitz, W. Membrane-bound hydrogenase I from the hyperthermophilic bacterium Aquifex aeolicus: Enzyme activation, redox intermediates and oxygen tolerance. *J. Am. Chem. Soc.* **2010**, *132*, 6991–7004. [CrossRef]
19. Parkin, A.; Goldet, G.; Cavazza, C.; Fontecilla-Camps, J.C.; Armstrong, F.A. The difference a Se makes? Oxygen-tolerant hydrogen production by the [NiFeSe]-hydrogenase from Desulfomicrobium baculatum. *J. Am. Chem. Soc.* **2008**, *130*, 13410–13416. [CrossRef]
20. Lenz, O.; Lauterbach, L.; Frielingsdorf, S. O_2-tolerant [NiFe]-hydrogenases of Ralstonia eutropha H16: Physiology, molecular biology, purification, and biochemical analysis. *Methods Enzymol.* **2018**, *613*, 117–151. [CrossRef]
21. Barilone, J.L.; Ogata, H.; Lubitz, W.; van Gastel, M. Structural differences between the active sites of the Ni-A and Ni-B states of the [NiFe] hydrogenase: An approach by quantum chemistry and single crystal ENDOR spectroscopy. *Phys. Chem. Chem. Phys.* **2015**, *17*, 16204–16212. [CrossRef] [PubMed]
22. Wulff, P.; Day, C.C.; Sargent, F.; Armstrong, F.A. How oxygen reacts with oxygen-tolerant respiratory [NiFe]-hydrogenases. *Proc. Natl. Acad. Sci. USA* **2014**, *111*, 6606–6611. [CrossRef] [PubMed]
23. Baltazar, C.S.A.; Marques, M.C.; Soares, C.M.; DeLacey, A.M.; Pereira, I.A.C.; Matias, P.M. Nickel–Iron–Selenium Hydrogenases—An Overview. *Eur. J. Inorg. Chem.* **2011**, *2011*, 948–962. [CrossRef]
24. Yang, X.; Elrod, L.C.; Reibenspies, J.H.; Hall, M.B.; Darensbourg, M.Y. Oxygen uptake in complexes related to [NiFeS]- and [NiFeSe]-hydrogenase active sites. *Chem. Sci.* **2019**, *10*, 1368–1373. [CrossRef] [PubMed]
25. Wombwell, C.; Caputo, C.A.; Reisner, E. [NiFeSe]-hydrogenase chemistry. *Acc. Chem. Res.* **2015**, *48*, 2858–2865. [CrossRef] [PubMed]
26. Reich, H.J.; Hondal, R.J. Why Nature Chose Selenium. *ACS Chem. Biol.* **2016**, *11*, 821–841. [CrossRef] [PubMed]
27. Rauchfuss, T.B. Diiron azadithiolates as models for the [FeFe]-hydrogenase active site and paradigm for the role of the second coordination sphere. *Acc. Chem. Res.* **2015**, *48*, 2107–2116. [CrossRef] [PubMed]

28. Liu, T.; Li, B.; Singleton, M.L.; Hall, M.B.; Darensbourg, M.Y. Sulfur oxygenates of biomimetics of the diiron subsite of the [FeFe]-hydrogenase active site: Properties and oxygen damage repair possibilities. *J. Am. Chem. Soc.* **2009**, *131*, 8296–8307. [CrossRef]
29. Ahmed, M.E.; Nayek, A.; Križan, A.; Coutard, N.; Morozan, A.; Ghosh Dey, S.; Lomoth, R.; Hammarström, L.; Artero, V.; Dey, A. A Bidirectional Bioinspired [FeFe]-Hydrogenase Model. *J. Am. Chem. Soc.* **2022**, *144*, 3614–3625. [CrossRef]
30. Leone, L.; Sgueglia, G.; La Gatta, S.; Chino, M.; Nastri, F.; Lombardi, A. Enzymatic and Bioinspired Systems for Hydrogen Production. *Int. J. Mol. Sci.* **2023**, *24*, 8605. [CrossRef]
31. Caputo, C.A.; Gross, M.A.; Lau, V.W.; Cavazza, C.; Lotsch, B.V.; Reisner, E. Photocatalytic hydrogen production using polymeric carbon nitride with a hydrogenase and a bioinspired synthetic Ni catalyst. *Angew. Chem. Int. Ed. Engl.* **2014**, *53*, 11538–11542. [CrossRef] [PubMed]
32. Ogo, S. H2 and O2 activation by [NiFe]hydrogenases—Insights from model complexes. *Coord. Chem. Rev.* **2017**, *334*, 43–53. [CrossRef]
33. Lindenmaier, N.J.; Wahlefeld, S.; Bill, E.; Szilvási, T.; Eberle, C.; Yao, S.; Hildebrandt, P.; Horch, M.; Zebger, I.; Driess, M. An S-Oxygenated [NiFe] Complex Modelling Sulfenate Intermediates of an O2-Tolerant Hydrogenase. *Angew. Chem. Int. Ed.* **2017**, *56*, 2208–2211. [CrossRef] [PubMed]
34. Song, L.-C.; Chen, W.; Feng, L. Two heterodinuclear NiFe-based sulfenate complexes mimicking an S-oxygenated intermediate of an O2-tolerant [NiFe]-H2ase: Synthesis, structures, and reactivity. *New J. Chem.* **2020**, *44*, 14015–14023. [CrossRef]
35. Yang, X.; Elrod, L.C.; Le, T.; Vega, V.S.; Naumann, H.; Rezenom, Y.; Reibenspies, J.H.; Hall, M.B.; Darensbourg, M.Y. Controlling O(2) Reactivity in Synthetic Analogues of [NiFeS]- and [NiFeSe]-Hydrogenase Active Sites. *J. Am. Chem. Soc.* **2019**, *141*, 15338–15347. [CrossRef] [PubMed]
36. Yang, X.; Darensbourg, M.Y. The roles of chalcogenides in O(2) protection of H(2)ase active sites. *Chem. Sci.* **2020**, *11*, 9366–9377. [CrossRef]
37. Denny, J.A.; Darensbourg, M.Y. Metallodithiolates as ligands in coordination, bioinorganic, and organometallic chemistry. *Chem. Rev.* **2015**, *115*, 5248–5273. [CrossRef]
38. Fogeron, T.; Todorova, T.K.; Porcher, J.-P.; Gomez-Mingot, M.; Chamoreau, L.-M.; Mellot-Draznieks, C.; Li, Y.; Fontecave, M. A Bioinspired Nickel(bis-dithiolene) Complex as a Homogeneous Catalyst for Carbon Dioxide Electroreduction. *ACS Catal.* **2018**, *8*, 2030–2038. [CrossRef]
39. Ghosh, P.; Quiroz, M.; Wang, N.; Bhuvanesh, N.; Darensbourg, M.Y. Complexes of MN2S2·Fe(η5-C5R5)(CO) as platform for exploring cooperative heterobimetallic effects in HER electrocatalysis. *Dalton Trans.* **2017**, *46*, 5617–5624. [CrossRef]
40. Jenkins, R.M.; Singleton, M.L.; Leamer, L.A.; Reibenspies, J.H.; Darensbourg, M.Y. Orientation and stereodynamic paths of planar monodentate ligands in square planar nickel N2S complexes. *Inorg. Chem.* **2010**, *49*, 5503–5514. [CrossRef]
41. Kishima, T.; Matsumoto, T.; Nakai, H.; Hayami, S.; Ohta, T.; Ogo, S. A High-Valent Iron(IV) Peroxo Core Derived from O2. *Angew. Chem. Int. Ed.* **2016**, *55*, 724–727. [CrossRef] [PubMed]
42. Isegawa, M.; Sharma, A.K.; Ogo, S.; Morokuma, K. DFT Study on Fe(IV)-Peroxo Formation and H Atom Transfer Triggered O2 Activation by NiFe Complex. *Organometallics* **2018**, *37*, 1534–1545. [CrossRef]

inorganics

MDPI

Article

Synthesis, Properties, and Electrochemistry of bis(iminophosphorane)pyridine Iron(II) Pincer Complexes

Nicolás Sánchez López [1], Erick Nuñez Bahena [1], Alexander D. Ryabov [2], Pierre Sutra [3], Alain Igau [3] and Ronan Le Lagadec [1,*]

[1] Instituto de Química, Universidad Nacional Autónoma de México, Circuito Exterior s/n, Ciudad Universitaria, Ciudad de México 04510, Mexico; nicosanlop@comunidad.unam.mx (N.S.L.); ericknb@comunidad.unam.mx (E.N.B.)

[2] Department of Chemistry, Carnegie Mellon University, 4400 Fifth Avenue, Pittsburgh, PA 15213, USA; ryabov@andrew.cmu.edu

[3] CNRS, LCC (Laboratoire de Chimie, de Coordination), Université de Toulouse, CEDEX 4, 31077 Toulouse, France; pierre.sutra@lcc-toulouse.fr (P.S.); alain.igau@lcc-toulouse.fr (A.I.)

* Correspondence: ronan@unam.mx

Abstract: Iron derivatives have emerged as valuable catalysts for a variety of transformations, as well as for biological and photophysical applications, and iminophosphorane can be considered an ideal ligand scaffold for modulating electronic and steric parameters in transition metal complexes. In this report, we aimed to synthesize dichloride and dibromide iron(II) complexes supported by symmetric bis(iminophosphorane)pyridine ligands, starting from readily available ferrous halides. The ease of synthesis of this class of ligands served to access several derivatives with distinct electronic and steric properties imparted by the phosphine moiety. The ligands and the resulting iron(II) complexes were characterized by ^{31}P and ^{1}H NMR spectroscopy and DART or ESI mass spectrometry. While none of these iron(II) complexes could be characterized by single-crystal X-ray diffraction, suitable crystals of a μ-O bridged dinuclear iron complex bearing an iminophosphorane ligand were obtained, confirming a κ^3 binding motif. The bis(iminophosphorane)pyridine ligands in the obtained iron(II) complexes are labile, as demonstrated by their facile substitution by terpyridine. Cyclic voltammetry studies revealed that the oxidation of bis(iminophosphorane)pyridine iron(II) complexes to iron(III) species is quasi-reversible, suggesting the strong thermodynamic stabilization of the iron(III) center imparted by the σ-donating iminophosphorane ligands.

Keywords: pincer ligands; iminophosphorane ligand; iron; cyclic voltammetry

1. Introduction

Iron complexes supported by 2,6-disubstituted pyridine pincer ligands have emerged as valuable catalysts for a variety of transformations, such as olefin polymerization and oligomerization, hydrogenation, and hydrofunctionalization reactions (Figure 1a) [1–3]. While, in some cases, the catalytic activity of these systems has been attributed to the redox-active nature of some of these ligands [4], it is generally accepted that adjusting the electronic properties at the metal center by ligand design is crucial for achieving optimal performance in catalysis [5]. Thus, developing new ligand sets in which ligand tuning can be easily accomplished is a challenging task. Diiminopyridine ligands are one of the most explored scaffolds for developing new transformations mediated or catalyzed by iron (Figure 1a, left). In considering this, we sought to access other types of easy-to-synthesize *N,N,N*-tridentate pincer ligands that could be suitable for tuning both the electronic and steric properties at the iron metal center [3,6].

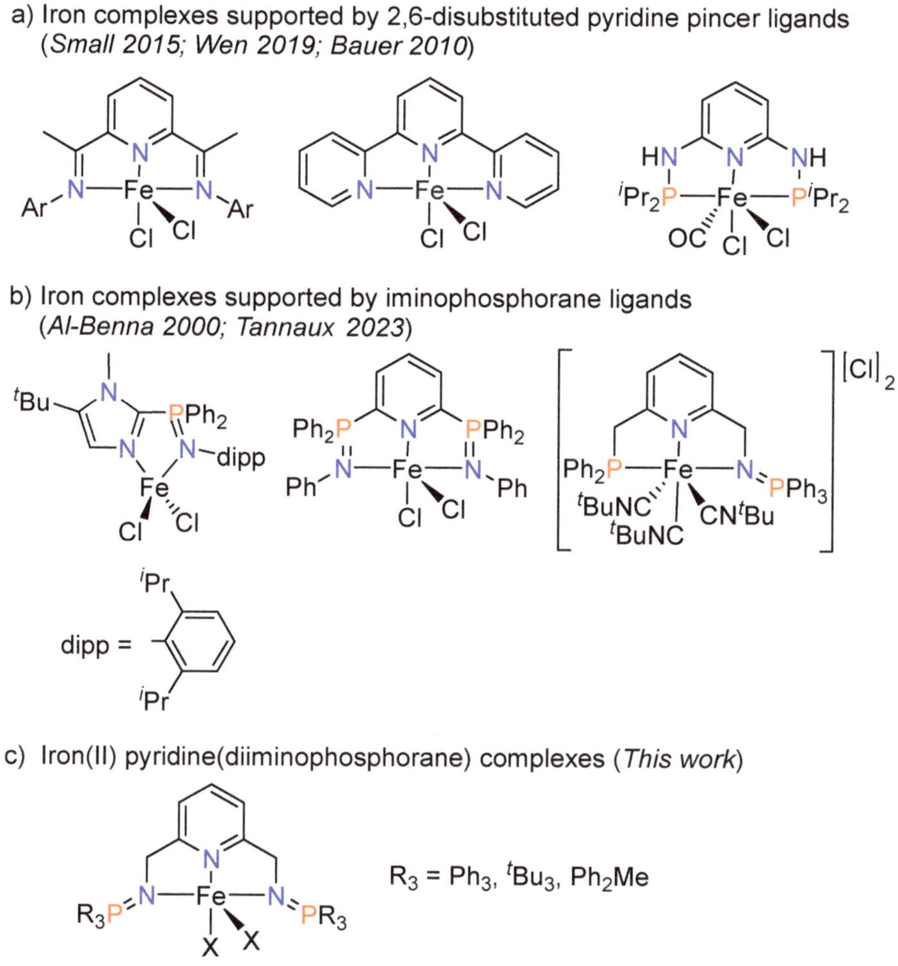

Figure 1. Examples of iron(II) pincer complexes and iminophosphorane iron(II) complexes [1–3,7,8].

Iminophosphoranes, the nitrogen analogs of phosphonium ylides, are suitable ligands for transition metal centers [6]. As shown in Scheme 1, the iminophosphorane motif is better represented by two canonical forms, one with phosphorus and nitrogen having a double bond $R_3P = NR'$ (ylene form) and the other with a single $R_3P^+{-}^-NR'$ bond with a positive charge on phosphorus and a negative charge on nitrogen (ylidic form). While experimental bond parameters support the depiction of iminophosphoranes as the $R_3P = NR'$ form (e.g., bond lengths of 1.54 Å–1.64 Å, and bond angles of 119°–143° in the solid-state) [9], DFT calculations indicate that the rotation energy of the P–N bond is only 2.1 kcal/mol, suggesting a significant contribution of the $R_3P^+{-}^-NR'$ form [10]. Due to the ylidic canonical form, iminophosphorane ligands are predominantly strong σ *N*-donors, resulting in convenient ligands to stabilize transition metal complexes in high oxidation states [11]. As the nature of the $R_3P = NR'$ partial double bonds involves negative hyperconjugation from a nitrogen lone pair p(N) to *s* *(P–C) orbitals [12,13], the σ-donor character of iminophosphoranes can be adjusted by the electronic properties of the phosphine moiety [11]. Consequently, iminophosphorane could be an ideal ligand scaffold for modulating electronic and steric parameters in transition metal complexes.

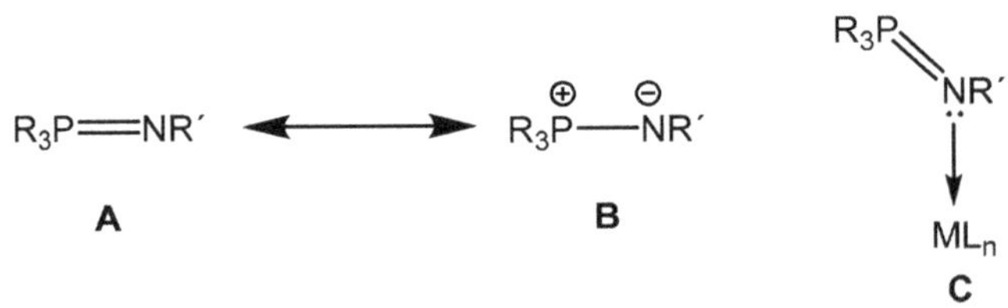

Scheme 1. Canonical forms of an iminophosphorane: ylene (**A**) and ylidic (**B**); its coordination to a metallic center (**C**).

While complexes supported by iminophosphorane ligands are common with lanthanides and early transition metals, examples of more electron-rich metal derivatives with iminophosphorane ligands have also been reported [6]. Iron catalysts bearing iminophosphoranes have been successfully employed for the activation of small molecules [14–16], as well as in ethylene oligomerization [17] and transfer hydrogenation [18].

As observed for other pincer ligands [19], tridentate ligands with one or two iminophosphorane fragments are less prone to engage in ligand substitution reactions than their bidentate and monodentate analogs, especially when bound to electron-rich metal centers [11]. In 2000, Bochmann and co-workers reported the synthesis of sterically hindered 2,6-bis(aryliminophosphoranyl)pyridine pincer complexes of vanadium, iron, cobalt, and nickel (e.g., Figure 1b, center) [7]. More recently, Auffrant and co-workers synthesized a related *P,N,N*-phosphine–pyridine–iminophosphorane iron pincer complex, which showed excellent catalytic activity in the hydrosilylation of acetophenones (Figure 1b) [8]. While these reports represent a significant advance toward using iminophosphoranes in the design of iron pincer complexes, their synthesis required the use of either $LiPR_2$ or Na/NH_3. Therefore, employing other pyridine iminophosphorane pincer ligands in designing iron coordination complexes that could be readily prepared from commercially available phosphines is highly desirable.

The *N,N,N*-pincer ligand bis(methyliminophosphoranyl)pyridine (**ImPh**, see Scheme 2a) is an ideal platform for stabilizing copper [20] and low-valent f elements [21,22]. In the case of copper, the coordination mode of **ImPh** depends on the oxidation state of the metal, with k^3-*N,N,N* and k^2-*N,N* coordination modes for Cu(II) and Cu(I) centers, respectively [20]. Due to the ease of synthesis of **ImPh**, which can be accessed from the corresponding organic diazide and triphenylphosphine, we hypothesized that other derivatives of this ligand scaffold could be obtained similarly using other phosphines. This would generate a set of ligands to modulate the electronic and steric properties of iron(II) pincer complexes. In addition, the potential of bis(iminophosphorane)pyridine ligands to exhibit hemilability could be advantageous for catalytic applications of the resulting iron complexes [20].

Herein, we report the synthesis of new bis(iminophosphorane)pyridine iron(II) pincer complexes. Their iminophosphorane ligands were effortlessly synthesized from commercially available phosphines. Reactivity studies of the complexes with terpyridine demonstrated the lability of the bis(iminophosphorane)pyridine ligands. The electronic properties of the new iron(II) complexes were evaluated by cyclic voltammetry.

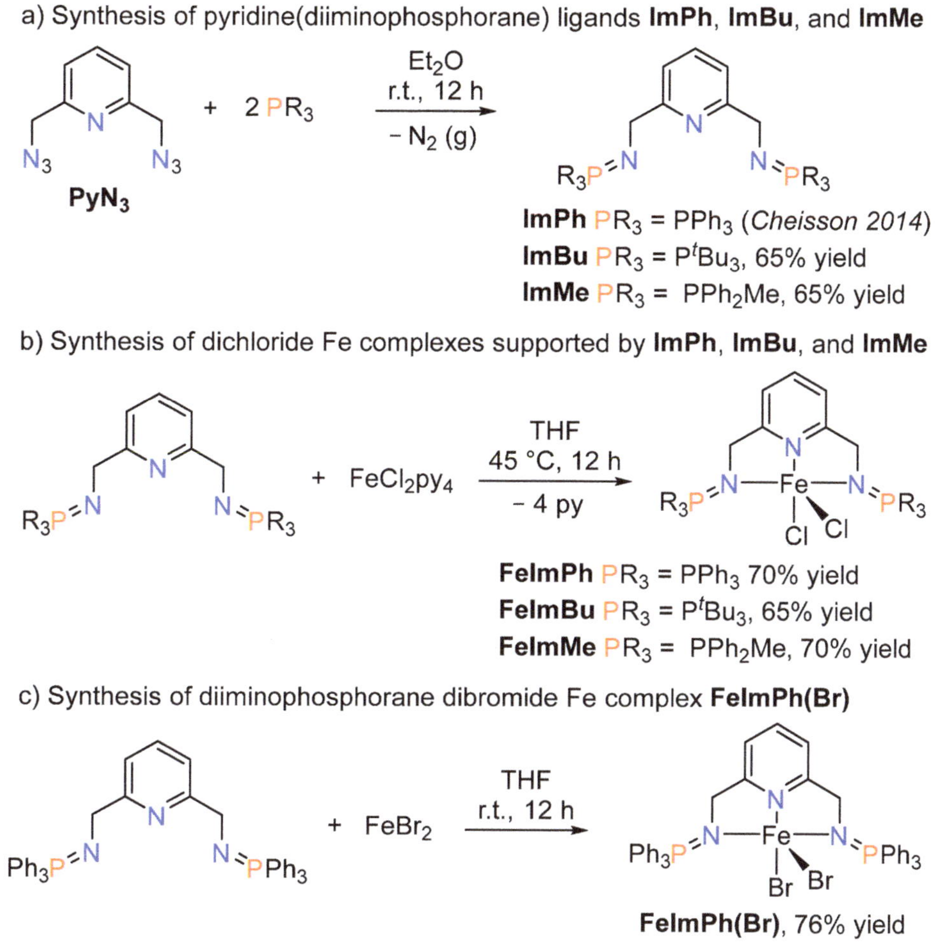

Scheme 2. Synthesis of bis(iminophosphorane)pyridine ligands and their iron(II) complexes [20].

2. Results and Discussion

2.1. Synthesis of Ligands and Iron Complexes

Iminophosphorane derivatives containing different phosphine moieties were synthesized analogously to the known **ImPh** ligand by the Staudinger reaction [20,23]. Accordingly, two equivalents of triphenylphosphine, tri-*tert*-butylphosphine, or diphenylmethylphosphine were added to a solution of 2,6-bis(azidomethyl)pyridine (**PyN3**) in diethyl ether and reacted overnight to obtain **ImPh** and new ligands **ImBu** and **ImMe**, respectively (Scheme 2a).

Iminophosphorane ligands **ImPh**, **ImBu**, and **ImMe** are white solids characterized by ^{1}H and ^{31}P NMR spectroscopy. In the ^{31}P NMR spectra, the resonances associated with the new ligands **ImBu** and **ImMe** appear as singlets at 53.65 and 12.85 ppm, respectively. In the ^{1}H NMR spectra, the methylene resonances of the iminophosphorane ligands appear as a singlet at 4.92 ppm for **ImBu** and as a broad doublet at 4.28 ppm for **ImMe**. Such differences observed in the multiplicity of the signal can be explained in terms of the electronic differences between the ligands **ImPh**, **ImMe**, and **ImBu**, which can lead to significant changes in the coupling constant values [24]. In the ATR-IR spectra, the band corresponding to N = P vibration is observed for **ImPh** at 1240 cm^{-1}, **ImBu** at 1109 cm^{-1}

and **ImMe at** 1223 cm^{-1}. The assignment of these structures was corroborated by MS (FAB+). The air and moisture-sensitive nature of ligands precluded their characterization by elemental analysis.

Next, the coordination of the iminophosphorane ligands to iron(II) was explored (Scheme 2b). The reaction of **ImPh** with $FeCl_2$ resulted in a complex reaction mixture, as observed by ^{31}P NMR spectroscopy. A yellow precipitate was formed when $FeCl_2py_4$ (py = pyridine) was employed as an iron precursor in the reaction with **ImPh** in THF at 45 °C for 12 h. In the ^{31}P NMR spectrum, the newly formed complex presented a characteristic singlet resonance at 38.64 ppm, which is significantly downfield shifted compared to that of the free ligand, thus supporting the coordination of **ImPh** to iron. The presence of only one signal in the ^{31}P NMR spectrum suggests a symmetric κ^3 binding mode in solution (Scheme 1b). When **ImBu** and **ImMe** independently reacted with $FeCl_2py_4$ under the same conditions, the formation of analog iron complexes was observed, as supported by the presence of singlet resonances in the ^{31}P NMR spectra (57.75 ppm for Fe**ImBu**, and 42.31 ppm for **FeImMe**). It should be noted that the presence of a small amount of paramagnetic impurity that could not be entirely removed causes the broadening of the ^{1}H NMR signals. In addition, the dibromide homolog **FeImPh(Br)** could be synthesized directly from $FeBr_2$ and **ImPh** at room temperature using a similar procedure (Scheme 2c). This is supported by the presence of a singlet signal at 42.31 ppm in the ^{31}P NMR spectrum of **FeImPh(Br)**, which is practically identical to that of the dichloride complex **FeImPh**. In the ATR-IR spectrum, the band corresponding to N = P vibration is observed at 1113 cm^{-1} for **FeImPh**, at 1173 cm^{-1} for **FeImBu**, at 1118 cm^{-1} for **FeImMe** and, at 1116 cm^{-1} for **FeImPh(Br)**. The structure of the synthesized bis(iminophosphorane)pyridine iron(II) complexes was corroborated by MS (DART) or MS (ESI). As mentioned for the free ligand, Fe**ImBu** is also highly sensitive to atmospheric conditions, precluding satisfactory elemental analysis [25]. Attempts to obtain MS data for Fe**ImBu** through different techniques such as DART+, FAB+, and ESI were unsuccessful.

Efforts to obtain crystalline material for X-ray diffraction of the bis(iminophosphorane) pyridine complexes under dry and air-free conditions were unsuccessful. Instead, suitable monocrystals of the μ-O-bridged dinuclear iron complex **Fe2ImPh** were obtained from a cold solution of **FeImPh** in THF (Scheme 3).

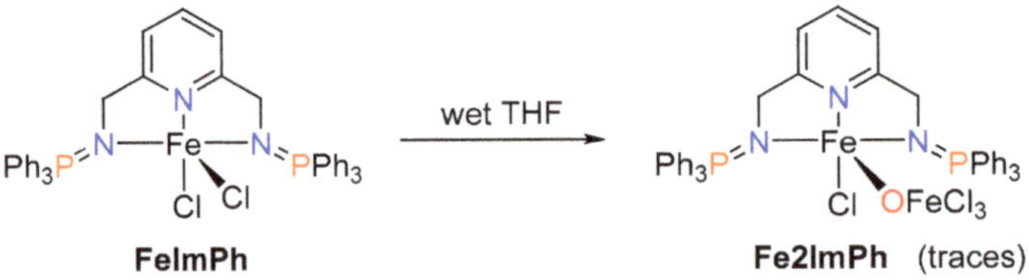

Scheme 3. Obtention of μ-oxo bridged dinuclear iron complex **Fe2ImPh** from **FeImPh**.

The **ImPh** ligand is part of a bimetallic unit with an oxygen bridge that links two iron atoms. This complex is obviously a minor byproduct. The bimetallic complex could form after the dissociation of **ImPh** from **FeImPh** to produce $FeCl_2$, followed by the oxidation of the latter by traces of dioxygen and the final association of a species produced with **FeImPh**. Bimetallic species **Fe2ImPh** contains a μ-oxo bridge that connects a $FeCl_3$ moiety with an iron center supported by the **ImPh** ligand. In the solid state of **Fe2ImPh** (Figure 2), the distances between the oxygen and the two iron centers are similar (Fe1−O1, 1.777(3) Å; Fe2−O1, 1.763(3) Å), suggesting a symmetrical binding mode of the *O*-donor. Analogous bimetallic μ-O-bridged iron species bearing (diimine)pyridine [26] and terpyridine [27] ligands have been reported. The three *N*-donor atoms of the bis(iminophosphorane)pyridine ligand are bound to Fe1 in a k^3 mode, as supported by bond distances (Fe1−N1, 2.098(4) Å; Fe1−N2, 2.134(4) Å; Fe1−N3, 2.132(4) Å). The parameter $\tau_5 = (\beta - \alpha)/60$, where β and α are the two greatest valence angles, was calculated [28]. The value $\tau_5 = 0.08$ confirmed that the complex's geometry around Fe1 could be described as a distorted square pyramidal, with the **ImPh** ligand and the chloride atom in the plane and the $OFeCl_3$ unit in the axial position.

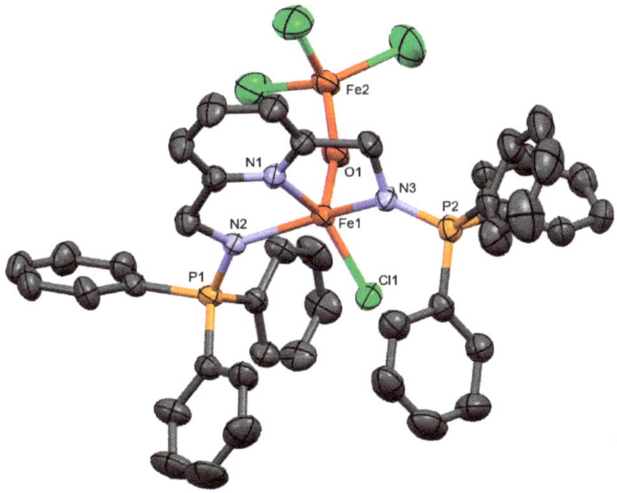

Figure 2. ORTEP representation of complex **Fe2ImPh** with ellipsoids is shown at 40% probability; hydrogen atoms are omitted for clarity. Selected bond lengths (Å) and angles (deg): Fe1−N1, 2.098(4); Fe1−N2, 2.134(4); Fe1−N3, 2.132(4); Fe1−O1, 1.777(3); Fe2−O1, 1.763(3); P1−N2, 1.610(4); P2−N3, 1.592(3); Fe1−O1−Fe2, 158.4(2).

The P−N bond distances in **Fe2ImPh** (P1−N2, 1.610(4) Å; P2−N3, 1.592(3) Å) are elongated with respect to the uncoordinated **ImPh** ligand (1.574(1) Å and 1.568(2) Å) [22]. This observation is consistent with the coordination of the nitrogen atoms of **ImPh** to the iron center.

2.2. Efforts toward the Synthesis of Heteroleptic Diiminophosphorane–Terpyridine Iron(II) Complexes

Octahedral ruthenium complexes with polypyridine ligands possess unique photophysical properties that have been broadly exploited [29–31]. While these complexes are the more studied group 8 metal compounds for this purpose, iron complexes have also been explored. Unfortunately, the metal-to-ligand charge transfer (MLCT) states of iron complexes are typically short-lived relative to those of the ruthenium analogs, thus translating into low quantum efficiencies [32]. Incorporating strong σ -donating ligands into an iron polypyridine fragment is an effective strategy for extending the lifetimes of MLCT states traditionally observed in polypyridine iron(II) complexes [32].

Within this context, several efforts were made to synthesize heteroleptic pyridine (diiminophosphorane)–terpyridine iron(II) complexes. When terpyridine (tpy) was introduced at room temperature to a solution or suspension of **FeImPh** in various organic solvents such as dichloromethane, methanol, acetonitrile, and THF, the bis(terpyridine)iron(II) complex $[Fe(tpy)_2]Cl_2$ was obtained as the major product, along with unreacted starting material, as observed by ^{31}P NMR. When this reaction was carried out under the same conditions with cationic iron(II) species formed in situ from **FeImPh** and different halogen abstracting agents such as $AgBF_4$, KPF_6, $TlPF_6$, and $NaBPh_4$, similar results were obtained (Scheme 4a). Starting from **FeImPh(Br)** led to similar results. For example, when terpyridine was reacted in THF with a cationic iron complex formed in situ from **FeImPh(Br)** and $NaBPh_4$ (room temperature, 12 h), $[Fe(tpy)_2][BPh_4]_2$ precipitated from the reaction mixture, while unreacted starting material remained in solution along with other uncharacterized iron-containing species (Scheme 4b). X-ray diffraction studies of crystals of the material obtained from the soluble portion of this mixture revealed the formation of a terpyridine-diamine iron(II) complex [Fe(pyNH)(tpy)][BPh_4]$_2$ (Scheme 4b). However, the quality of the crystals was only sufficient for connectivity determination. We hypothesize that terpyridine heteroleptic complexes could be generated when the dissociated diiminophosphorane ligand is hydrolyzed by traces of water to form the corresponding primary amine, which could coordinate to a terpyridine-iron(II) fragment. However, more complex mechanisms involving reactive unsaturated iron intermediates cannot be ruled out.

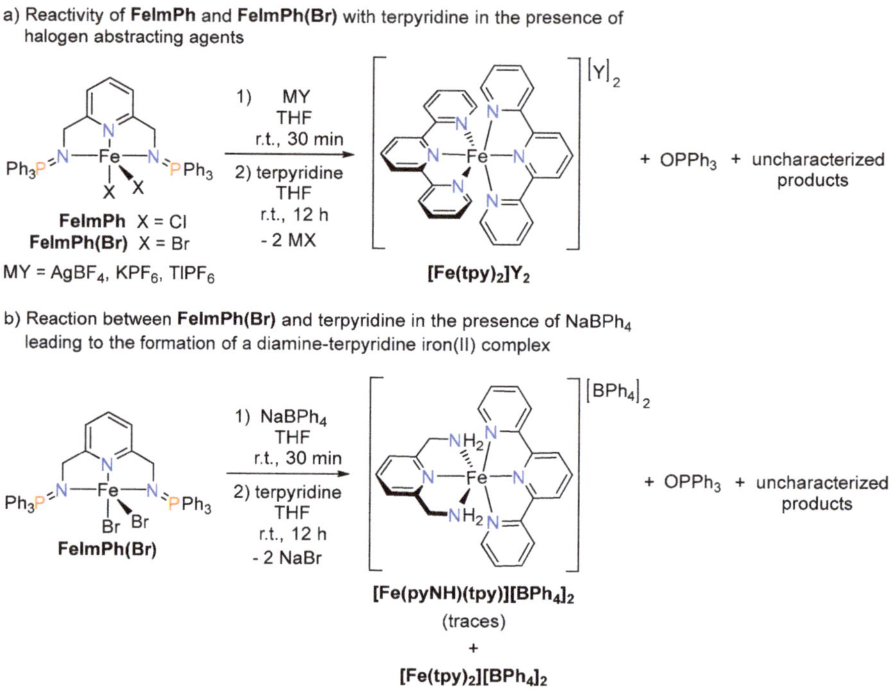

Scheme 4. Efforts toward the synthesis of a heteroleptic diiminophosphorane–terpyridine iron(II) complex.

Overall, these results highlight the lability of the diiminophosphorane ligand in these iron(II) complexes, which facilitates the formation of bis(terpyridine)-iron(II) complexes regardless of the reaction stoichiometry. Decoordination of the iminophosphorane ligand was similarly observed when bipyridine or monodentate ligands such as PPh_3 and CO were reacted with **FeImPh**. Such lability of our pincer ligands was not expected. However, the lability and hemilability of related ligands has previously been discussed [6,20,33].

2.3. Cyclic Voltammetry Studies

Redox properties of the synthesized bis(iminophosphorane)pyridine iron complexes were explored by cyclic voltammetry (CV). A quasi-reversible one-electron wave centered at $E_{1/2}$ = 0.103 V vs. Ag/AgCl was observed for **FeImPh** (Figure 3, right). This feature is tentatively assigned to a Fe(II)/Fe(III) couple. A similar quasi-reversible redox process centered at $E_{1/2}$ = 0.270 V vs. Ag/AgCl was observed for **FeImPh(Br)**. A slight potential shift was attributed to the more electron-donating nature of bromide with respect to chloride. For complexes **FeImBu** and **FeImMe**, almost irreversible redox waves centered around 0.62 and 0.05 V vs. Ag/AgCl, respectively, were observed. Other waves, probably due to pyridine-centered events, were observed in the studied potential window of −2 to 2 V (Figure 3) [34].

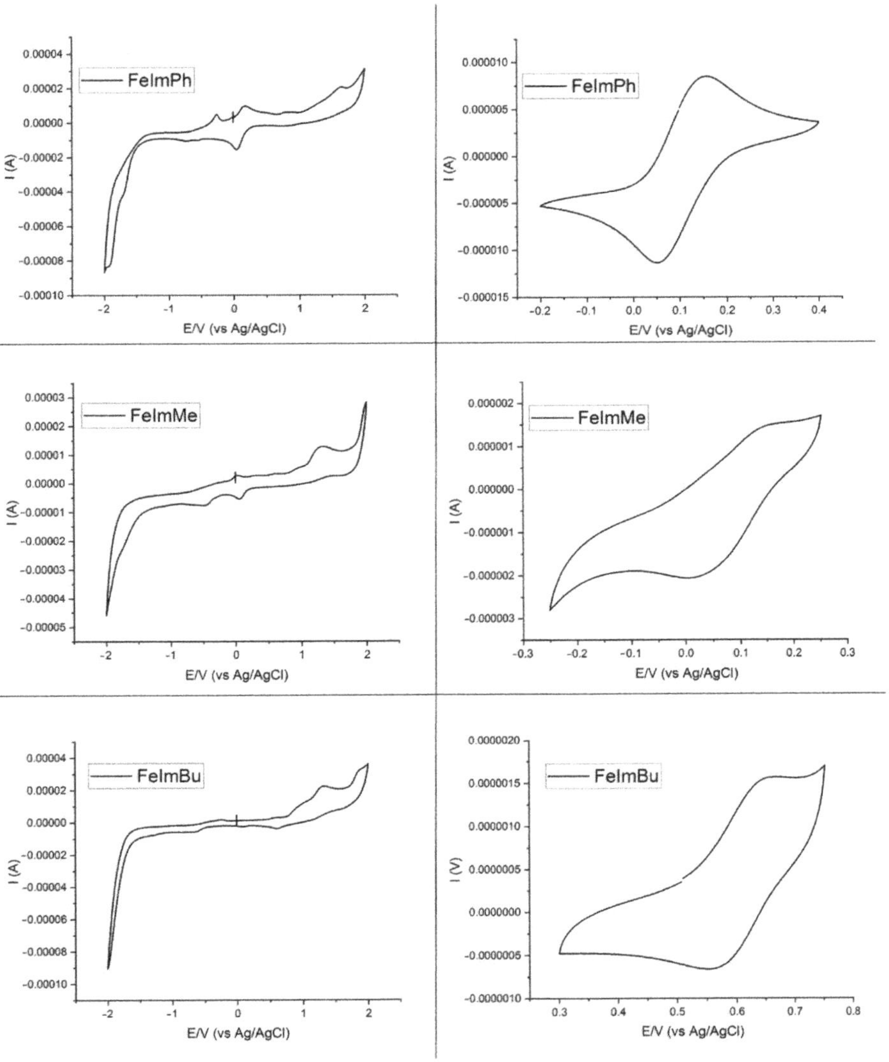

Figure 3. Cyclic voltammograms of complexes **FeImPh**, **FeImMe**, and **FeImBu** (0.1 M $n\text{Bu}_4\text{NPF}_6$, 100 mVs^{-1}, glassy carbon, Ag/AgCl, 25 °C) in CH_2Cl_2 at 1×10^{-4} M.

3. Materials and Methods

3.1. Materials and Reagents

All reactions were carried out under a nitrogen atmosphere using standard Schlenk techniques. THF and Et_2O were distilled over sodium prior to use. Pyridine-2,6-dicarboxylic acid, $FeCl_2$, $FeBr_2$ Na_2CO_3, $NaBH_4$, PBr_3, PPh_3, PCH_3Ph_2, and tri-*tert*-butylphosphine solution (1.0 M in THF) were purchased from Sigma-Aldrich (St. Louis, MO, USA), and NaN_3 was purchased from MCF Productos Científicos (Hermosillo, Mexico). All purchased reagents were used as received. 2,6-bis(azidomethyl)pyridine and **ImPh** were prepared according to the literature procedures [20]. The precursor $FeCl_2py_4$ was prepared following the literature procedures [35].

^{1}H, ^{13}C NMR, and ^{31}P NMR spectra were carried out on a JEOL GX 300 spectrometer (300.5296 MHz for ^{1}H, 75.5682 MHz for ^{13}C, and 121.5 MHz for ^{31}P). The δ scale was used throughout; chemical shifts were in ppm, and the coupling constants were in Hz. The samples of the iron complexes were prepared in an inert atmosphere and transferred to an NMR tube coupled to a J Young valve. FAB+ mass spectra were obtained using a JEOL JMS-SX102A instrument with *m*-nitrobenzyl alcohol as a matrix. DART mass spectra were obtained using the Joel AccuTOF JMS-T100LC instrument. Infrared spectra were performed on a Bruker Alpha ATR spectrometer. A three-electrode configuration was used with a BAS working glassy carbon electrode, Ag/AgCl reference electrode, and auxiliary Pt electrode. Before each measurement, the working electrode was polished with a diamond paste and rinsed with acetone and distilled water. All potential scans were carried out at a scanning rate of 100 mVs^{-1} in dry CH_2Cl_2 at a concentration of 1×10^{-4} M and 0.1 M of tetra-*n*butylammonium hexafluorophosphate. Under those conditions, $E_{1/2}$ = 0.494 V (vs. Ag/AgCl) for the ferrocene/ferrocenium redox couple.

Crystallography. Dark brown crystals of **Fe2ImPh** were obtained by cooling a THF solution of **FeImPh** to 0 °C for two days. The X-ray intensity data were measured at 298(2) K on a Bruker Smart Apex CCD diffractometer using standard MoKα radiation (λ = 0.71073 Å). A multi-scan absorption correction procedure was applied. The integration of the data was done using a triclinic unit cell to yield a total of 13654 reflections to a maximum 2θ angle of 51.06°, of which 9594 [R(int) = 0.0739] were independent. The integration structure solution was performed using SHELXS-2012, and refinement (full-matrix least squares) was performed using the SHELXS-2014/7 program [36]. Hydrogen atoms were placed in calculated positions and were allowed to ride on the atoms to which they were attached. Crystal structure parameters and experimental data on the structure solution and refinement are given in Table S10 in Supporting Information.

3.2. Synthesis of ImBu

In a Schlenk flask, 0.214 g (1.056 mmol) of tri-*tert*-butylphosphine were slowly added to a solution of 0.10 g (0.528 mmol) of 2,6-bis(azidomethyl)pyridine in 10 mL of dry diethyl ether, and the reaction mixture was stirred at room temperature for 12 h. The solvent was concentrated under a vacuum to about 5 mL, and the white precipitate was filtered off through a cannula fitted with filter paper, then washed with 30 mL of cold hexane and dried under a vacuum. The white solid corresponding to **ImBu** was obtained in 65% yield (0.225 g, 0.343 mmol).

$^{31}P\{^{1}H\}$ NMR ($CDCl_3$, 121 MHz) 53.65 (s). ^{1}H NMR ($CDCl_3$, 300 MHz) 7.44 (t, 1H), 7.17 (d, J_{HH} = 9.0, 2H), 4.92 (s, 4H), 1.43 (d, J_{PH} = 12.0, 54H). IR (ν cm^{-1}): 1109 ($\nu_{N=P}$).

3.3. Synthesis of ImMe

In a Schlenk flask, 0.211 g (1.056 mmol) of methyldiphenylphosphine were slowly added to a solution of 0.10 g (0.528 mmol) of 2,6-bis(azidomethyl)pyridine in 10 mL of diethyl ether, and the reaction mixture was stirred at room temperature for 12 h. The solvent was concentrated under a vacuum to about 5 mL, and the white precipitate was filtered off through a cannula fitted with filter paper, then washed with 30 mL of cold diethyl ether and dried under a vacuum. The white solid corresponding to **ImMe** was obtained in 65% yield (0.168 g, 0.316 mmol).

$^{31}P\{^{1}H\}$ NMR ($CDCl_3$, 121 MHz) 12.85 (s). ^{1}H NMR ($CDCl_3$, 300 MHz) 7.66–7.54 (m, 12H, H1, H2, H6), 7.42–7.31 (m, 13H, H7, H8), 4.28 (d, J_{PH} = 21.0, 4H, H4) 1.91 (d, J_{PH} = 12.0, 6H, H9). MS(DART) *m/z* Calcd for $C_{33}H_{33}N_3P_2$ $[M+H]^+$: 533.60; found: 534. See Figure S5 for compound numeration.

3.4. *Synthesis of **FeImPh***

In a Schlenk flask, 0.067 g, (0.152 mmol) of $FeCl_2py_4$ were added to a solution of 0.10 g (0.152 mmol) of **ImPh** in 20 mL of dry THF. The reaction mixture was stirred for 12 h at 45 °C. The brown precipitate was obtained and filtered through a cannula fitted with filter paper under a nitrogen atmosphere. The brown solid was dissolved in 15 mL of CH_2Cl_2 and filtered through a cannula fitted with filter paper. The solvent was evaporated to dryness, and the dark residue was dissolved in 100 mL of acetone, filtered through a short plug of Celite®, and the solvent was evaporated to dryness. The yellow solid was washed with 30 mL of diethyl ether and dried under vacuum to give the compound **FeImPh** in 70% yield (0.083 g, 0.106 mmol).

$^{31}P\{^1H\}$ NMR ($CDCl_3$, 121 MHz) 38.64 (s). ^1H NMR ($CDCl_3$, 300 MHz) 7.45–7.90 (m, 34H, H1, H2, H6, H7, H8), 4.15 (d, J_{PH} = 18, 4H, H4). See Figure S6 for compound numeration. MS(ESI+) *m/z* Calcd for $C_{43}H_{37}C_{12}FeN_3P_2$ [M]+: 783.12; found: 783.60. IR (ν cm^{-1}): 1113 ($\nu_{N=P}$). Anal Calcd for $C_{43}H_{37}C_{12}FeN_3P_2 \cdot 0.7\ CH_2Cl_2$: N, 4.98; C, 62.19; H, 4.59. Found: N, 5.21; C, 62.03; H, 4.69.

3.5. *Synthesis of **FeImBu***

In a Schlenk flask, 0.10 g, (0.186 mmol) of $FeCl_2py_4$ were added to a solution of 0.10 g (0.152 mmol) of **ImBu** in 20 mL of dry THF. The reaction mixture was stirred for 12 h at 45 °C. The brown precipitate was obtained and filtered through a cannula fitted with filter paper. The brown solid was dissolved in 15 mL of CH_2Cl_2 and filtered through a cannula fitted with filter paper. The solvent was evaporated to dryness, and the dark residue was dissolved in 100 mL of acetone, filtered through a short plug of Celite®, and the solvent was evaporated to dryness. The yellow solid was washed with 30 mL of diethyl ether and dried under vacuum to give the compound **FeImBu** in 65% yield (0.080 g, 0.120 mmol).

$^{31}P\{^1H\}$ NMR ($CDCl_3$, 121 MHz) 57.75 (s). ^1H NMR ($CDCl_3$, 300 MHz) 7.70–7.41 (m, 3H), 5.36 (m, 4H), 1.68–1.04 (m, 54 H). IR (ν cm^{-1}): 1144 ($\nu_{N=P}$). See Figure S7 for compound numeration. Satisfactory elemental analysis could not be obtained due to the high air and moisture-sensitive nature of this complex.

3.6. *Synthesis of **FeImMe***

In a Schlenk flask, 0.083 g (0.187 mmol) of $FeCl_2py_4$ were added to a solution of 0.10 g (0.187 mmol) of **ImMe** in 20 mL of dry THF. The reaction mixture was stirred for 12 h at 45 °C. The brown precipitate was obtained and filtered through a cannula fitted with filter paper. The brown solid was dissolved in 15 mL of CH_2Cl_2 and filtered through a cannula fitted with filter paper. The solvent was evaporated to dryness, and the dark residue was dissolved in 100 mL acetone, filtered through a short plug of Celite®, and the solvent was evaporated to dryness. The yellow solid was washed with 30 mL of diethyl ether and dried under vacuum to give the compound **FeImMe** in 70% yield (0.086 g, 0.130 mmol).

$^{31}P\{^1H\}$ NMR ($CDCl_3$, 121 MHz) 42.31 (s). ^1H NMR ($CDCl_3$, 300 MHz) 7.81–7.47 (m, 23H, H1, H2, H3, H6, H7, H8), 4.19 (m, 4H, H4), 2.66 (m, 6H, CH_3). See Figure S8 for compound numeration. MS(DART) *m/z* Calcd for $C_{33}H_{33}C_{12}FeN_3P_2$ $[M+H]^+$: 660.09; found: 660. IR (ν cm^{-1}): 1118 ($\nu_{N=P}$). Anal Calcd for $C_{33}H_{33}Cl_2FeN_3P_2 \cdot 0.8\ CH_2Cl_2$: N, 5.77; C, 55.74; H, 4.79. Found: N, 6.57; C, 55.06; H, 5.00.

3.7. *Synthesis of **FeImPh(Br)***

In a Schlenk flask, 0.039 g (0.182 mmol) of $FeBr_2$ were added to a solution of 0.12 g (0.182 mmol) of **ImPh** in 20 mL of dry THF. The reaction mixture was stirred for 12 **h** at room temperature. After this time, a yellow precipitate was obtained and filtered through a cannula fitted with filter paper. This solid was washed with 30 mL of cold THF and dried under vacuum to give the compound **FeImPh(Br)** as a yellow solid in 76% yield (0.121 g, 0.138 mmol).

$^{31}P\{^1H\}$ NMR ($CDCl_3$, 121 MHz) 42.31 (s). ^1H NMR ($CDCl_3$, 300 MHz) 8.61 (s, 4H) 7.73–7.28 (m, 27H), 6.24 (s, 2H), 4.22 (s, 4H). MS(ESI+) *m/z* Calcd for $C_{43}H_{37}Br_2FeN_3P_2$ $[M+H]^+$: 874.02; found: 874.2. IR (ν cm^{-1}): 1116 ($\nu_{N=P}$).

*3.8. Reaction between **FeImPh** and terpyridine*

In a typical reaction, 0.1 g (0.127 mmol) of **FeImPh** were added to a solution of 0.029 g (0.127 mmol) of terpyridine in 20 mL of dry solvent in a Schlenk flask. The reaction mixture was stirred at room temperature. The purple solution was evaporated to dryness under vacuum, and the residue was dissolved in 15 mL of CH_2Cl_2. Upon chromatography through a short column of alumina, two fractions were obtained. A yellow fraction contained mainly unreacted **FeImPh**, while the purple fractions contained $[Fe(tpy)_2]Cl_2$. After the evaporation of the solvent, the purple solid was washed with 20 mL of diethyl ether and dried under vacuum, leading to $[Fe(tpy)_2]Cl_2$ in yields from 15% to 25%, depending on the reaction time and solvent.

^1H NMR ($CDCl_3$, 300 MHz) 7.69–7.62 (m, 8H), 7.57–7.51 (m, 5H), 7.48–7,42 (m, 9H). See Figure S12 in Supporting Information.

4. Conclusions

Pyridine(diiminophosphoranes) are convenient ligands for accessing iron(II) coordination complexes. The methodology developed here allows for the easy preparation of various bis(iminophosphorane)pyridine complexes of iron(II). Their electronic and steric properties can be tuned by the phosphine ligand, which can be incorporated into the iminophosphorane framework through a straightforward synthetic procedure. We hypothesize that such iron(II) bis(iminophosphorane)pyridine complexes are candidates for catalytic processes known to be catalyzed by other iron(II) pincer complexes.

Supplementary Materials: The following supporting information can be downloaded at: https://www.mdpi.com/article/10.3390/inorganics12040115/s1: Synthesis and characterization of **PyN3** and **ImPh** (S2–S3); IR, ^1H- and ^{31}P-NMR spectra, IR, M/S data, and cyclic voltammogram for **IMBu, ImMe, FeImPh, FeImBu, FeImMe**, and **FeImPh(Br)** (S4–S9); Crystallographic details for **Fe2ImPh** (S10); Studies of the stability of **FeImPh** and **FeImBu** in solution (S11); NMR spectra for $[Fe(tpy)_2]Cl_2$ and $OPPh_3$ (S12); CV of ferrocene in CH_2Cl_2 (S13). CCDC 2332976 for **Fe2ImPh** contains the supplementary crystallographic data for this paper. These data can be obtained free of charge from The Cambridge Crystallographic Data Centre.

Author Contributions: Conceptualization, N.S.L., A.I. and R.L.L.; methodology, N.S.L., E.N.B. and R.L.L.; investigation, N.S.L., E.N.B. and R.L.L.; writing—original draft preparation, N.S.L., E.N.B. and R.L.L.; writing—review and editing, P.S., A.I., A.D.R. and R.L.L.; funding acquisition, R.L.L. All authors have read and agreed to the published version of the manuscript.

Funding: This research was funded by DGAPA—UNAM (PAPIIT projects IN-207419 and IN-211522) and Consejo Nacional de Humanidades, Ciencia y Tecnología CONAHCYT (Project A1-S-15068, and grant to N. Sánchez López).

Data Availability Statement: The data presented in this study are available in the article and Supplementary Materials.

Acknowledgments: Rubén A. Toscano and Marcos Flores-Álamo are thanked for X-rya diffraction studies. We thank M. P. Orta Perez, E. Huerta Salazar, E. García Ríos, R. L. Gaviño Ramirez, J. D. Vázquez Cuevas, M. M. Aguilar Araiza, and G. E. Cortés Romero for obtaining analytical data and for technical assistance. J. F. Zambrano Bedoya is thanked for his help in evaluating the stability of the compounds.

Conflicts of Interest: The authors declare no conflicts of interest.

References

1. Small, B.L. Discovery and Development of Pyridine-bis(imine) and Related Catalysts for Olefin Polymerization and Oligomerization. *Acc. Chem. Res.* **2015**, *48*, 2599–2611. [CrossRef] [PubMed]
2. Wen, H.; Liu, G.; Huang, Z. Recent advances in tridentate iron and cobalt complexes for alkene and alkyne hydrofunctionalizations. *Coord. Chem. Rev.* **2019**, *386*, 138–153. [CrossRef]
3. Bauer, G.; Hu, X. Recent developments of iron pincer complexes for catalytic applications. *Inorg. Chem. Front.* **2016**, *3*, 741–765. [CrossRef]
4. Tondreau, A.M.; Milsmann, C.; Patrick, A.D.; Hoyt, H.M.; Lobkovsky, E.; Wieghardt, K.; Chirik, P.J. Synthesis and Electronic Structure of Cationic, Neutral, and Anionic Bis(imino)pyridine Iron Alkyl Complexes: Evaluation of Redox Activity in Single-Component Ethylene Polymerization Catalysts. *J. Am. Chem. Soc.* **2010**, *132*, 15046–15059. [CrossRef] [PubMed]
5. Arevalo, R.; Chirik, P.J. Enabling Two-Electron Pathways with Iron and Cobalt: From Ligand Design to Catalytic Applications. *J. Am. Chem. Soc.* **2019**, *141*, 9106–9123. [CrossRef] [PubMed]
6. García-Álvarez, J.; García-Garrido, S.E.; Cadierno, V. Iminophosphorane–phosphines: Versatile ligands for homogeneous catalysis. *J. Organomet. Chem.* **2014**, *751*, 792–808. [CrossRef]
7. Al-Benna, S.; Sarsfield, M.J.; Thornton-Pett, M.; Ormsby, D.L.; Maddox, P.J.; Brès, P.; Bochmann, M. Sterically hindered iminophosphorane complexes of vanadium, iron, cobalt and nickel: A synthetic, structural and catalytic study †. *J. Chem. Soc. Dalton Trans.* **2000**, 4247–4257. [CrossRef]
8. Tannoux, T.; Mazaud, L.; Cheisson, T.; Casaretto, N.; Auffrant, A. Fe(II) complexes supported by an iminophosphorane ligand: Synthesis and reactivity. *Dalton Trans.* **2023**, *52*, 12010–12019. [CrossRef] [PubMed]
9. Sudhakar, P.V.; Lammertsma, K. Nature of bonding in phosphazoylides. A comparative study of N2H4, NPH4, and P2H4. *J. Am. Chem. Soc.* **2002**, *113*, 1899–1906. [CrossRef]
10. Molina, P.; Alajarin, M.; Lopez Leonardo, C.; Claramunt, R.M.; Foces-Foces, M.d.l.C.; Hernandez Cano, F.; Catalan, J.; De Paz, J.L.G.; Elguero, J. Experimental and theoretical study of the R3P+-X- bond. Case of betaines derived from N-iminophosphoranes and alkyl isocyanates. *J. Am. Chem. Soc.* **2002**, *111*, 355–363. [CrossRef]
11. Tannoux, T.; Auffrant, A. Complexes featuring tridentate iminophosphorane ligands: Synthesis, reactivity, and catalysis. *Coord. Chem. Rev.* **2023**, *474*, 214845. [CrossRef]
12. Chaplin, A.B.; Harrison, J.A.; Dyson, P.J. Revisiting the Electronic Structure of Phosphazenes. *Inorg. Chem.* **2005**, *44*, 8407–8417. [CrossRef] [PubMed]
13. Reed, A.E.; Schleyer, P.v.R. Chemical bonding in hypervalent molecules. The dominance of ionic bonding and negative hyperconjugation over d-orbital participation. *J. Am. Chem. Soc.* **2002**, *112*, 1434–1445. [CrossRef]
14. Winslow, C.; Lee, H.B.; Field, M.J.; Teat, S.J.; Rittle, J. Structure and Reactivity of a High-Spin, Nonheme Iron(III)- Superoxo Complex Supported by Phosphinimide Ligands. *J. Am. Chem. Soc.* **2021**, *143*, 13686–13693. [CrossRef] [PubMed]
15. Hein, N.M.; Pick, F.S.; Fryzuk, M.D. Synthesis and Reactivity of a Low-Coordinate Iron(II) Hydride Complex: Applications in Catalytic Hydrodefluorination. *Inorg. Chem.* **2017**, *56*, 14513–14523. [CrossRef] [PubMed]
16. Spentzos, A.Z.; May, S.R.; Confer, A.M.; Gau, M.R.; Carroll, P.J.; Goldberg, D.P.; Tomson, N.C. Investigating Metal-Metal Bond Polarization in a Heteroleptic Tris-Ylide Diiron System. *Inorg. Chem.* **2023**, *62*, 11487–11499. [CrossRef] [PubMed]
17. Spencer, L.P.; Altwer, R.; Wei, P.; Gelmini, L.; Gauld, J.; Stephan, D.W. Pyridine– and Imidazole–Phosphinimine Bidentate Ligand Complexes: Considerations for Ethylene Oligomerization Catalysts. *Organometallics* **2003**, *22*, 3841–3854. [CrossRef]
18. Oheix, E.; Herrero, C.; Moutet, J.; Rebilly, J.N.; Cordier, M.; Guillot, R.; Bourcier, S.; Banse, F.; Senechal-David, K.; Auffrant, A. Fe(III) and Fe(II) Phosphasalen Complexes: Synthesis, Characterization, and Catalytic Application for 2-Naphthol Oxidative Coupling. *Chemistry* **2020**, *26*, 13634–13643. [CrossRef]
19. Peris, E.; Crabtree, R.H. Key factors in pincer ligand design. *Chem. Soc. Rev.* **2018**, *47*, 1959–1968. [CrossRef]
20. Cheisson, T.; Auffrant, A. Versatile coordination chemistry of a bis(methyliminophosphoranyl)pyridine ligand on copper centres. *Dalton Trans.* **2014**, *43*, 13399–13409. [CrossRef]
21. Cheisson, T.; Ricard, L.; Heinemann, F.W.; Meyer, K.; Auffrant, A.; Nocton, G. Synthesis and Reactivity of Low-Valent f-Element Iodide Complexes with Neutral Iminophosphorane Ligands. *Inorg. Chem.* **2018**, *57*, 9230–9240. [CrossRef]
22. Cheisson, T.; Auffrant, A.; Nocton, G. η5–η1 Switch in Divalent Phosphaytterbocene Complexes with Neutral Iminophosphoranyl Pincer Ligands: Solid-State Structures and Solution NMR 1JYb–P Coupling Constants. *Organometallics* **2015**, *34*, 5470–5478. [CrossRef]
23. Staudinger, H.; Meyer, J. Über neue organische phosphorverbindungen III. Phosphinmethylenderivate und phosphinimine. *Helv. Chim. Acta* **1919**, *2*, 635–646. [CrossRef]
24. Gorenstein, D.G. Conformation and Dynamics of DNA and Protein-DNA Complexes by 31P NMR. *Chem. Rev.* **1994**, *94*, 1315–1338. [CrossRef]
25. Kuveke, R.E.H.; Barwise, L.; van Ingen, Y.; Vashisth, K.; Roberts, N.; Chitnis, S.S.; Dutton, J.L.; Martin, C.D.; Melen, R.L. An International Study Evaluating Elemental Analysis. *ACS Cent. Sci.* **2022**, *8*, 855–863. [CrossRef]
26. Small, B.L.; Brookhart, M. Polymerization of propylene by a new generation of iron catalysts: Mechanisms of chain initiation, propagation, and termination. *Macromolecules* **1999**, *32*, 2120–2130. [CrossRef]
27. Bocian, A.; Napierała, S.; Gorczyński, A.; Kubicki, M.; Wałęsa-Chorab, M.; Patroniak, V. The first example of an asymmetrical μ-oxo bridged dinuclear iron complex with a terpyridine ligand. *New J. Chem.* **2019**, *43*, 12650–12656. [CrossRef]

28. Addison, A.W.; Rao, T.N.; Reedijk, J.; van Rijn, J.; Verschoor, G.C. Synthesis, structure, and spectroscopic properties of copper(II) compounds containing nitrogen–sulphur donor ligands; the crystal and molecular structure of aqua [1,7-bis(N-methylbenzimidazol-2′-yl)-2,6-dithiaheptane]copper(II) perchlorate. *J. Chem. Soc. Dalton Trans.* **1984**, *7*, 1349–1356. [CrossRef]
29. Hagfeldt, A.; Boschloo, G.; Sun, L.; Kloo, L.; Pettersson, H. Dye-sensitized solar cells. *Chem. Rev.* **2010**, *110*, 6595–6663. [CrossRef]
30. Prier, C.K.; Rankic, D.A.; MacMillan, D.W. Visible light photoredox catalysis with transition metal complexes: Applications in organic synthesis. *Chem. Rev.* **2013**, *113*, 5322–5363. [CrossRef]
31. Lebon, E.; Bastin, S.; Sutra, P.; Vendier, L.; Piau, R.E.; Dixon, I.M.; Boggio-Pasqua, M.; Alary, F.; Heully, J.-L.; Igau, A. Can a functionalized phosphine ligand promote room temperature luminescence of the [Ru (bpy)(tpy)] 2+ core? *Chem. Commun.* **2012**, *48*, 741–743. [CrossRef]
32. Liu, Y.; Persson, P.; Sundstrom, V.; Warnmark, K. Fe N-heterocyclic carbene complexes as promising photosensitizers. *Acc. Chem. Res.* **2016**, *49*, 1477–1485. [CrossRef]
33. Fukui, M.; Itoh, K.; Ishii, Y. Iminophosphorane Complexes of Palladium(II). *Bull. Chem. Soc. Jpn.* **1975**, *48*, 2044–2046. [CrossRef]
34. Martin, D.J.; McCarthy, B.D.; Piro, N.A.; Dempsey, J.L. Synthesis and electrochemical characterization of a tridentate Schiff-base ligated Fe(II) complex. *Polyhedron* **2016**, *114*, 200–204. [CrossRef]
35. Wu, J.Y.; Stanzl, B.N.; Ritter, T. A strategy for the synthesis of well-defined iron catalysts and application to regioselective diene hydrosilylation. *J. Am. Chem. Soc.* **2010**, *132*, 13214–13216. [CrossRef]
36. Sheldrick, G.M. Crystal structure refinement with SHELXL. *Acta Crystallogr. Sect. C Struct. Chem.* **2015**, *71*, 3–8. [CrossRef]

inorganics

MDPI

Article

Synthesis, Spectral Characterization, and Structural Modelling of Di- and Trinuclear Iron(III) Monensinates with Different Bridging Patterns

Nikolay Petkov [1,*], Alia Tadjer [1], Svetlana Simova [2], Zara Cherkezova-Zheleva [3], Daniela Paneva [3], Radostina Stoyanova [4], Rositsa Kukeva [4], Petar Dorkov [5] and Ivayla Pantcheva [1,*]

1 Faculty of Chemistry and Pharmacy, Sofia University St. Kliment Ohridski, 1164 Sofia, Bulgaria; tadjer@chem.uni-sofia.bg
2 Institute of Organic Chemistry with Centre of Phytochemistry, Bulgarian Academy of Sciences, 1113 Sofia, Bulgaria; svetlana.simova@orgchm.bas.bg
3 Institute of Catalysis, Bulgarian Academy of Sciences, 1113 Sofia, Bulgaria; zzhel@ic.bas.bg (Z.C.-Z.); daniela@ic.bas.bg (D.P.)
4 Institute of General and Inorganic Chemistry, Bulgarian Academy of Sciences, 1113 Sofia, Bulgaria; radstoy@svr.igic.bas.bg (R.S.); rositsakukeva@yahoo.com (R.K.)
5 Research and Development Department, Biovet Ltd., 4550 Peshtera, Bulgaria; p_dorkov@biovet.com
* Correspondence: ahnp@chem.uni-sofia.bg (N.P.); ahip@chem.uni-sofia.bg (I.P.); Tel.: +359-2-8161446 (N.P. & I.P.)

Abstract: In the present study, we report the solid-state isolation and structural characterization of novel iron(III) complexes of the veterinary antibiotic monensin. Monensic acid ($MonH \times H_2O$) forms a dinuclear complex of composition with $FeCl_3$ $[FeCl(Mon)_2]_2$ (**1**), while its interaction with $FeSO_4$ leads to the isolation of a triangular oxo-ferric coordination species $[Fe_3O(Mon \times H_2O)_6(H_2O)_2(OH)]$ (**2**). During the procedure resulting in **2**, oxidation of the Fe(II) ions by atmospheric oxygen was observed. In the presence of organic bases, both complexation reactions proceeded to successfully deprotonate the carboxylic function of the ligand. Iron(III) complexes **1** and **2** were characterized by IR, EPR, NMR, and Mössbauer spectroscopies as well as with thermal (TG-DTA/MS) and elemental analyses. In addition, the structures of the two coordination compounds were modelled and selected calculated parameters were compared with the experimental results. The biological assay revealed the enhanced antibacterial potential of the newly obtained complexes against the Gram-positive aerobic microorganisms *Bacillus cereus* and *Bacillus subtilis*.

Keywords: polyether ionophore; antibacterial iron(III) complexes; antiferromagnetism; DFT modelling

1. Introduction

Iron ions play a significant role in almost all living organisms [1]. They are vital for transporting oxygen, activating various substrates, and assisting in electron transfer reactions. Most of these processes occur due to the versatile coordination ability of both Fe(II) and Fe(III). The chemistry and biochemistry of ferric complexes are well known and have been subject to intensive reviewing over the years [2–5]. Iron(III) can form a wide variety of structural arrangements ranging from mono- to polynuclear coordination species with ligands involving diverse donor atoms and exhibiting different denticity. The literature overview reveals that a number of iron(III) complexes contain carboxylate ligands and exist as di- or trinuclear coordination species. In the first category, the carboxylate group participates as a terminal ligand in the formation of diamond-core $Fe_2(\mu_2\text{-}O)_2$ complexes with hydroxide/alkoxide bridges or acts as a $(\mu\text{-}\eta_1{:}\eta_1)$-link in both symmetric and asymmetric di-iron species [6–16]. Although rare, acetate complexes of various metal ions containing μ_2-aqua or μ_2-alcohol linkers are also known [17–25]. The trinuclear iron(III) carboxylates can be found mainly as oxo-ferric complexes adopting the typical triangular acetate structure [26–29]. In the above-mentioned constructs, three types of carboxylate coordination modes with respect to the isolated iron centre can be realised (Figure 1), excluding the pure metal–ligand ionic bond: monodentate, bidentate, and bridging [30,31]. Which configuration will prevail in the corresponding coordination species depends on many factors such as the overall ligand structure (i.e., the presence of other functional groups), the reaction conditions, the solvent effect, the metal salt counterion, etc.

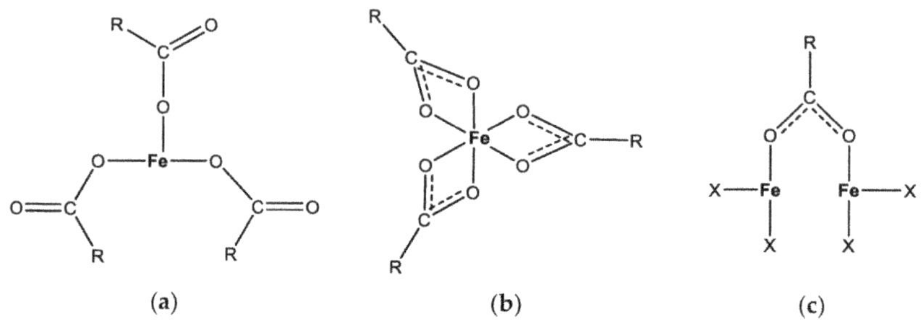

Figure 1. The possible (**a**) monodentate, (**b**) bidentate, and (**c**) bridging coordination patterns of the carboxylate group to iron(III) cation(s).

In addition to the rich and diverse coordination chemistry of iron that makes it suitable for a large variety of functions within biological systems [32], its metabolism was found to be crucial also for the life cycle of cancer cells, triggering ferroptosis—a recently described form of regulated cell death caused by iron-dependent lipid peroxidation [33]. The ferroptosis is driven by various mechanisms, including the following:

(i) glutathione level reduction and decreased activity of glutathione peroxidase, which results in the deposition of harmful lipid-reactive oxygen species (L-ROS) from polyunsaturated fatty acids in the presence of high concentrations of iron ions, thus promoting cell death [34];
(ii) overexpression of transferrin receptor 1 (TFR1) and decrease in ferritin levels, since the upregulation of TFR1 is detected in many malformations such as glioblastoma, leukaemia, breast cancer, ovarian cancer, hepatic cancer, thyroid cancer, and colorectal cancer [35];
(iii) Fenton reaction that strongly depends on the intracellular iron concentration and can be a possible mechanism of ROS generation.

Ferroptosis has been suggested to be an endogenous anticancer mechanism providing new opportunities in the treatment of drug-resistant tumours. In this sense, any natural compound that can induce such regulated cell death can be treated as a potential anticancer agent [36,37]. Recently, it was shown that some members of the natural carboxylic polyether ionophores—salinomycin and ironomycin—accumulate in lysosomes, sequester the lysosomal iron, and produce ROS in this organelle via Fenton reaction [38–40].

To gain deeper insight into the possible chemical interactions between polyether ionophores and Fe(II)/Fe(III), we initiated targeted research on the coordination ability of monensin and salinomycin (HL) to bind iron. Monensic acid (MonH, Figure S1) was selected as the most widely applied antibiotic in veterinary medicine, and salinomycinic acid (SalH) was deemed as a promising anticancer agent. Our first findings led to the isolation and characterization of iron(III) monensinate and salinomycinate of composition $[Fe_3(\mu_3\text{–}O)L_3(OH)_4]$ [41]. Then, under completely different reaction conditions, two new Fe(III) coordination compounds of monensin were obtained with the involvement of the antibiotic carboxylate function in the complex formation. The reported iron(III) complexes of monensin can be described as dinuclear chloro-containing (**1**) and trinuclear oxo-ferric (**2**) coordination species. The experimental data reveal that the antibiotic ligand serves in a pure monodentate/bidentate or in a bridged bidentate coordination mode, with the involvement of a terminal alcohol group in the bidentate ones. Employing the experimental and theoretical chemistry tools, we were able to derive reliable structures of these new iron(III) monensinates, combining various spectroscopies and the computational DFT method.

2. Results and Discussion

2.1. General Remarks

In the presence of organic bases (Et_4NOH or Et_3N), monensic acid effectively deprotonates and acts as a carboxylate monoanion in complexation reactions with metal ions of different oxidation states [41–45]. Monensinate reaction with $FeCl_3$ or $FeSO_4$ leads to the formation of new complexes which exhibit spectral properties characteristic of di- (**1**) and trinuclear (**2**) coordination species of iron(III), respectively. In addition, the formation of **1** is accompanied by co-precipitation of the mono-complex **1a** (Figure S2). The complete set of experimental data is significantly different from that previously reported for iron(III) monensinate and salinomycinate [41], a fact indicating that we have prepared new ferric coordination compounds of the polyether ionophore monensin. The isolated solids **1** and **2** are amorphous without any sign of crystallinity, making their precise characterization difficult. The research methodology for structure elucidation described below is based on a series of spectroscopic techniques along with an appropriate computational chemistry protocol.

2.2. Physicochemical Properties of Complexes 1 and 2

2.2.1. Vibrational and Thermal Analysis

The IR spectra of **1** and **2** (Figure 2) reveal the deprotonation of monensin during its reaction with iron ions. The band of MonH × H_2O at 1710 cm^{-1} assigned to the carboxylic group is replaced by two new bands at 1592 and 1419 cm^{-1} in the spectrum of **1**, attributed to the corresponding asymmetric and symmetric stretching vibrations of the carboxylate function. The multicomponent band at 1592 cm^{-1} was fitted with three Lorentzians (inset in Figure 1) thus assuming different coordination modes of the carboxylate moiety in the solid sample of complex **1** [46]. The areas of the sub-bands (1) at 1594 cm^{-1} (FWHM = 66 cm^{-1}) and (2) at 1554 cm^{-1} (FWHM = 46 cm^{-1}) are almost equal (ca. 45% for each component), whereas the area of sub-band (3) is noticeably smaller. It is supposed that **1** comprises two main types of carboxylates whose ligation varies to some extent due to the different values of $\Delta = \nu_{asym} - \nu_{sym}$, 175, and 135 cm^{-1}, respectively. The weak band at 1643 cm^{-1} (FWHM = 75 cm^{-1}, Δ = 224 cm^{-1}) is attributed to the presence of monensinate bound in another type of complex species. The latter probably exists as an impurity in the solid sample rather than as a constituent of the intrinsic structure of **1**. As will be described further, this assumption is supported by the magnetic studies used for structural characterization of **1**.

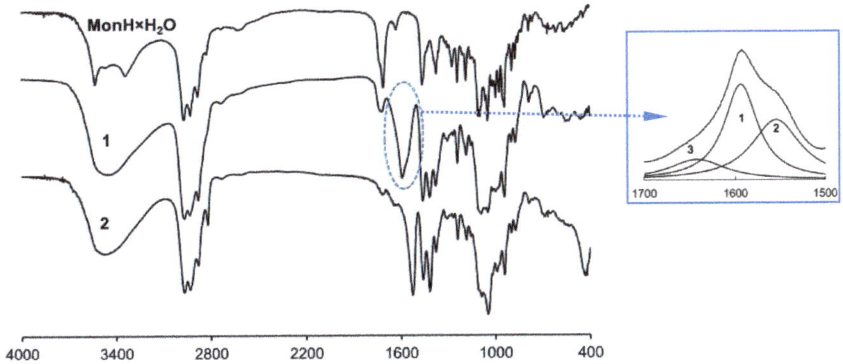

Figure 2. IR spectra of MonH × H_2O and complexes **1–2** in KBr pellets.

The spectrum of **2** also consists of two bands at 1527 and 1417 cm^{-1}, but the significantly smaller value of Δ = 110 cm^{-1} points to a bridging coordination mode of the carboxylate moiety. Complex **2** exhibits intense characteristic bands below 1000 cm^{-1} which are absent in **1**. These vibrations are assigned to the formation of Fe-OH (δ_{FeOH} = 1047 cm^{-1}) and Fe-O (ν_{FeO} = 430 cm^{-1}) bonds, respectively. The bands in the range 4000–3000 cm^{-1}, related to asymmetric and symmetric stretching vibrations of OH-bonds in the water molecule (3530–3460 cm^{-1}), and ν_{OH} (3336 cm^{-1}) in MonH $\times$ H_2O, broaden in the spectra of **1** and **2** (3450–3470 cm^{-1}) due to the OH-group's engagement in various intramolecular interactions.

No intense endothermic peaks below 185 °C were observed in the TG-DTA/TG-MS curves of complex **1** (Figure S3a), ruling out the presence of coordinated water molecule(s). In contrast, the endothermic peak at 104 °C in **2** refers to a two-step water loss, which is an indication of different types of water molecules present in the studied sample (Figure S3b).

Based on the IR and thermal data, combined with microanalysis results, it can be concluded that (i) monensin is bound as a monoanion in the structures of **1** and **2**; (ii) the molar metal-to-antibiotic ratio is 1:2 in both complexes; (iii) species **1** contains additional chloride ions, whereas a hydroxide anion and water ligands participate in the composition of **2**.

2.2.2. Magnetic Studies

The EPR spectra of complexes **1–2** are registered in the temperature range from 77 K (100 K) to 295 K (Figure 3). A general characteristic of the spectra at r. t. is the broad signal with g = 2.02, attributable to iron(III) ions, confirming the oxidation of Fe(II) under atmospheric conditions during the preparation procedure of **2**.

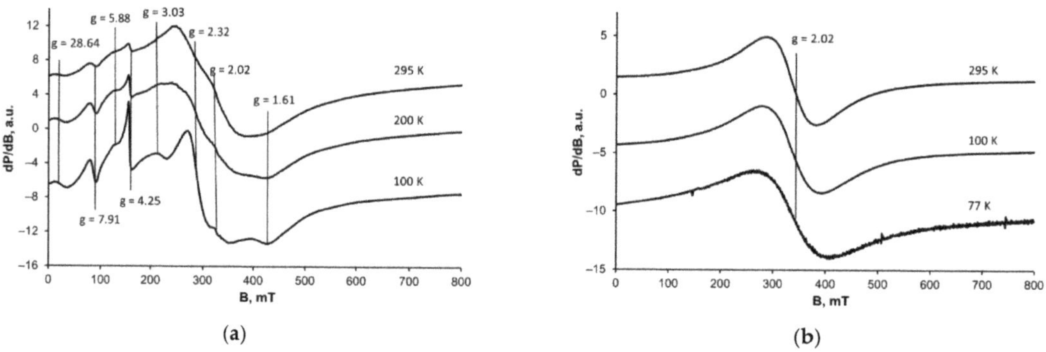

Figure 3. X-band solid-state EPR spectra of **1** (**a**) and **2** (**b**) registered in the range 77–293 K.

The peak-to-peak linewidth and intensity of the central signal (g = 2.02) in **1–2** show a well-defined tendency towards signal broadening with temperature decrease. Such a temperature dependence of the linewidth relates to an antiferromagnetic spin interaction of the metal centres, which becomes stronger at low temperatures.

The relatively broad signal with g = 2.02 observed for **1** is attributed to magnetically coupled Fe(III) ions [47]. It is accompanied by several extra signals which can be seen even at room temperature with effective g-factor values ranging from 28.64 to 1.61, labelled on Figure 3a. As the temperature decreases, the spectrum of **1** undergoes a transformation in which the less pronounced set of signals turns narrower at 100 K and becomes the main feature of the low temperature spectrum. On the other hand, the temperature behaviour of the signal with g = 2.02 is reversed. The result is that at 100 K its intensity is negligible compared with the intensities of the narrower lines set. Such a central field signal broadening can be attributed to the presence of closely spaced iron ions bound by a suitable ligand. Based on the composition data, we assume that chloride anions or hydroxyl groups of the monensinate may play the role of such a linker between two metal centres to witness the observed EPR behaviour of **1**.

The set of lines that dominates the low-temperature spectrum of sample **1** refers to the fine structure of Fe^{3+} ions and its origin can be explained by allowed transitions with $\Delta m_s = 1$, although the existence of forbidden transition lines cannot be excluded either. These signals are ascribed to isolated Fe(III) ions placed in an axial symmetry field [48] and their presence can be attributed to the formation of the mono-complex **1a**, whose composition and structure will be discussed later. We could not eliminate **1a** despite our numerous attempts to purify the dinuclear complex. The presence of the assumed mono-species **1a** does not affect the overall elemental composition of **1**. The results are consistent with the IR-data, where the third weak band attributable to the asymmetric stretching of COO− is assigned to the formation of a second type of coordination species present in a minor quantity.

For sample **2** the signal at g = 2.02 is the main spectral characteristic over the entire temperature range. The g-factor value remains constant in the whole temperature span (Figure 3b). The negative Curie–Weiss constant (−608 ± 12 K) and the signal broadening at low temperatures (95 mT at 295 K and 142 mT at 77 K) indicate the occurrence of exchange-coupled iron(III) ions.

To obtain additional information on the binding mode of monensin in **1–2**, we also performed NMR analysis, which might be strongly affected by the presence of paramagnetic ions, but in some cases can contribute to a deeper understanding of the properties of the metal complexes. We were unable to adjust the NMR settings and record any spectra of complex **2** using conventional NMR techniques, which may serve as indirect evidence that the studied monensinate sample contains paramagnetic iron(III) cations.

To our surprise, we recorded r. t. NMR spectra of **1** in $CDCl_3$ to observe a negligible shift of the ^{13}C signals compared with those of the uncoordinated monensic acid (Table S1) at low sample concentration, while augmenting the concentration resulted in a significant signal broadening. We hypothesized that a possible dissociation of the dinuclear iron(III) complex may occur, inflicting structure breakdown under the solvent action [49]. To confirm this, we also measured the EPR spectrum of the same solution at 120 K (Figure 4). Experimental results reveal that there are relatively narrow signals (*ca.* 15–20 mT) with g = 4.33 and 2.01, which can be attributed to isolated Fe^{3+} ions placed in a low and in an octahedral symmetry, respectively [50]. The hump detected at 240 mT (as part of a broad signal with g = 2.02) is attributed to the presence of exchange-coupled Fe(III) ions. EPR data recorded both in solution and in the solid state confirm that **1** comprises different types of coordination species. At the current stage of research, we cannot explain the negligible effect of Fe(III) on monensin NMR signals, but it may be related to the lability/inertness of the coordination species formed—a phenomenon that deserves further investigation, which is beyond the scope of the present study.

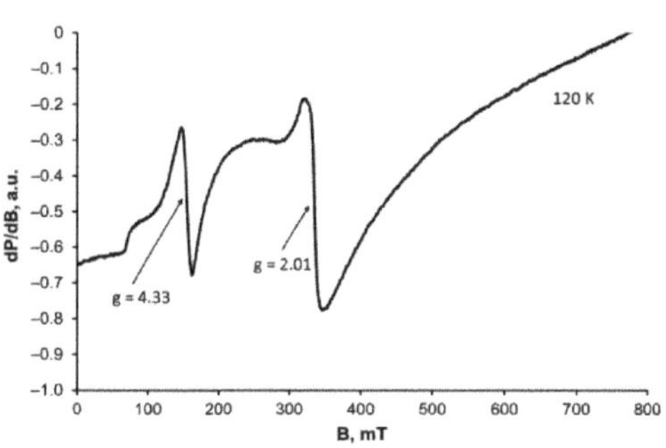

Figure 4. X-band EPR spectrum of frozen solution of **1** in $CDCl_3$.

2.2.3. Mössbauer Studies

Monensin, as a representative of natural polyether ionophorous antibiotics, contains only O-donor atoms, so we suppose that the iron ions involved in the composition of complexes **1** and **2** will retain the high-spin configuration. Our hypothesis is confirmed by the subsequent Mössbauer studies performed at both room and liquid nitrogen temperatures.

The Mössbauer spectrum of **1** at 293 K consists of two asymmetric quadrupole signals that can be assigned to a minimum of two different types of iron ions. Consistent with the EPR data suggesting the presence of di- and mono-species, we decomposed the observed signals into three components by the least-squares fitting procedure (Figure 5a). Calculations reveal that two of the doublets (in 1:1 ratio, sub-spectra (1) and (2)) exhibit very close isomer shifts (IS, δ = ca. 0.4 mm/s) and quadrupole splittings (QS, Δ = ca. 0.7 mm/s) (Table 1). The results are consistent with the presence of high-spin iron(III) ions placed in a nearly identical octahedral environment [51] and corroborate previously reported data for ferric dimers containing two OH-bridges [52,53]. The minor sub-spectrum (3) differs from the rest, especially in the IS value. The doublet is assigned to the presence of mono-complex **1a** and its low intensity agrees well with the IR and EPR data recorded. It can be concluded that Fe(III) in **1a** is placed in a ligand environment similar to that of **1**, as no significant change in its QS value is observed.

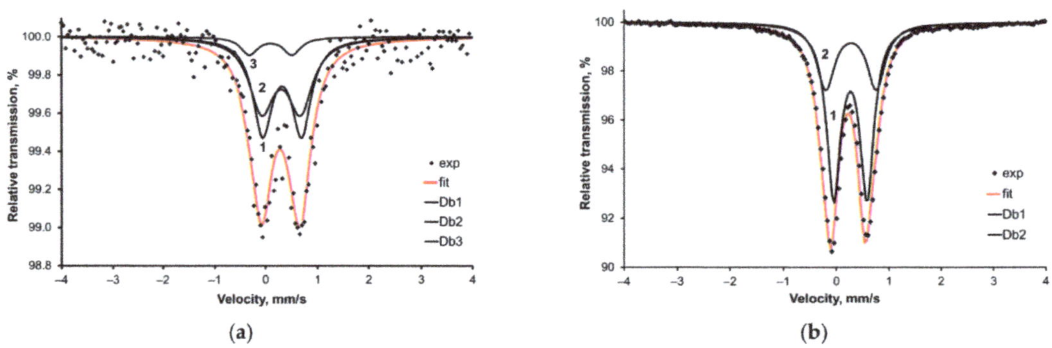

Figure 5. Mössbauer spectra at 293 K: (**a**) complex **1**, (**b**) complex **2**.

Table 1. Mössbauer parameters for solid samples **1** and **2** at 293 K (and 77 K in parentheses).

Complex	Component	δ, mm/s	Δ, mm/s	Γ, mm/s	A, %	M, %
1	Db1	0.43 (0.60)	0.75 (0.75)	0.45 (0.45)	47 (48)	0.49 (1.48)
	Db2	0.41 (0.61)	0.74 (0.79)	0.57 (0.52)	45 (47)	0.37 (1.29)
	Db3	0.21 (0.40)	0.82 (0.75)	0.40 (0.40)	8 (5)	0.09 (0.19)
2	Db1	0.37 (0.57)	0.63 (0.62)	0.32 (0.31)	66 (64)	6.96 (8.41)
	Db2	0.39 (0.58)	0.96 (1.00)	0.42 (0.38)	34 (36)	2.99 (3.91)

The Mössbauer spectrum of **2** at 293 K (Figure 5b) can be fitted in two doublets with similar IS values and 2:1 area, revealing the presence of three high-spin iron(III) ions bound in an octahedral crystal field. The QS values show that one of the metal centres is placed in a less symmetric environment than the other two. When this is compared with complex **1**, and relying on Mössbauer/EPR studies performed, it can be assumed that coordination species **2** probably belongs to the triangular iron(III) complexes of the "acetate" type, where the carboxylate groups of the antibiotic are the "main characters" in the structure formed. The formation of mixed-valence Fe(II)-Fe(III) coordination species in the case of complex **2** is ruled out due to the absence of Mössbauer signals with high IS-values (>1.2 mm/s), thus directly confirming the full oxidation of Fe(II) ions under atmospheric conditions [54,55].

The increase in the isomer shift of both complexes at liquid nitrogen temperature is due to the second-order Doppler effect [56] (Figure S4). The quadrupole splitting, as expected for high-spin iron(III), does not exhibit an important temperature dependence. The average linewidth of all doublets (*ca.* 0.43 mm/s at 293 K and ca. 0.41 mm/s at 77 K) confirms the amorphous nature of the studied complex species.

2.2.4. Proposed Structures of Complexes **1–2**

Based on the collected experimental data, we suggest that the isolated complex **1** represents a dinuclear iron(III) complex of monensin. Taking into account its composition (Fe^{3+}:Cl^-:Mon^- = 1:1:2) and the observed spectral results, viable architectures can be constructed as follows:

(i) two monensinates are bound to each iron(III) ion in a bidentate manner through their terminal carboxylate and hydroxyl functions, and two chloride anions link the metal centres (Figure 6a);
(ii) two bidentate monensinate ligands bridge the iron ions via tail hydroxyl groups and each chloride anion is terminally bound to the metal cation (Figure 6b);
(iii) the presence of a carboxylate linker as a structural motif in **1** is excluded due to the higher Δ-value(s) detected in the IR-spectra of the solid complex.

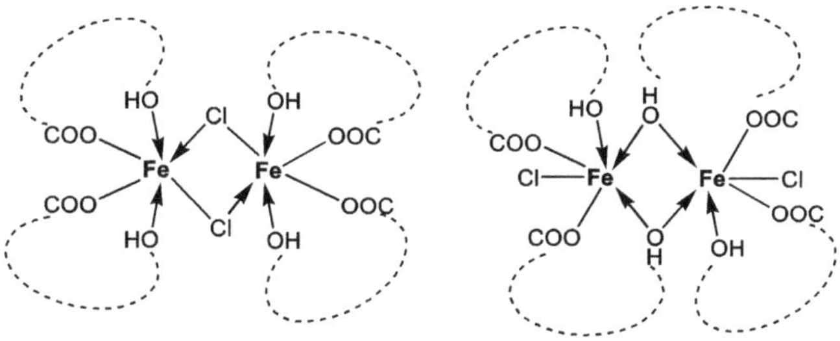

Figure 6. Proposed structure of complex **1** as dinuclear chloro-bridged (**a**) or hydroxyl-bridged (**b**) species. The monensinate skeleton is represented schematically by the dashed line connecting the terminal ligating groups.

The formation of either construct (Figure 6) can explain the observed magnetic behaviour of **1**, but to determine its most reliable structure, we applied a computational approach as described in Section 3.3.

Using the structures proposed above, we also assume that the isolation of **1** in solid state is accompanied by the presence of another type of coordination species, where the monensinate anion is bound in a similar fashion. This compound is deemed to be the mono-complex **1a** of composition [FeCl(Mon)$_2$] (Figure S2) which subsists in equilibrium with the dinuclear parent complex [FeCl(Mon)$_2$]$_2$. Its co-precipitation cannot be controlled, but obviously its formation corroborates the observed experimental EPR and Mössbauer data, especially at low temperatures.

The spectral and microanalysis results reveal that complex **2** most likely belongs to the group of the trinuclear oxo-ferric-monocarboxylates with composition [Fe_3(μ_3-O)(Mon × H_2O)$_6$(H_2O)$_2$(OH)] (Figure 7). We infer that each pair of ionophores serve as bridges between each pair of metal centres through the carboxylate function in a similar way to the known structure of ferric acetate. Binding of a hydroxide anion ensures the overall neutral character of the species formed. The magnetic data disclose the antiferromagnetic properties of **2**, which may arise in an indirect manner, i.e., through an oxo-anion placed into the core of a trinuclear iron(III) cluster.

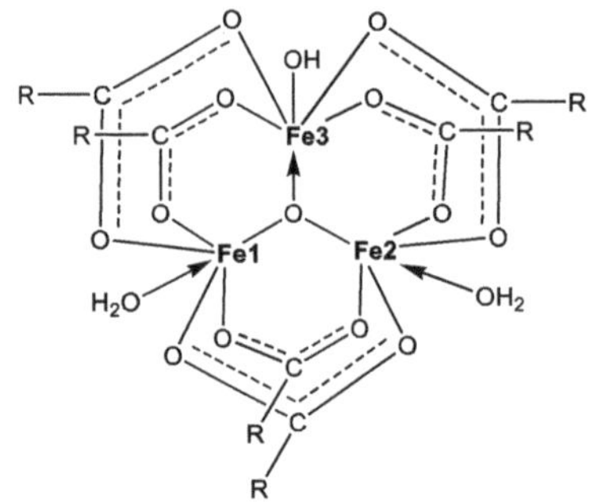

Figure 7. Proposed structure of the triangular oxo-complex **2** (R represents the polyether cavity hosting a water molecule).

To elucidate the architecture of the first coordination sphere of species **1–2**, we conducted an additional molecular modelling. As will be seen, the computed structural parameters are consistent with those observed experimentally, thus confirming the plausibility of the hypothesized constructs.

2.3. Theoretical Studies

2.3.1. Molecular Modelling of 1

To understand the intrinsic properties of complex **1**, we built four dinuclear constructs with C_i point group symmetry as shown in Figure 8, where the monensinate binding was modelled by four acetates (representing its "head" carboxylate functions) and four ethanol molecules (as avatars of the "tail" segment hydroxyl groups). The chloride position was either bridging (**1A**) or terminal (**1B–D**). Structure **1B** comprises chloride anions which are perpendicular to the Fe_2O_6-chromophore and occupy the axial positions in the primary coordination shell. In the other two constructs the inorganic ions lie in the plane of the diamond-core, but the carboxylate groups are parallel (**1C**) and antiparallel (**1D**) to each other.

An indication of the preferred topology of **1** can be derived from the comparison between the calculated IR spectra for **1A–D** and the experimental one. A specific feature of the experimental spectrum is the splitting of the asymmetric C=O vibration into two bands with frequencies 1594 and 1554 cm^{-1} (Figure 2), which are a good match for the corresponding computed unscaled bands for **1B–D** (Table 2). In contrast, no splitting is found in the calculated vibrational spectrum of **1A**—just a narrow band at 1613 cm^{-1}. This is a sign that the dichloro-bridged structure **1A** does not correspond to the experimentally observed vibrational behaviour of complex **1**.

Table 2. Frequencies of asymmetric stretches of the carboxylate function in monensin and similarity factors—SF (the ratio of the experimental to the calculated value for each pair of results).

Complex	Band Position, cm^{-1}		
	Exp.	Calc.	SF
1A	–	1613	–
1B (sub-band 1)	1594	1647	0.97
1B (sub-band 2)	1554	1578	0.98
1C (sub-band 1)	1594	1626	0.98
1C (sub-band 2)	1554	1604	0.97
1D (sub-band 1)	1594	1655	0.96
1D (sub-band 2)	1554	1607	0.97
2	1527	1649	0.93

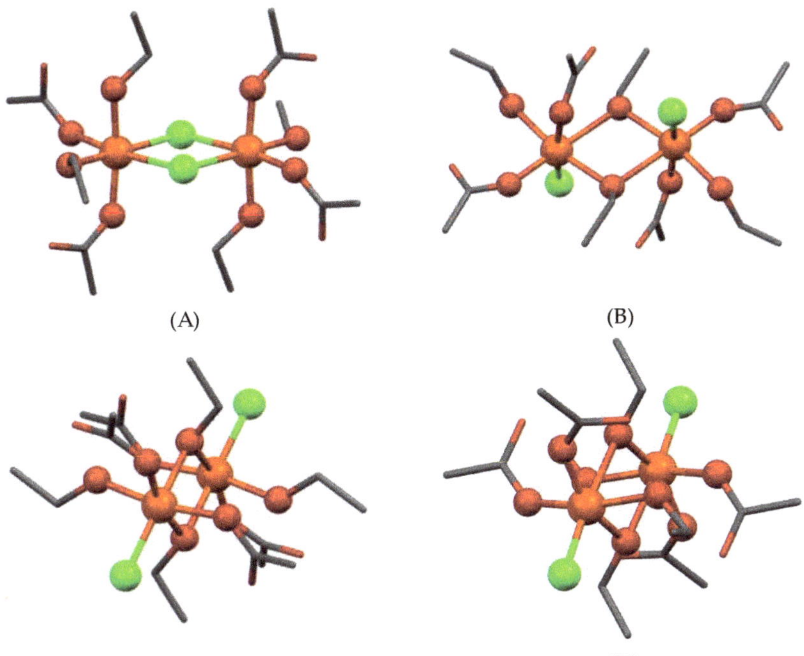

Figure 8. Modelled structures of ferric dimers **1** (**A–D**). Hydrogens are omitted for clarity. Colour code: C—grey, O—red, Cl—green, Fe—orange.

In order to evaluate the impact of the spin state on the EPR properties of target complex **1**, the possible states for the two high-spin Fe(III) ions ($S_{5/2}$) are taken into account, allowing for the complex to have either a ferromagnetic high spin (HS, S_t = 10/2) or an antiferromagnetic singlet (AFMS, S_t = 0) coupling (Scheme S1a). For **1A** (chloride-bridged) and **1B** (hydroxyl-bridged with axial chlorides) constructs the HS state is slightly more stable. The temperature decrease would lead to a population increase in the HS state, leading to an increased intensity of the experimental EPR signal. In the remaining models with hydroxyl oxygen linking (**1C–D**, equatorially placed chlorides), the calculated energy difference favours the low spin state at 293 K. At lower temperatures, the population of the LS state should increase and the intensity of the EPR signal should decrease with a tendency to disappear (EPR-silent state). In line with the experimental EPR data obtained (decrease in the signal at g = 2.02) we can conclude that complex **1** can be described as an alcohol-bridged dimer with terminally ligated chloride anions **1C–D** [57].

The comparison between the anticipated and observed properties of all modelled structures discloses that those with equatorial chlorides are the most likely. The calculated Boltzmann distribution curves show 99.80% at 293 K and 100% at 77 K in favour of **1D** over **1C**. Thus, it can be summarized that the alcohol-bridged dimer with antiparallel orientation of the carboxylates (**1D**) is the most feasible model matching the structure of complex **1**.

2.3.2. Molecular Design of **2**

As a starting construct to design the primary coordination shell of complex **2**, the crystal structure of ferric acetate [29] was used by replacing one of the water molecules with a hydroxide anion (Figure 9). The model consists of three high-spin iron ions which can interact with each other in different ways. We performed a full geometry optimization of the possible multiplicities—high spin (HS, S_t = 15/2) and antiferromagnetic sextet (AFMSx, S_t = 5/2); the spin alignments are presented in Scheme S1b. The modelled structure belongs to the C_{2v} symmetry point group, where two of the iron ions are equivalent with respect to the principal symmetry axis and the corresponding plane containing it. Considering the energy dependence of the spin-flip position (Table 3), the predominantly populated AFMSx is the one in which the spins flip on the iron lying on C_2 (*ca.* 123 kcal/mol). The state population at 293 and 77 K is calculated to be 82.9% of AFMSx (17.1% of HS) at 293 K and increases to 99.8% at 77 K.

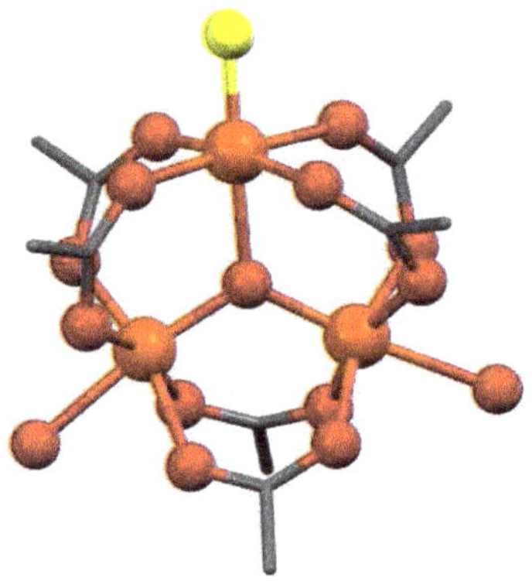

Figure 9. Modelled structure of complex **2**. Hydrogens are omitted for clarity. Colour code: C—grey, O—red, Fe—orange. The hydroxide oxygen is presented in yellow.

Table 3. Enthalpy difference and calculated g-factors for the states of the proposed construct of complex **2**.

Multiplicity	ΔH, kcal/mol, 293 K	g-Factor, 0 K
HS	0.91	2.005
AFMSx	0.00	2.004
AFMSx-ooa *	122.71	2.087

* Spin-flip is performed on the iron ion which does not lie on the principal axis.

To gain a deeper insight into the intimate properties of complex **2**, an additional g-factor calculation was performed (Table 3), which matches well with the experimental findings (Figure 3b). The calculated value of the asymmetric C=O vibration (Table 2) is also in good agreement with the observed one.

2.4. Antibacterial Activity

The biological activity of MonH × H_2O and complexes **1**–**2** is evaluated in terms of their minimum inhibitory concentration (MIC), at which the tested compounds effectively inhibit the visible growth of the target Gram-positive microorganisms. In the currently applied protocol conditions, *B. cereus* (BC) appears to be more sensitive to monensic acid compared with the *B. subtilis* strain (BS) (Table 4), while the parent salts $FeCl_3 \times 6H_2O$ and $FeSO_4 \times 7H_2O$ are ineffective below 3.6–3.7 mM against both bacterial strains and the same holds for the solvent used (methanol). The studied iron(III) complexes have four- (**1**) and six-fold (**2**) increased antibacterial efficacy against *B. subtilis,* which may be due to the presence of four and six antibiotic ligands in the composition of species **1**–**2**, respectively. On the other hand, the new coordination species are eight (**1**) and twelve (**2**) times more potent in the case of *B. cereus*—an activity that cannot be explained in terms of a simple additive effect of metal cations and ligands, linked together in the complex structures. The observed enhanced bioactivity calls for further dedicated investigation of the efficacy of the newly obtained iron(III) monensinates against different target bacterial strains/cell lines to explore their potential as suitable bioactive metal-based drugs.

Table 4. MIC of the polyether ionophore and its complexes **1–2**.

Bacteria Compound	MW, g/mol	BC		BS	
		μg/mL	μM	μg/mL	μM
MonH × H_2O	688.90	3.91	5.67	15.63	22.69
Complex **1**	2862.08	1.95	0.68	15.63	5.46
Complex **2**	4363.90	1.95	0.47	15.63	3.58

3. Experimental Section

3.1. Materials and Methods

Sodium monensinate (MonNa) was generously provided by Biovet Ltd. (Huvepharma, Peshtera, Bulgaria) in a chemically pure form and was used without further purification. Monensic acid monohydrate (MonH × H_2O) was prepared by treating MonNa with HCl [58]. $FeCl_3$ × $6H_2O$, $FeSO_4$ × $7H_2O$, Et_4NOH, Et_3N, acetonitrile (MeCN), methanol (MeOH) p.a. grade, and $CDCl_3$ were delivered by local suppliers. Deionized water was used in all experiments when necessary.

The following approaches and devices were utilized in the present study: infrared spectroscopy (IR) on a Nicolet 6700 FT-IR spectrometer (Thermo Scientific, Madison, WI, USA); thermogravimetry (TG-DTA, TG-MS) on a Setaram Labsys Evo 1600 (Caluire-et-Cuire, France); electron paramagnetic resonance (EPR) on a Bruker BioSpin EMXplus10/12 EPR spectrometer (Karlsruhe, Germany); nuclear magnetic resonance (NMR) on a Bruker NEO 600 spectrometer (Karlsruhe, Germany); ^{57}Fe Mössbauer measurements on a Wissel spectrometer (Wissenschaftliche Elektronik GmbH, Starnberg, Germany); microanalysis on a Vario MACRO cube Elementar analysensysteme Gmbh (Stuttgart, Germany) (C, H) and Perkin-Elmer SCIEX-ELAN DRC-e ICP-MS (Massachusetts, USA) (Fe). Details of the technical parameters of the spectrometers and the corresponding sample preparation procedures are described in [41,59].

3.2. Synthesis of Complexes **1–2**

Complex **1**: To a solution of MonH × H_2O (0.5 mmol, 344.45 mg in 15 mL MeCN), Et_4NOH (0.5 mmol, 180 μL, 40% in H_2O) was added. The reaction mixture was stirred for 15 min to ensure the deprotonation of the antibiotic and $FeCl_3$ × $6H_2O$ (0.17 mmol, 45.00 mg in 5 MeCN) was gradually added to spontaneously form an ochre solid phase. The precipitate was filtered off, washed with MeCN, and dried in a desiccator. [$Fe_2Cl_2(Mon)_4$]: $C_{144}H_{244}Cl_2Fe_2O_{44}$, MW 2862.08 g/mol. Calc. H, 8.89; C, 60.43; Cl, 2.48; Fe, 3.90%. Found: H, 8.19; C, 60.64; Cl, 2.16; Fe, 3.07%. Yield: 153.50 mg, 63%.

Complex **2**: To a solution containing MonH × H_2O (0.5 mmol, 344.45 mg in 20 mL MeCN/MeOH) and Et_3N (1.0 mmol, 139.5 μL), solid $FeSO_4$ × $7H_2O$ (0.5 mmol, 139 mg) was added. The reaction mixture was stirred for 30 min until the iron salt was completely dissolved to turn the colourless solution into a yellow mixture. Subsequent addition of water afforded the formation of dark green precipitates which changed colour to tile red/rusty brown within 10–15 min, indicating Fe(II) oxidation. The solid phase was filtered off, washed with water, and dried in a desiccator. [$Fe_3O(Mon \times H_2O)_6(H_2O)_2(OH)$]: $C_{216}H_{383}Fe_3O_{76}$, MW 4363.90 g/mol. Calc. H, 8.85; C, 59.45; Fe, 3.84%. Found: H, 9.68; C, 59.01; Fe, 3.94%. Yield: 327.35 mg, 90%.

3.3. Computational Protocol

The quantum chemical calculations were performed according to a previously published protocol [41]. Briefly, the geometries of all constructs were optimized with Becke's three-parameter hybrid-exchange functional combined with the Lee-Yang-Parr correlation functional [60], and with Grimme D3 correction [61] for the dispersion interactions and the 6–31G(d) basis set. Vibrational frequencies analysis was performed to verify the minima of all structures. The optimization and the vibrational spectra calculations were carried out with the software package Gaussian 16 [62]. The magnetic properties and EPR parameters for each EPR-active spin state were computed using ORCA 5.0.3 [63] with the BHandHLYP functional [64–66] which has a higher percentage of exact Hartree–Fock exchange, the latter being of great importance for the proper estimation of the spin exchange interactions in the studied systems. The TZVP basis set of Ahlrichs and co-workers [67,68] was implemented for the ligands and the CP(PPP) basis set for the iron ions [69]. All calculations were performed in vacuo.

3.4. Antibacterial Assay

The double-layer agar hole diffusion method [70] was applied to evaluate the effect of the compounds of interest towards Gram-positive microorganisms. Two aerobic bacterial strains were used in the present study—B. subtilis (NBIMCC 1709) and B. cereus (NBIMCC 1085), supplied by the National Bank for Industrial Microorganisms and Cell Cultures (NBIMCC, Bulgaria). The antibacterial efficiency of MonH × H_2O, complexes **1–2**, and parent iron salts was assessed according to the protocol described in [41], replacing DMSO with MeOH as a solvent where necessary.

4. Conclusions

Two new iron(III) monensinates were synthesized and characterized via a variety of experimental methods, supplemented by molecular modelling. Common features of the obtained octahedral complexes are that (i) the monensinate ligands are bound to the iron ions via their carboxylate termini, and (ii) the ratio Fe:ligand is 1:2. However, the pattern of coordination differs. With the Fe(III) salt, an antiferromagnetic dinuclear complex is formed with metal ions also exchange-coupled by hydroxyl groups of the ligand, whereas with the Fe(II) salt a trinuclear oxo-ferric complex results, in which the iron ions, being fully oxidized by the atmospheric oxygen to Fe(III), are linked solely by the carboxylate ligand functions to form an antiferromagnetic sextet. The structure of the complexes was decoded by means of comparison between the computed characteristics of models corresponding to the elemental analysis data and the results measured with the employed experimental approaches. The antibacterial activity tests reveal that the complexes exhibit equal (against *B. subtilis*) or higher (against *B. cereus*) bioactivity, which indicates that the newly synthesized complexes may find a useful implementation as medication in the veterinary practice.

Supplementary Materials: The following supporting information can be downloaded at https://www.mdpi.com/article/10.3390/inorganics12040114/s1: Figure S1: Structure of monensin; Figure S2: Proposed structure of mono-complex 1a; Table S1: ^{13}C-NMR chemical shift (δ, ppm) of monensic acid and complex **1** in $CDCl_3$ ($\Delta = \delta_1 - \delta_{MonH}$); Figure S3: TG-DTA/MS curves of (a) complex **1** and (b) complex **2**; Figure S4: Mössbauer spectra at 77 K: (a) complex **1**, (b) complex **2**; Scheme S1. Electron configurations of the iron ions in (a) complex **1** and (b) complex **2**.

Author Contributions: Conceptualization, N.P. and I.P.; methodology and analysis, N.P., A.T., S.S., Z.C.-Z., D.P., R.S. and R.K.; resources, P.D.; writing—original draft preparation, N.P., I.P. and A.T.; writing—review and editing, I.P. and A.T. All authors have read and agreed to the published version of the manuscript.

Funding: This research was funded by the European Union—NextGenerationEU, through the National Recovery and Resilience Plan of the Republic of Bulgaria, project SUMMIT BG-RRP-2.004-0008-C01 (№ 70-123-186).

Data Availability Statement: Data are available from the authors upon request.

Conflicts of Interest: Author Petar Dorkov is employed by the company Biovet Ltd. The remaining authors declare that the research was conducted in the absence of any commercial or financial relationships that could be construed as a potential conflict of interest. Biovet Ltd. provided the material sodium monensinate (MonNa), but the company was not involved in the study design, collection, analysis, interpretation of data, the writing of this article or the decision to submit it for publication.

References

1. Crichton, R.R.; Pierre, J.-L. Old iron, young copper: From Mars to Venus. *BioMetals* **2001**, *14*, 99–112. [CrossRef] [PubMed]
2. Crichton, R. *Iron Metabolism: From Molecular Mechanisms to Clinical Consequences*, 4th ed.; John Wiley & Sons, Ltd.: Hoboken, NJ, USA, 2016; ISBN 978-1-118-92561-4.
3. Ward, R.J.; Crichton, R.R. *Ironing Out the Brain. Metal Ions in Life Science*; De Gruyter: Berlin, Germany, 2019. [CrossRef]
4. Ward, R.J.; Dexter, D.T.; Crichton, R.R. Neurodegenerative diseases and therapeutic strategies using iron chelators. *J. Trace Elem. Med. Biol.* **2015**, *31*, 267–273. [CrossRef] [PubMed]
5. Roemhild, K.; von Maltzahn, F.; Weiskirchen, R.; Knüchel, R.; von Stillfried, S.; Lammers, T. Iron metabolism: Pathophysiology and pharmacology. *Trends Pharmacol. Sci.* **2021**, *48*, 640–656. [CrossRef]
6. Thich, J.A.; Ou, C.-C.; Powers, D.; Vasilious, B.; Mastropaolo, D.; Potenza, J.A.; Schugar, H.J. Molecular structure and magnetic properties of mu-dihydroxo-bis[2,6-pyridine dicarboxylato aquoiron(III)] and mu-dihydroxo-bis[4-hydroxo-2,6-pyridine dicarboxylato aquoiron(III)] tetrahydrate. *J. Am. Chem. Soc.* **1976**, *98*, 1425–1433. [CrossRef]
7. Ou, C.-C.; Lalancette, R.A.; Potenza, J.A.; Schugar, H.J. Molecular structure and magnetic properties of mu-dihydroxo-bis[4-dimethylamino-2,6-pyridinedicarboxylatoaquo iron(III)] dihydrate, $[(CH_3)_2NC_7H_2NO_4(H_2O)FeOH]_2.2H_2O$. *J. Am. Chem. Soc.* **1978**, *100*, 2053–2057. [CrossRef]
8. Kato, M.; Yamada, T.; Inagaki, T.; Mori, W.; Sakai, K.; Tsubomura, T.; Sato, M.; Yano, S. Structures and magnetic properties of iron(III) dinuclear complexes with alkoxo and carboxylato bridges. *Inorg. Chem.* **1995**, *34*, 2645–2651. [CrossRef]
9. Norman, R.E.; Yan, S.; Que, L., Jr.; Backes, G.; Ling, J.; Sanders-Loehr, J.; Zhang, J.H.; O'Connor, C.J. (mu-Oxo)(mu-carboxylato) diiron(III) complexes with distinct iron sites. Consequences of the inequivalence and its relevance to dinuclear iron-oxo proteins. *J. Am. Chem. Soc.* **1990**, *112*, 1554–1562. [CrossRef]
10. Trukhan, V.M.; Gritsenko, O.N.; Nordlander, E.; Shteinman, A.A. Design and synthesis of new models for diiron biosites. *J. Inorg. Biochem.* **2000**, *79*, 41–46. [CrossRef]
11. Pardo, E.; Lloret, F.; Carrasco, R.; Muñoz, M.C.; Temporal-Sánchez, T.; Ruiz-García, R. Chemistry and reactivity of dinuclear iron oxamate complexes: Alkane oxidation with hydrogen peroxide catalysed by an oxo-bridged diiron(III) complex with amide and carboxylate ligation. *Inorg. Chim. Acta* **2004**, *357*, 2713–2720. [CrossRef]
12. Romakh, V.B.; Therrien, B.; Labat, G.; Stoekli-Evans, H.; Shul'pin, G.B.; Süss-Fink, G. Dinuclear iron, ruthenium and cobalt complexes containing 1,4-dimethyl-1,4,7-triazacyclononane ligands as well as carboxylato and oxo or hydroxo bridges. *Inorg. Chim. Acta* **2006**, *359*, 3297–3051. [CrossRef]
13. Do, L.H.; Lippard, S.J. Toward functional carboxylate-bridged diiron protein mimics: Achieving structural stability and conformational flexibility using a macrocyclic ligand framework. *J. Am. Chem. Soc.* **2011**, *133*, 10568–105681. [CrossRef] [PubMed]
14. Abe, K.; Sakiyama, S.; Nishida, T. Structural investigation of four dinuclear iron(III) complexes relevant to renal injuries. *J. Comp. Chem.* **2015**, *14*, 23–29. [CrossRef]
15. Comba, P.; Daumann, L.; Lefebvre, J.; Linti, G.; Martin, B.; Straub, J.; Zessin, T. Mono- and dinuclear copper(II) and iron(III) complexes of a tetradentate bispidine-diacetate ligand. *Aust. J. Chem.* **2009**, *62*, 1238–1245. [CrossRef]
16. Paredes-García, V.; Latorre, R.O.; Spodine, E. Electronic and magnetic properties of iron(III) dinuclear complexes with carbox-ylate bridges. *Polyhedron* **2014**, *23*, 1869–1876. [CrossRef]
17. Bell, M.; Edwards, A.J.; Hoskins, B.F.; Kachab, E.H.; Robson, R. Synthesis and crystal structure of a complex of a tetranucleating macrocyclic ligand which binds four NiII ions in a square arrangement with a μ_4-hydroxo group at the centre. *J. Chem. Soc. Chem. Commun.* **1987**, *24*, 1852–1854. [CrossRef]
18. Starosta, W.; Ptasiewicz-Bak, H.; Leciejewicz, J. The crystal and molecular structure of a new calcium(II) complex with pyri-dine-2,6-dicarboxylate, water and nitrate ligands. *J. Coord. Chem.* **2002**, *55*, 1147–1153. [CrossRef]
19. Chen, Z.-Q.; Zhang, W.-Z.; Xu, Q. Poly[[mu]2-aqua-[[mu]2-1,1′-(butane-1,4-di yl) diimidazole] bis ([mu]4-naphthalene-1,4-dicarboxyl ato) dimanganese(II)]. *Acta Cryst. Sect. E Struct. Rep. Online* **2008**, *64*, m1543. [CrossRef] [PubMed]
20. Shen, X.-Q.; Yao, H.-C.; Yang, R.; Li, Z.-J.; Zhang, H.-Y.; Wu, B.-L.; Hou, H.-W. Synthesis, structure and thermal properties of polynuclear complexes with a new multidentate ligand, N,N′-bis(5-ethyl-1,3,4-thiadiazol-2-yl)-2,6-pyridinedicarboxamide. *Polyhedron* **2008**, *27*, 203–209. [CrossRef]
21. Azadmeher, A.; Amini, M.M.; Hadipour, N.; Khavasi, H.R.; Fun, H.-K.; Chen, C.-J. Synthesis and structural characterization of diorganotin(IV) complexes with 2,6-pyridinedicarboxylic acid. *Appl. Organomet. Chem.* **2008**, *22*, 19–24. [CrossRef]
22. Su, Q.-J.; Li, S.-H.; Wang, L.; Xie, C.-Z.; Ouyang, Y.; Xu, J.-Y. Synthesis, structure and characterization of 1D threefold-bridging copper(II) chain with strong ferromagnetic coupling. *Inorg. Chem. Comm.* **2010**, *13*, 1210–1212. [CrossRef]
23. White, N.G.; Kitchen, J.A.; Joule, J.A.; Brooker, S. Copper-induced N–N bond cleavage results in an octanuclear expanded-core grid-like complex. *Chem. Commun.* **2012**, *48*, 6229–6231. [CrossRef] [PubMed]
24. Chen, M.; Sanudo, E.C.; Hu, J.-Y.; Hu, M.; Fang, S.-M.; Liu, C.-S. Structural diversity and magnetic properties of five Cu(II) complexes with mixed naphthalene-based dicarboxyl tecton and different N-donor co-ligands. *Aust. J. Chem.* **2013**, *66*, 963–971. [CrossRef]
25. Gou, L.; Zhang, H.-X.; Fan, X.-Y.; Li, D.-L. Lithium based coordination polymer as anode for Li-ion battery. *Inorg. Chim. Acta* **2013**, *394*, 10–14. [CrossRef]
26. Long, G.J.; Robinson, W.T.; Tappmeyer, W.P.; Bridges, D.L. The magnetic, electronic, and Mössbauer spectral properties of several trinuclear iron(III) carboxylate complexes. *J. Chem. Soc. Dalton Trans.* **1973**, *6*, 573–579. [CrossRef]

27. Boudalis, A.K.; Sanakis, Y.; Raptopoulou, C.P.; Terzis, A.; Tuchagues, J.-P.; Perlepes, S.P. A trinuclear cluster containing the {$Fe_3(\mu_3$-O)}7+ core: Structural, magnetic and spectroscopic (IR, Mössbauer, EPR) studies. *Polyhedron* **2005**, *24*, 1540–1548. [CrossRef]
28. Kantouri, M.-L.; Papadopoulos, C.D.; Hatzidimitriou, A.G.; Bakas, T.; Pachini, S. A trinuclear iron(III) complex containing the semi-cubane [$Fe_3(\mu_3$-O)]7+ core: Structural, spectroscopic, magnetic and electrochemical study. *Z. Für Anorg. Und Allg. Chem.* **2010**, *636*, 531–538. [CrossRef]
29. Balić, T.; Jagličić, Z.; Sadrollah, E.; Litterst, F.J.; Počkaj, M.; Baabe, B.; Kovač-Andrić, E.; Bijelić, J.; Gašo-Sokač, D.; Djerdj, I. Single crystal growth, structural characterization and magnetic properties study of an antiferromagnetic trinuclear iron(III) acetate complex with uncoordinated hexamine. *Inorg. Chim. Acta* **2021**, *520*, 120292. [CrossRef]
30. Bronstein, L.M.; Huang, X.; Retrum, J.; Schmucker, A.; Pink, M.; Stein, B.D.; Dragnea, B. Influence of iron oleate complex structure on iron oxide nanoparticle formation. *Chem. Mater.* **2007**, *19*, 3624–3632. [CrossRef]
31. Lin, M.M.; Kim, D.K. In situ thermolysis of magnetic nanoparticles using non-hydrated iron oleate complex. *J. Nanopart. Res.* **2012**, *14*, 688. [CrossRef]
32. Sánchez, M.; Sabio, L.; Gálvez, N.; Capdevila, M.; Dominguez-Vera, J.M. Iron chemistry at the service of life. *IUBMB Life* **2017**, *69*, 382–388. [CrossRef]
33. Stockwell, B.R.; Jiang, X. The chemistry and biology of ferroptosis. *Cell Chem. Biol.* **2020**, *27*, 365–375. [CrossRef] [PubMed]
34. Qiu, Y.; Cao, Y.; Cao, W.; Jia, Y.; Lu, N. The application of ferroptosis in diseases. *Pharmacol. Res.* **2020**, *159*, 104919. [CrossRef] [PubMed]
35. Guo, Q.; Li, L.; Hou, S.; Yuan, Z.; Li, C.; Zhang, W.; Zheng, L.; Li, X. The role of iron in cancer progression. *Front. Oncol.* **2021**, *11*, 778492. [CrossRef] [PubMed]
36. Wu, Z.; Zhong, M.; Liu, Y.; Xiong, Y.; Gao, Z.; Ma, J.; Zhuang, G.; Hong, X. Application of natural products for inducing ferroptosis in tumor cells. *Biotechnol. Appl. Biochem.* **2022**, *69*, 190–197. [CrossRef] [PubMed]
37. Su, Y.; Zhao, B.; Zhou, L.; Zhang, Z.; Shen, Y.; Lv, H.; Al Qudsy, L.H.H.; Shang, P. Ferroptosis, a novel pharmacological mechanism of anti-cancer drugs. *Cancer Lett.* **2020**, *483*, 127–136. [CrossRef] [PubMed]
38. Mai, T.T.; Hamaï, A.; Hienzsch, A.; Cañeque, T.; Müller, S.; Wicinski, J.; Cabaud, O.; Leroy, C.; David, A.; Acevedo, V.; et al. Salinomycin kills cancer stem cells by sequestering iron in lysosomes. *Nat. Chem.* **2017**, *9*, 1025–1033. [CrossRef] [PubMed]
39. Versini, A.; Colombeau, L.; Hienzch, A.; Gaillet, C.; Retailleau, P.; Debieu, S.; Müller, S.; Cañeque, T.; Rodriguez, R. Salinomycin derivatives kill breast cancer stem cells by lysosomal iron targeting. *Chem.-Eur. J.* **2020**, *26*, 7416–7424. [CrossRef] [PubMed]
40. Antoszczak, M.; Müller, S.; Cañeque, T.; Colombeau, L.; Dusetti, N.; Santofimia-Castaño, P.; Gaillet, G.; Puisieux, A.; Iovanna, J.L.; Rodriguez, R. Iron-sensitive prodrugs that trigger active ferroptosis in drug-tolerant pancreatic cancer cells. *J. Am. Chem. Soc.* **2022**, *144*, 11536–11545. [CrossRef] [PubMed]
41. Petkov, N.; Tadjer, A.; Encheva, E.; Cherkezova-Zheleva, Z.; Paneva, D.; Stoyanova, R.; Kukeva, R.; Dorkov, P.; Pantcheva, I. Experimental and DFT study of monensinate and salinomycinate complexes containing {$Fe_3(\mu_3$–O)}7+ core. *Molecules* **2024**, *26*, 364. [CrossRef]
42. Pantcheva, I.N.; Ivanova, J.; Zhorova, R.; Mitewa, M.; Simova, S.; Mayer-Figge, H.; Sheldrick, W.S. Nickel(II) and zinc(II) dim-onensinates: Crystal structure, spectral properties and bactericidal activity. *Inorg. Chim. Acta* **2010**, *363*, 1879–1886. [CrossRef]
43. Ivanova, J.; Pantcheva, I.N.; Mitewa, M.; Simova, S.; Mayer-Figge, H.; Sheldrick, W.S. Crystal structures and spectral properties of new Cd(II) and Hg(II) complexes of monensic acid with different coordination modes of the ligand. *Centr. Eur. J. Chem.* **2010**, *8*, 852–860. [CrossRef]
44. Pantcheva, I.; Dimitrova, R.; Nedzhib, A.; Dorkov, P.; Stoyanova, R.; Zhecheva, E. Coordination compounds of polyether ionophore Monensin with gadolinium(III) ions. In *Nanoscience & Nanotechnology (Sofia)*; Balabanova, E., Mileva, E., Eds.; Acad. Evgeni Budevski Institute of Electrochemistry and Energy Systems (IEES-BAS), Bulgarian Academy of Sciences (BAS): Sofia, Bulgaria, 2019; Volume 19, pp. 40–46, ISSN 1313-8995 (print), ISSN 2738-8743 (online).
45. Pantcheva, I.; Dimitrova, R.; Ivanova, V.; Nedzhib, A.; Dorkov, P.; Dinev, D.; Spasov, R.; Alexandrova, R. Spectral properties and biological activity of La(III) and Nd(III) monensinates. *Open Chem.* **2019**, *17*, 1423–1434. [CrossRef]
46. Wojdyr, M. Fityk: A general-purpose peak fitting program. *J. Appl. Crystallogr.* **2010**, *43*, 1126–1128. [CrossRef]
47. Muralidhara, R.S.; Kesavulu, C.R.; Rao, J.L.; Anavekar, R.V.; Chakradhar, R.P.S. EPR and optical absorption studies of Fe3+ ions in sodium borophosphate glasses. *J. Phys. Chem. Solids* **2010**, *71*, 1651–1655. [CrossRef]
48. Kakazey, M.; Vlasova, M.; Gonzalez-Rodriguez, G.; Salazar-Hernández, B. Fine structure of EPR spectra of Fe3+ in α-Al_2O_3 crystal, powders, and textured ceramics. *Mat. Sci. Engin.* **2002**, *90*, 114–119. [CrossRef]
49. Borer, L.; Thalken, L.; Zhang, J.H.; Reiff, W.M. Characterization of a dimeric iron(III) complex of N,N′-ethylenebis(salicylamine). *Inorg. Chem.* **1983**, *22*, 3174–3176. [CrossRef]
50. Vercamer, V.; Lelong, G.; Hijiya, H.; Kondo, Y.; Galoisy, L.; Calas, G. Diluted Fe3+ in silicate glasses: Structural effects of Fe-redox state and matrix composition. An optical absorption and X-band/Q-band EPR study. *J. Non-Cryst. Solids* **2015**, *428*, 138–145. [CrossRef]
51. Murray, K.S. Binuclear oxo-bridged iron(III) complexes. *Coord. Chem. Rev.* **1974**, *12*, 1–35. [CrossRef]
52. Rodriguez, F.; Moreno, M. The influence of water addition on solutions of $FeCl_3$ in isopropanol. *Z. Für Naturforschung* **1983**, *38*, 701–702. [CrossRef]
53. Vertes, A.; Korecz, L.; Burger, K. *Mössbauer Spectroscopy*; Elsevier: Amsterdam, The Netherlands, 1979.

54. Reedijk, J.; van Der Kraan, A.M. Mössbauer effect of octahedral iron(II) solvates. *Recl. Trav. Chim. Pays-Bas.* **1969**, *88*, 828–832. [CrossRef]
55. Goldanskii, V.I.; Herber, R.H. *Chemical Applications of Mossbauer Spectroscopy*; Academic Press: Cambridge, MA, USA, 1968.
56. Greenwood, N.N.; Gibb, T.C. *Mössbauer Spectroscopy*; Chapman and Hall: London, UK, 1971.
57. You, Z.-L.; Zhu, H.-L. A dinuclear Schiff base iron(III) complex with the ligand N,N′-bis(2-oxidophenylmethyleneimino) propane-1,2-diamine. *Acta Cryst. Sect. E: Struct. Rep. Online* **2004**, *60*, m1046–m1048. [CrossRef]
58. Gertenbach, P.G.; Popov, A.I. Solution chemistry of monensin and its alkali metal ion complexes. Potentiometric and spec-troscopic studies. *J. Am. Chem. Soc.* **1975**, *97*, 4738–4744. [CrossRef]
59. Petkov, N.; Pantcheva, I.; Ivanova, A.; Stoyanova, R.; Kukeva, R.; Alexandrova, R.; Abudalleh, A.; Dorkov, P. Novel cerium(IV) coordination compounds of monensin and salinomycin. *Molecules* **2023**, *28*, 4676. [CrossRef] [PubMed]
60. Stephens, P.J.; Devlin, F.J.; Chabalowski, C.F.; Frisch, M.J. Ab initio calculation of vibrational absorption and circular dichroism spectra using density functional force fields. *J. Phys. Chem.* **1994**, *98*, 11623–11627. [CrossRef]
61. Grimme, S.; Antony, J.; Ehrlich, S.; Krieg, H. A consistent and accurate ab initio parametrization of density functional dispersion correction (DFT-D) for the 94 elements H-Pu. *J. Chem. Phys.* **2010**, *132*, 154104. [CrossRef] [PubMed]
62. Frisch, M.J.; Trucks, G.W.; Schlegel, H.B.; Scuseria, G.E.; Robb, M.A.; Cheeseman, J.R.; Scalmani, G.; Barone, V.; Petersson, G.A.; Nakatsuji, H.; et al. *Gaussian 16, Revision B.01*; GaussView 5.0.; Gaussian, Inc.: Wallingford, CT, USA, 2016.
63. Neese, F. Software update: The ORCA program system—Version 5.0. *WIREs Comp. Mol. Sci.* **2022**, *12*, e1606. [CrossRef]
64. Becke, A.D. A new mixing of Hartree-Fock and local density-functional theories. *J. Chem. Phys.* **1993**, *98*, 1372–1377. [CrossRef]
65. Yang, Y.; Ratner, M.A.; Schatz, G.C. Computational modeling of octahedral iron oxide clusters: Hexaaquairon(III) and its dimers. *J. Phys. Chem. C* **2013**, *117*, 21706–21717. [CrossRef]
66. Zhao, H.; Fang, C.; Gao, J.; Liu, C. Spin-state energies of heme-related models from spin-flip TDDFT calculations. *Phys. Chem. Chem. Phys.* **2016**, *18*, 29486–29494. [CrossRef]
67. Schaefer, A.; Horn, H.; Ahlrichs, R. Fully optimized contracted Gaussian-basis sets for atoms Li to Kr. *J. Chem. Phys.* **1992**, *97*, 2571–2577. [CrossRef]
68. Schaefer, A.; Huber, C.; Ahlrichs, R. Fully optimized contracted Gaussian-basis sets of triple zeta valence quality for atoms Li to Kr. *J. Chem. Phys.* **1994**, *100*, 5829–5835. [CrossRef]
69. Neese, F. Prediction and interpretation of the 57Fe isomer shift in Mössbauer spectra by density functional theory. *Inorg. Chim. Acta* **2002**, *337*, 181–192. [CrossRef]
70. Andrews, J.M. Determination of minimum inhibitory concentrations. *J. Antimicrob. Chemother.* **2001**, *48* (Suppl. S1), 5–16. [CrossRef] [PubMed]

inorganics

MDPI

Review

Homo- and Hetero-Multinuclear Iridium(III) Complexes with Cytotoxic Activity

Irena Kostova

Department of Chemistry, Faculty of Pharmacy, Medical University, 2 Dunav St., 1000 Sofia, Bulgaria; irenakostova@yahoo.com

Abstract: Towards the efforts to expand the bioactivity and to reduce toxic and adverse properties of known metal-based drugs, various multinuclear complexes have recently been studied. They have shown enhancement of target specificity and selectivity. Different from small organic compounds and traditional metal-based complexes with anticancer activity, iridium(III) multinuclear or heteronuclear metallodrugs have confirmed potential advantages due to their unique biological and chemical diversities, better activity and different anticancer mechanisms. Ir(III) coordination compounds, similar to most Pt group compounds, are of excessive interest because of their potential cytotoxic activity, effective cellular uptake and tolerance by healthy cells. Although mononuclear Ir(III) complex compounds have been extensively studied as promising candidates for antitumor application, the research on the antineoplastic potential of homo- or hetero-multinuclear iridium(III) complexes is not as abundant; nevertheless, intensive investigations have been conducted in the recent years towards developing complexes that are anticipated to have improved therapeutic potential and biotarget selectivity. Multimetallic iridium(III) frameworks have offered interesting possibilities for designing new antitumor agents by exploiting the action of different metal cations at the same time. This method was very successful in the design of homo- and hetero-multinuclear cyclometalated and half-sandwich organometallic Ir(III) compounds. In the described background, many homonuclear and heteronuclear Ir(III) complexes have been estimated and have exposed promising advantages in cancer therapy. This review intends to summarize newly reported innovative and promising multinuclear Ir(III)-based complexes and to afford a wide-ranging overview of current development and perspectives for the practical impact of these complexes in the tumor therapy field. It is anticipated that this analysis will provide significant direction for the further progress of active homonuclear and heteronuclear iridium-based anticancer agents.

Keywords: homonuclear; heteronuclear; iridium(III) complexes; cytotoxic activity

1. Introduction

Metal-based anticancer compounds, particularly the complexes of platinum group metals, are amongst the most extensively applied antineoplastic agents in clinics. The healing potential and possible application of such compounds are at a disadvantage because of resistance problems, toxic and severe dose-limiting side effects, or unsatisfactory antimetastatic and diagnostic properties. All this has provoked the consideration of scientists for the discovery of suitable alternatives. Among the best antitumor metal-based drugs are the complexes of the Pt group, including Ir(III) complexes [1–4]. They have been proven to be effective antineoplastic agents with good cytotoxicity [5,6], high selectivity [7], minor adverse effects and unique target specificity [8,9], offering a wide spectrum of bioactivities. Their structural and electronic properties have been mostly controlled by the type of involved biologically active ligands since the ligands influence significantly the target site recognition. Compared to classical metallodrugs, numerous Ir(III) cyclometalated compounds have displayed enhanced cytotoxicity, inherent luminescence properties [10–12] and specificity in targeting organelles [13–17]. Consequently, the physicochemical characteristics of organoiridium complex compounds gave them substantial bioactivity together with notable properties to be proposed as potential antineoplastic agents. Additionally, they have been exploited for imaging usage as they have exceptional luminescence features such as high quantum yields photostability, long excited-state lifetime, small photobleaching and large Stokes shift. In this context, the progress of Ir(III) compounds, which possessed high cytotoxicity, has directed these complexes to be applied as theranostics [18]. Iridium(III) complex compounds have suitable coordination numbers and geometry, obtainable by their different oxidation states, adjustable chemical interactions, ligand exchange varieties and flexible physical features. As target-specific, iridium(III) complex compounds have been observed to accumulate in various organelles for instance mitochondria [19–21], endoplasmic reticulum [22], nucleus [23], lysosomes [24,25], etc., where mitochondria and lysosomes were the key targets [26]. They can generate ROS and decrease mitochondrial membrane potentials (MMPs) by creating mitochondrial dysfunction or triggering mitochondrial apoptotic functions. By acquiring in-depth knowledge about the chemicophysical properties of iridium(III) complexes and their outstanding antitumor activity in various mechanistic pathways, many research groups have designed different types of Ir(III) organometallic complexes, including multinuclear ones [27–29], as appropriate chemotherapeutics with diverse from Pt(II) and Pt(IV) agents antineoplastic mechanism, and high cytotoxic potency against resistant to cisplatin tumor cells [30]. In many cases, their cytotoxicity was better than that of the classical platinum drugs [31,32].

Iridium is a rare third-row transition metal belonging to the platinum group. It is often considered to be one of the most inert metals. In this group, the reactivity of Rh, Ir, Pd and Pt are much closer to each other than to the chemical activity of Ru and Os. Rh and Ir readily form complexes with a +3 oxidation state, that, similar to Co analogs, have kinetical inertness [2]. The typical stability of Ir(III) complex compounds makes them potential drug candidates which are capable of reaching the targets without any modifications. The variety of biologically active (N^N, C^N, O^N, etc.) ligands for coordination with iridium is an additional reason for its application, because the bioligands regulate the targets' recognition. Iridium(III) complex compounds represent two main classes, viz., half-sandwich and cyclometalated Ir(III) complexes. In comparison with half-sandwich iridium(III) complexes, the cyclometalated ones possess improved optical characteristics [33] and could inhibit tumor cells by light exciting, making them appropriate possible agents for phototherapeutic use. Numerous cyclometalated [34–37] and half-sandwich organoiridium(III) compounds with cyclopentadienyls [38–41] have been obtained and utilized as antineoplastic agents [42,43].

In recent years, polynuclear metal-based compounds have attracted the consideration of medicinal chemistry researchers because of their exceptional antineoplastic activity compared to the activity of mononuclear corresponding complexes [44,45]. Ma et al.l. have recently reviewed Pt-containing heterometallic complexes with different transition metals, such as Ru, Au, Pd, Rh, Ti, Gd, Eu, Ir, Re, Fe, Cu, Tc, etc. for the treatment of malignant diseases [46]. The synthesis, antitumor activity, mechanisms of action, and interactions between platinum and additional metallic centers have been comprehensively discussed. The multinuclear complex compounds have shown improved cytotoxicity, selectivity and

DNA-binding ability. Most of them were able to overcome cisplatin-acquired resistance, which can be explained by the synergistic effects of various metal centers, such as Ru [47], Pt [48], Au [49,50], Re [51] and Ir [52]. Concerning heterometallic anticancer complexes, platinum-based complexes constitute the most extensively studied compounds. Numerous examples of platinum-containing heterometallic complexes have been designated, displaying great potential for overcoming the limitations of traditional clinically used platinum drugs, such as resistance, toxic effects, and diagnostic issues. These multinuclear metal complexes represent combinations of platinum with gold [53], ruthenium [54,55], rhodium [56], palladium [57], and undoubtedly iridium [58], affording theragnostic heterometallic complexes [59]. Multinuclear homometallic or heterometallic complexes have recently been reviewed [60] with an emphasis on multinuclear ruthenium(II), rhodium(III), osmium(II) and iridium(III) derived compounds with various metal-containing species in a framework of homometallic or heterometallic units with cytotoxic and antimicrobial activity. Specific consideration has been paid to the evaluation of structural analogs with diverse numbers or types of metal centers, coordinated with the biologically active ligands [60]. The multinuclear complexes, reported in the literature have shown enhanced permeability and retention, a greater degree of selectivity and lower adverse effects than their mononuclear counterparts [61]. In many cases, malignant cells retained the resulting macromolecules because of their damaged lymphatic drainage, whereas healthy cells readily excreted them [61]. The choice of the metal centers, forming the respective hetero-multinuclear complex, is crucial as it determines the possible therapeutic and synergistic effects [62].

With the recognized features of iridium(III) complexes for anticancer application, their utilization as biomolecular recognition probes for monitoring cellular changes and their potential for bioimaging applications, the majority of active iridium(III) complexes have been combined with other cytotoxic metals during the last decades. The incorporation of a second metallic center with distinct biotargets and advantageous physicochemical characteristics into iridium compounds is a good approach to improve the antineoplastic efficiency of iridium-based agents and to overcome the possible disadvantages. Many diiridium(III) [63] and iridium(III) heterometallic complexes have exposed a great antiproliferative potential and their photo- and bioimaging assets have also been improved. The design of such compounds offers growth in the resulting biological activity through the multiple metal centers or by combining different metal centers with different mechanisms of action and bioeffects. This review provides a wide-ranging analysis and outlook of recently reported representatives of iridium-containing homonuclear and heteronuclear complexes examined for anticancer treatment and might guide investigators in discovering the challenges and prospects in this developing field.

2. Homonuclear Iridium(III) Complexes

2.1. *Iridium(III) Cyclopentadienyl Dinuclear Complexes*

Half-sandwich Cp*-Ir(III) organometallic complexes with pseudo-octahedral geometry, where Cp* is a derivative of 1,2,3,4,5-pentamethylcyclopenta-1,3-dienyl, have been demonstrated to be potent anticancer metal-based drugs. Dinuclear Ir(III) complexes of bioactive ligands with different structures have shown significant activity and have exhibited different modes of activity as compared to the respective mononuclear corresponding complexes or to similar homonuclear analogous complexes.

A series of homologous dinuclear iridium complexes, presented in Figure 1, $[Ir(\eta^5\text{-}C_5Me_4R)Cl(\mu\text{-}Cl)]_2$ (R = Me, Figure 1a; R = H, Figure 1b; R = Pr, Figure 1c; R = 4-C_6H_4F, Figure 1d; R = 4-C_6H_4OH, Figure 1e) have been prepared, characterized, and assessed for their cytotoxic action on rodent and human tumor cells, such as mouse melanoma B16, rat glioma C6, colorectal carcinoma SW620 and HCT116, breast adenocarcinoma MCF-7, ovarian A2780 carcinoma, and human fetal lung MRC5 fibroblast cells [64]. The evaluation of properties of the adapted Cp* ring by substitution of the Me group with a variety of additional groups and their influence on the cytotoxic activity has also been presented. Predominantly, a substitution of one Me group with H has been found to be satisfactory for improved cytotoxicity.

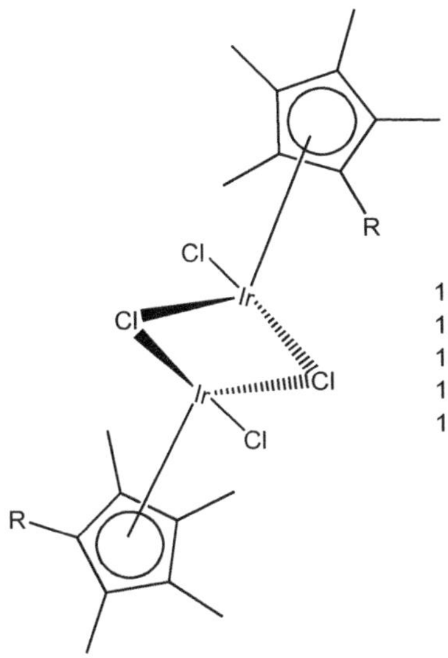

1a R=Me
1b R=H
1c R=Pr
1d R=4-C_6H_4F
1e R=4-C_6H_4OH

Figure 1. Structural formula of diiridium complexes bearing variable tetramethylcyclopentadienyl ligands [64].

Half-sandwiched Ir(III) complexes are attractive organometallic antineoplastic agents. The half-sandwich Cp* iridium complex $[Ir_2(m\text{-pzpy})(\eta^5\text{-Cp*})_2Cl_3]PF_6$ (Figure 2), containing the bridging pzpy ligand (Hpzpy = N′-[(1Z)-1-(pyridin-2-yl)ethylidene]pyrazine-2-carbohydrazonamide) as the tridentate one has been synthesized [65]. In the dinuclear iridium complex, the ligands coordinate to one Ir(III) center bidentately and to the other ligand in a monodentate fashion. The cytotoxicity of Ir(III) complex has been estimated against the colorectal carcinoma cell lines HT 29, HCT116 p53þ/þ, HCT116 p53/ and the normal human epithelial resulting from the retina ARPE-19 cells. However, the Ir(III) complex exhibited only moderate cytotoxic activity ($IC_{50} > 100$ μM), which was lesser than that of cisplatin and its Rh(III) analogs, Table 1.

Related studies have been extended to other Cp* iridium polypyridyl complexes. The best result has been found for bis-intercalators comprising rigid linkers which favor slow kinetical dissociation and probably further groove reactions. Potential bimetallic bis-intercalators, for instance, the tetracations $[\{(\eta^5\text{-Cp*})Ir(dppz)\}_2(\mu\text{-pyz})]^{4+}$, and $[\{(\eta^5\text{-}C_5Me_5)Ir(pp)\}_2(\mu\text{-}4,4'\text{-bpy})]^{4+}$, where pp is dipyrido[3,2-d:2′,3′-f]quinoxaline (dpq), dipyrido[2,3-a:2′,3′-c]phenazine (dppz) or benzo[i]dipyrido[3,2-a:2′,3′-c]phenazine (dppn)), which comprise $\{(\eta^5\text{-Cp*})Ir(dppz)\}^{2+}$ groups bridged by rigid bipyridinyl ligands, such as pyrazine (pyz) or 4,4′-bipyridyl (4,4′-bpy), have been designed by Nazif et al. [66]. The DNA-binding characteristics and in vitro cytotoxicity of the binuclear complexes with the formula $[\{Ir(N^\wedge N)(\eta^5\text{-}C_5Me_5)\}_2L]^{4+}$ (Figure 3a–c) have been studied by the same group [66]. While, the pyz-bridged complex ion appeared to be only a mono-intercalator, NOESY- NMR spectra revealed strong bis-intercalation for the 4,4′-bpy compound, that sandwiched two DNA base pairs through its parallel dppz ligands. As the binuclear complexes displayed inhibitory effects analogous to these of $[(\eta^5\text{-Cp*})IrCl(dppz)]^+$ where the covalent nucleobase binding was more stable, it could be hypothesized that the character and strength of the possible binding mechanisms were not of principal significance for the resulting potency. The antiproliferative effects of the compounds have been determined against colon HT-29 and breast MCF-7 carcinoma cells. IC_{50} values of 3.5/1.8 and 3.1/3.8 μM, respectively, were calculated for the pyz and 4,4′-bpy complexes against MCF-7/HT-29 cell line, Table 1. The dipyrido[3,2-a:2′,3′-c]phenazine (dppz) complex, shown in Figure 3b, bonded through a bis-intercalation manner to DNA including the involvement of both the dppz ligands in the base pairs. In contrast, complexes in Figure 3a,c with smaller dipyridoquinoxaline (dpq) and bigger benzodipyridophenazine (dppn) bioligands did not intercalate into DNA base pairs. Therefore, the size of the hypothetically intercalating diimine bioligands is of significance to the design of active bis-intercalators. The outcomes of the lengths of rigid bridging ligands on DNA-binding modes have been examined. For instance, the distance between the effectively parallel dppz bioligands in the complex, shown in Figure 3b, is around 10.2 Å, which is the perfect distance (3 × 3.4 Å) for two base sandwiching pairs.

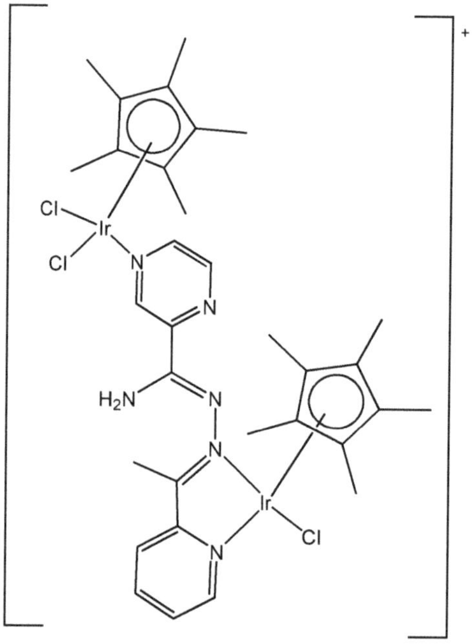

Figure 2. Half-sandwich Cp*iridium complex $[Ir_2(m\text{-pzpy})(\eta^5\text{-Cp*})_2Cl_3]PF_6$ [65].

Improved cytotoxicity has been realized by the corresponding complexes comprising larger dipyridinyl bridging bioligands, precisely 4-[(*E*)-2-(4-pyridinyl)ethenyl]pyridine (dpee) in Figure 4a, 4-(2-pyridin-4-ylethynyl)pyridine (dpey) in Figure 4b or 1,4-di(2-pyridin-4-ylethynyl)benzene (dpeb) in Figure 4c. Engaging longer rigid bridge ligands (Figure 4) led to improved potency (IC_{50} = 0.61 and 0.49 μM) for the complexes $[\{(\eta^5\text{-Cp*})Ir(dppz)\}_2(\mu\text{-B})](CF_3SO_3)_4$, where B = dpee, dpey, Figure 4a,b, against the less quickly proliferating MCF-7 cell line, Table 1. Interestingly, complexes dpee (Figure 4a) and dpey (Figure 4b) fix to DNA only in a mono-intercalation manner, although their iridium–

iridium distances (13.1 and 13.5 Å, respectively), were close to the perfect distance for 3-base sandwiching pairs, which is 13.6 Å [67]. The reported diiridium complexes have been found to be only mono-intercalators but they strongly cleaved the supercoiled DNA of pBR322 plasmid to produce the nicked forms in dark conditions [67]. The addition of DNA, complex dpeb, Figure 4c, (with iridium–iridium distance of 20.6 Å) quickly formed kinetically-preferred monointercalated adducts, which were gradually transformed to thermodynamically stable intertwined bis-intercalation modes. UV/Vis and CD spectral investigations were in agreement with a stable intertwined bis-intercalation mode for ion pairs in Figure 4c with 1,4-di(2-pyridin-4-ylethynyl)benzene), whose much longer metal–metal distance allowed a stack of five aromatic chromophores to be sandwiched between its effective parallel dppz bioligands. The obtained IC_{50} values of the nuclease complexes in Figure 4a,b were smaller than those of the bis-intercalator Figure 4c by one order of magnitude. Additionally, the complex in Figure 4c possesses photoinduced nuclease action. The dpeb correspondent (Figure 4c) was much less active (IC_{50} = 2.2 μM) in the inhibition of MCF-7 proliferation, in spite of its capability to bind to DNA in the more stable bis-intercalative manner, Table 1, [67]. The results suggested that the nuclease characteristics of the dpee and dpey complex compounds could probably be responsible for their better cytotoxic activity. It is also notable that the inhibiting properties of the non-intercalating complexes $[\{(\eta^5\text{-Cp}^*)\text{Ir}(5,6\text{-Me}_2\text{phen})\}_2(\mu\text{-B})](\text{CF}_3\text{SO}_3)_4$ with identical bridging ligands (dpey, dpeb) were similar to those of the corresponding dppz complex compounds against HT-29 and MCF-7 cells [67]. Noticeable antileukemic action (IC_{50} = 6–7 μM) connected with bigger ROS levels and apoptotic induction has been documented for the complexes towards Jurkat cells. It has been shown that the nature of polypyridyl chelating N,N-ligand appeared to be more significant for the consequential antiproliferative action than the bridging ligand, while the bridging ligands regulated the type of nuclease action, since dpee- and dpey derivatives cleaved DNA in dark conditions, which was detected for the dpeb derivative only when irradiated.

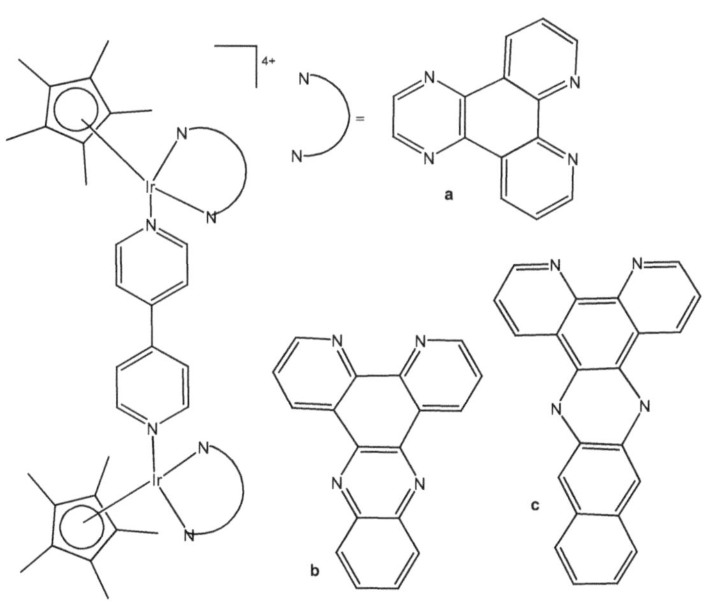

Figure 3. Structures of $[\{(\eta^5\text{-C}_5\text{Me}_5)\text{Ir}(\text{pp})\}_2(\mu\text{-}4,4'\text{-bpy})]^{4+}$, where pp = dpq (**a**), dppz (**b**), dppn (**c**) [66].

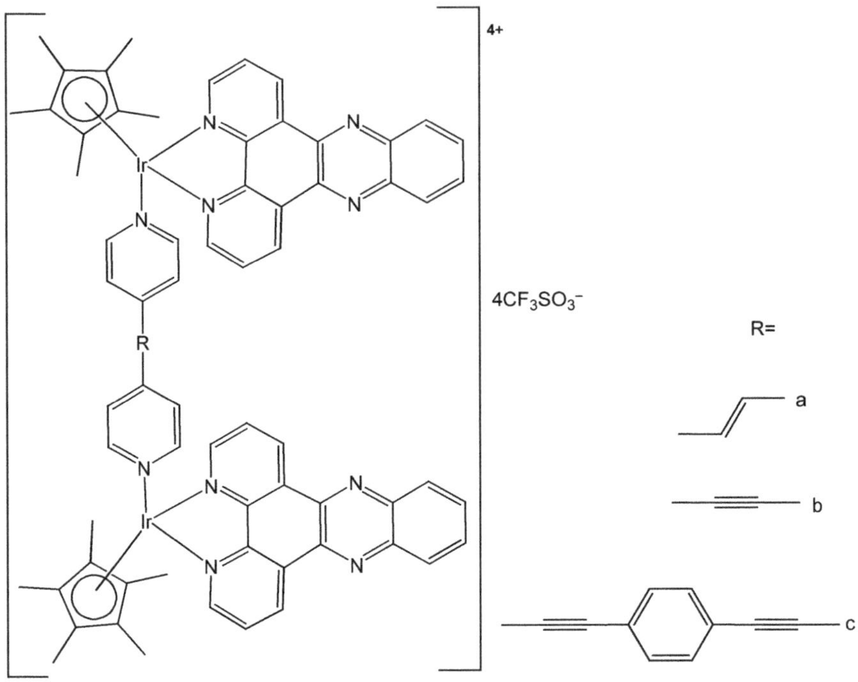

Figure 4. Complexes [{(η^5-Cp*)Ir(dppz)}$_2$(μ-B)](CF_3SO_3)$_4$, where R = dpee (**a**), dpey (**b**) and dpeb (**c**) bridging bioligands [67].

Dinuclear chloridoiridium(III)-Cp* complexes, shown in Figure 5, with higher in vitro antineoplastic activity than cisplatin and analogous Ru derivatives have been discovered by Parveen et al. [68]. Organometallic iridium complexes have been prepared by stirring a methanolic solution of [(Cp*)IrCl$_2$]$_2$ with deprotonated 1,n-bis[3-hydroxy-2-methyl-4(1H)-pyridinon-1-yl]alkane (n = 4, 6, 8, 12) and the products have been extracted with diethyl ether and dichloromethane to obtain the respective complexes. The cytotoxicity of the complexes has been assessed against colon HCT116, non-small cell lung NCI-H460, cervix SiHa and colon SW480 cells. The IC_{50} of the complex, presented in Figure 5d, against the tested cells was less than 1 μM (one order of magnitude more potent than cisplatin in chemo-resistant SW480 colon cancer cell line). The iridium(III) complex, shown in Figure 5d, has induced the formation of ROS in a concentration-dependent manner. To get an idea of the overall toxicity of the compounds, the most active derivative, shown in Figure 5d, was chosen for hemolytic investigations with mouse red blood cells. The complex was non-hemolytic up to 125 μM and less noxious in comparison with Pt-based antitumor agents. All the studied dinuclear iridium(III)-Cp* complexes (Figure 5) have shown in vitro capability superior to cisplatin. The mechanism of action seemed to be associated with DNA damage and ROS-mediated stress pathways. The most cytotoxic complex, shown in Figure 5d, was well tolerated without any mortality or detectable abnormalities observed in the developing zebrafish embryos. This compound has not demonstrated any disturbance in vascular vessel formation, showing the anticipated low toxic effects, despite its marked cytotoxicity in vitro in tumor cell lines.

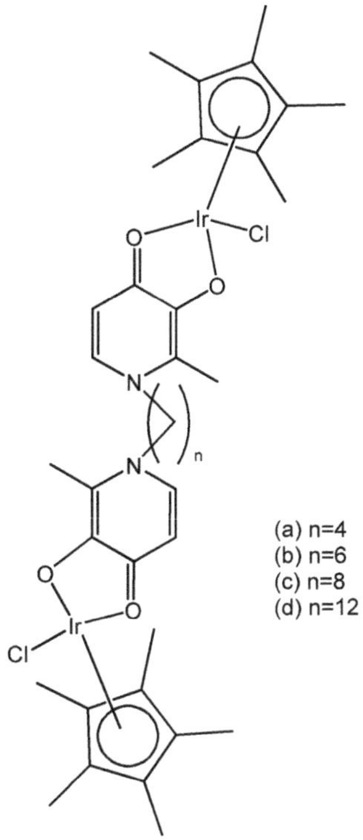

Figure 5. Dinuclear chloridoiridium(III) complexes [68].

A series of binuclear half-sandwich Ir(III) complexes $[Ir_2(\mu\text{-}L^n)(\eta^5\text{-}Cp^*)_2Cl_2](PF_6)_2$ with 4,4′-biphenyl-based bridging Schiff base bioligands, such as N,N′-(biphenyl-4,4′-diyldimethylidyne)bis-2-(pyridin-2-yl) methanamine and N,N′-(biphenyl-4,4′-diyldimethylidyne)bis-2-(pyridin-2-yl)ethanamine, Figure 6, have been reported by Štarha et al. [69]. These complexes have been tested against human ovarian A2780 carcinoma, cisplatin-resistant ovarian A2780R carcinoma, breast MCF-7 carcinoma, osteosarcoma HOS, colon HT-29 carcinoma, pancreatic PANC-1 carcinoma cells and against two healthy human cell lines. Complex, containing N,N′-(biphenyl-4,4′-diyldimethylidyne)bis-2-(pyridin-2-yl)ethanamine bridging ligand having a six-membered MN_2C_3 chelate ring, exposed higher activity towards A2780 (IC_{50} = 3.1 μM) and MCF-7 (IC_{50} = 6.0 μM) cells, as compared with that of N,N′-(biphenyl-4,4′-diyldimethylidyne)bis-2-(pyridin-2-yl) methanamine bridging ligand with a five-membered MN_2C_2 chelate rings, Table 1. The more active complex exceeded twice the in vitro cytotoxicity of cisplatin and displayed higher selective effects towards A2780 and MCF-7 cells over healthy human cells. Contrary to cisplatin, the complex has not induced cell cycle modifications with high superoxide anion levels at the human ovarian A2780 carcinoma cell line. The investigated iridium(III) dinuclear complexes were the first binuclear compounds with two {Ir(η-ar)Cl} moieties linked by nitrogen-donor bridging ligands chelating two metallic ions.

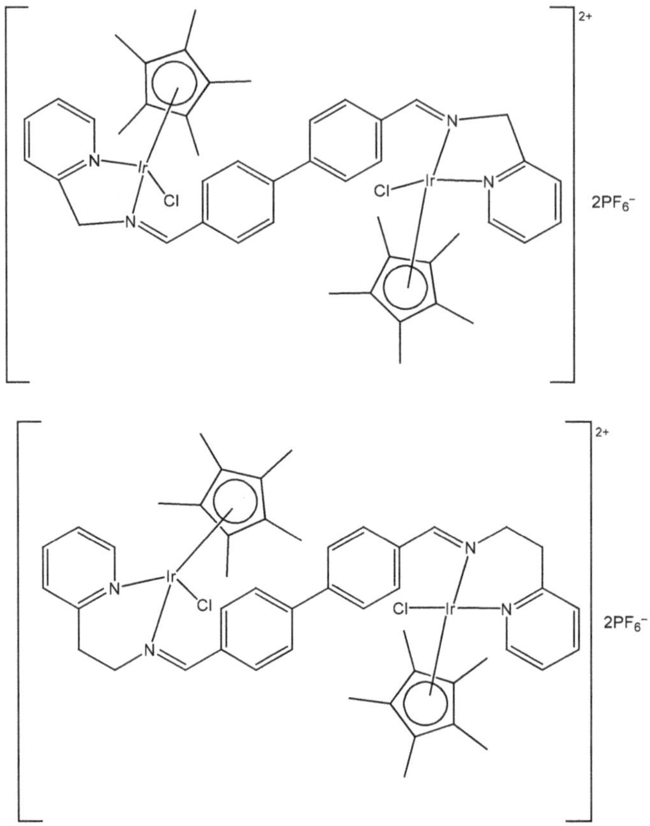

Figure 6. Binuclear half-sandwich monochloridoiridium(III) complexes $[Ir_2(\mu\text{-}L^n)(\eta^5\text{-}Cp^*)_2Cl_2](PF_6)_2$, where L = N,N′-(biphenyl-4,4′-diyldimethylidyne)bis-2-(pyridin-2-yl) methanamine and N,N′-(biphenyl-4,4′-diyldimethylidyne)bis-2-(pyridin-2-yl)ethanamine [69].

The above work has been followed by the synthesis of less active corresponding monochloridoiridium(III) complexes with analogous tetradentate nitrogen-donor ligands, produced from 4,4′–methylenedianiline [70] and benzene-1,4-diamine [71]. The binuclear half-sandwich Ir(III) complexes $[Ir_2(\mu\text{-}L1)(\eta^5\text{-}Cp^*)_2Cl_2](PF_6)_2$ (Figure 7a), $[Ir_2(\mu\text{-}L2)(\eta^5\text{-}Cp^*)_2Cl_2](PF_6)_2$ (Figure 7b) and $[Ir_2(\mu\text{-}L3)(\eta^5\text{-}Cp^*)_2Cl_2](PF_6)_2$ (Figure 7c) with tetradentate 4,4-methylenedianiline-based nitrogen-donor ligands (L1–L3) have been synthesized and identified by elemental analysis, mass spectral analysis, NMR and FTIR spectroscopy by Masaryk et al. [70]. The complexes have been tested against human prostate cancer DU-145, melanoma A375, human liver hepatocellular carcinoma HepG2, human lung cancer A549 and breast cancer MCF-7 cells by MTT assay. The most potent complex of the series, shown in Figure 7b, has shown comparable cytotoxic activity to cisplatin in the HepG2 cell line with IC_{50} of 12.6 μM, and acceptable IC_{50} for A549 cells (16.5 μM), Table 1. The IC_{50} of clinically used cisplatin against HepG2 and A549 cell lines were 12.0 μM and 12.3 μM, correspondingly. For the other tumor cells, the same complex showed lower cytotoxicity. The complexes in Figure 7a,c have shown comparatively weak cytotoxicity in the tested cells. Additionally, the studied complexes (Figure 7) partially oxidized NADH to NAD^+, that could be connected with the redox-mediated mode of cytotoxicity.

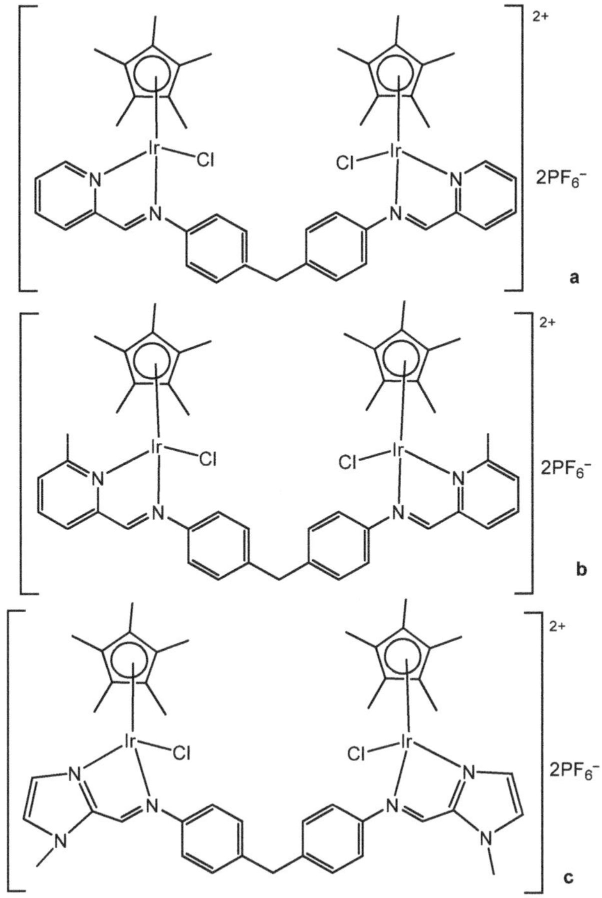

Figure 7. The structures of $[Ir_2(\mu\text{-}L^n)(\eta^5\text{-}Cp^*)_2Cl_2](PF_6)_2$ complexes with (1E)-pyridin-2-yl-N-(4-(4-(((E)-pyridin-2-ylmethylidene)amino)benzyl)phenyl)methanimine) ((**a**); L1) and its derivatives (1E)-(6-methylpyridin-2-yl)-N-(4-(4-(((E)-(6-methylpyridin2-yl)methylidene)amino)benzyl)phenyl)methanimine ((**b**); L2) and (1E)-(1-methyl-1 H-imidazol-2-yl)-N-(4-(4-(((E)-(1-methyl-1H-imidazol-2-yl)methylidene)amino)benzyl)phenyl) methan-imine ((**c**); L3) [70].

Kavukcu et al. have described the preparation of a series of monometallic and dimetallic iridium(III)–arene compounds with various aromatic and aliphatic groups and examined their activity on Vero and HepG2 cells through cell death mechanisms [72]. The electro-neutral $[Ir_2(\mu\text{-mpa})_2(\eta^5\text{-Cp}^*)_2Cl_2]$ complex (Figure 8) involved 2-mercapto-N-phenylacetamide (Hmpa), which possessed diverse coordination modes (as a S-donor bridging ligand) whereas its Ru correspondent $[Ru_2(\mu\text{-mpa})_2(\eta^6\text{-pcym})_2]$ with mpa ligand coordinated as a N,S-chelating ligand bridging the two metallic centers through the S atom of mpa. The complex was synthesized by using $[IrCl_2(Cp^*)]_2$ in dry methyl cyanide at room temperature. The bimetallic Ir(III)–pentamethylcyclopentadienyl complex with S,S-double coordination, shown in Figure 8, has exhibited a reduction of tumor cell viability in mg mL^{-1}, demonstrating a higher cytotoxic activity, comparable with the potency of its ruthenium analog. The complex has been shown to exhibit an increase in RIPK1 immunoreactivity in both cell lines.

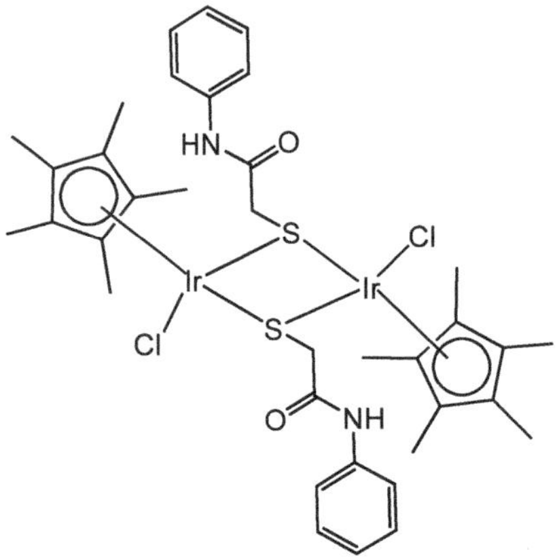

Figure 8. $[Ir_2(\mu\text{-mpa})_2(\eta^5\text{-Cp}^*)_2Cl_2]$ complex of the ligand 2-mercapto-N-phenylacetamide (Hmpa) [72].

The complexes with triphenylamine-modified thiosemicarbazone (TSC) have been reported by Shao et al. [73]. Due to the "enol" configuration of TSC, the studied compounds produced unique dimeric configurations. The most active complex is shown in Figure 9. In comparison with cisplatin, the antineoplastic activity in vivo and in vitro has confirmed that this complex might efficiently inhibit cancer growth, displaying better cytotoxic activity. The complex was active against A549 and HeLa tumor cell lines, with low selectivity against the same cell lines over the normal BEAS-2B cells tested. Meanwhile, it could circumvent adverse effects in the trials of safety estimation. Aided by the suitable fluorescence characteristics, the studied compounds can enter cancer cells in an energy-dependent manner, can be accumulated in lysosomes, and produce lysosome integrity damage. Ir(III) complexes blocked the cell cycle and enhanced the levels of ROS, which led to apoptosis. In particular, half-sandwich Ir(III)-thiosemicarbazone complexes were projected to be hopeful candidates for antitumor therapy.

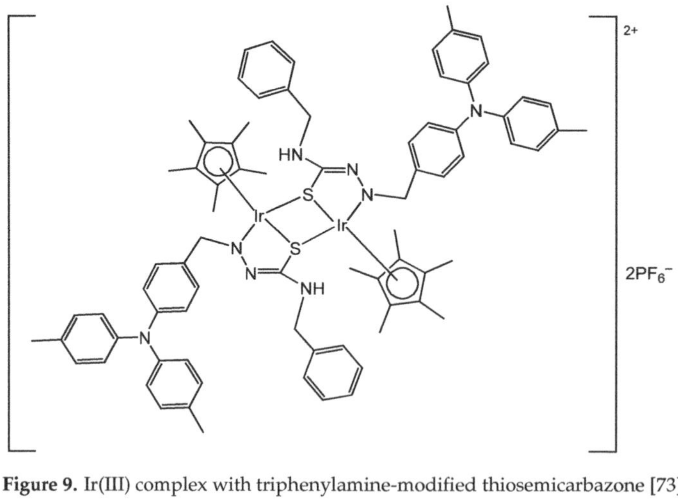

Figure 9. Ir(III) complex with triphenylamine-modified thiosemicarbazone [73].

2.2. Iridium Cyclometalated Dinuclear Complexes

Coordination cyclometalated Ir(III) complexes have exposed promising activity as novel modules of antitumor metal-based drugs. These metal coordination compounds with exceptional luminescent characteristics and the ability to photosensitize O_2-forming ROS have been exploited in imaging diagnostic methods and photodynamic anticancer therapy.

The binuclear organo-Ir(I) complex $[IrCl(cod)]_2$ with square-planar geometry, where cod = diene 1,5-cyclooctadiene (C_8H_{12}), has displayed strong antimetastatic action in the Lewis lung model nonetheless it has not shown primary tumors' inhibition (Figure 10) [74,75].

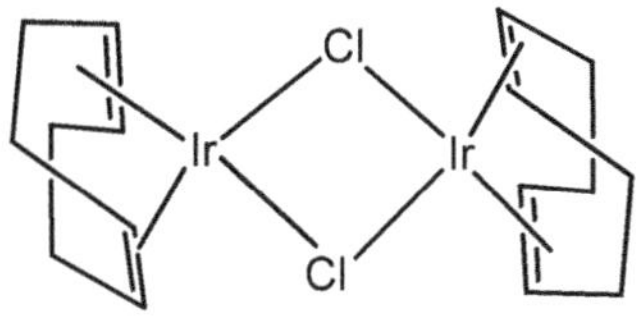

Figure 10. Organo-Ir(I) square-planar complex $[IrCl(cod)]_2$ [74].

Wang et al. have reported new N-substituted carbazole mono- and binuclear Ir(I) complexes, Figure 11, [76]. To confirm an extended π-conjugated ligand structure, acetylene moiety was incorporated between the bipyridine units and the N-substituted carbazoles. The studied homonuclear complex has demonstrated higher absorbance in the visible spectrum and higher $^1O_2$ quantum yields because the complex possessed a conjugated carbazole bridge ligand. Thus, the newly synthesized bimetallic complex could be a potential candidate for numerous applications like triplet photosensitizer for Triplet-Triplet Annihilation Up-conversion (TTA-UC).

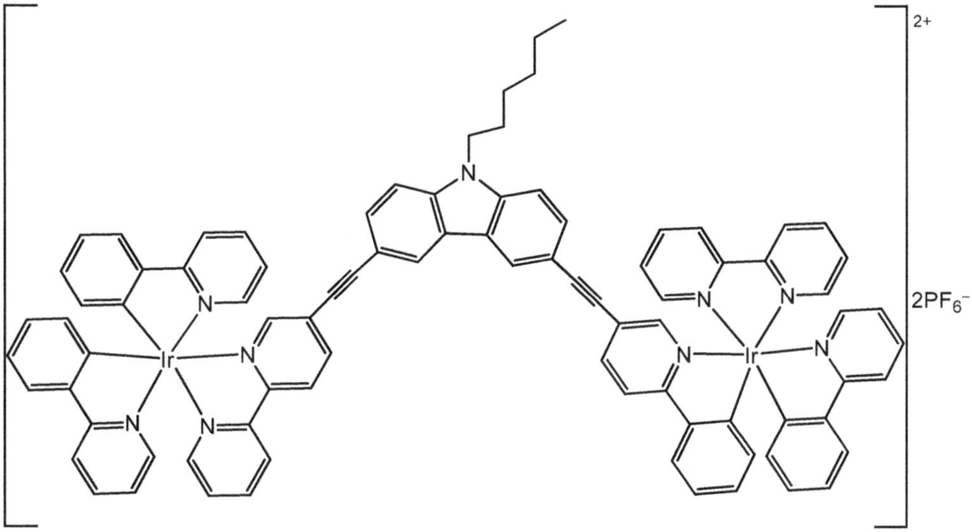

Figure 11. Structure of N-substituted carbazole dinuclear Ir(I) complex [76].

A series of highly charged binuclear Ir(III) cationic complexes have been studied for their antineoplastic effects against breast cancer MCF-7 and MDA-MB-231 cells [77]. The alkane linker between the two iridium(III) centers has been varied from 7 to 16 carbon atoms

(Figure 12) to investigate its effect on the bioactivity of the compounds. The potency of the complexes improved with increasing the chain lengths. The most active complex, Figure 12, comprised a 16-carbon chain linker. However, while the complex, shown in Figure 12, was cytotoxic against the metastatic MDA-MB-231 cell line (3 μM), it was comparatively non-active against the non-metastatic MCF-7 cells (29 μM). The anticancer activity, lipophilicity (log *P*) and DNA-binding ability have been studied. Because of their cationic nature, the compounds bonded to ct-DNA with high affinity as demonstrated by circular dichroism spectroscopy. The dinuclear Ir(III) cationic complexes in Figure 12 induced significant variations in the circular dichroism spectral intensity of DNA samples, signifying that they bonded to DNA with high affinity and caused substantial structural distortion. Additional studies might help to understand the mechanisms of cell death for instance apoptotic induction. Further investigations covering wild-type and cisplatinum-resistant cells can afford more insights into the effectiveness of the cytotoxic complexes, as shown in Figure 12.

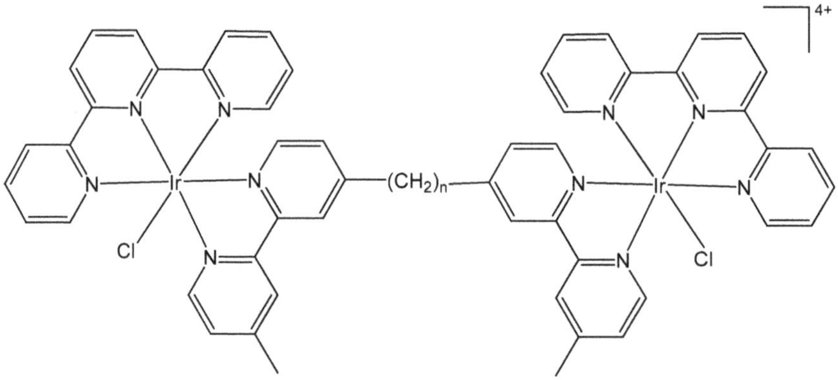

Figure 12. Structure of binuclear Ir(III) complexes $[\{Ir(tpy)Cl\}_2\{\mu\text{-}bb_n\}]^{4+}$\{Cl-Irbb$_n$, where tpy = 2,2′:6′,2″-terpyridine, bb$_n$ = bis[4(4′-methyl-2,2′-bipyridyl)]-1,*n*-alkane and *n* = 7, 12 and 16 [77].

Unique Ir(III) luminescent coordination compounds also consist of dinuclear bis-tridentate Ir(III) complexes, described by Liu et al., which include analogous cyclometalated ligands that were deliberated above [78]. Dinuclear Ir(III) complexes having terpyridine-capped fluorenyl bridge ligands and a variety of polypyridyl- or cyclometalating terminal tridentate ligands displayed photodynamic beneficial in vitro effects against melanoma SK-MEL-28 cell line upon visible light irradiation, with sub-micromolar photoinduced cytotoxicity and phototherapeutic indices. The reported complexes incorporated 9,9-dioctyl-2,7-di(terpyridyl)-9*H*-fluorene as a bridging ligand, with variation at the terminal tridentate bioligands, such as 4′-phenyl-2,2′:6′,2″-terpyridine (N^N^N), 1,3-dipyridyl-4,6-dimethylbenzene (N^C^N), 4,6-diphenyl-2,2′-bipyridine (C^N^N), or 2,4,6-triphenylpyridine (C^N^C). The complexes phosphoresced brightly in compromised cells and displayed photoactivated cell uptake, confirming the theranostic properties of the studied compounds. Together, both bridging and terminal ligands influenced the photophysical and biological activity significantly. Their properties depended on the bridging linker length. Even though the compounds have exposed very potent photophysical characteristics, their cytotoxic activity has not been correlated with the ROS production quantum yield or their photocleaving ability. Based on photophysical properties, the complex with 1,3-dipyridyl-4,6-dimethylbenzene (N^C^N), shown in Figure 13, was projected to be the most potent in vitro PDT candidate. For an in vitro PDT application, there has been a clear suggestion that terminal tridentate N^C^N ligands performed best in combination with Me substituents on the central cyclometalating rings.

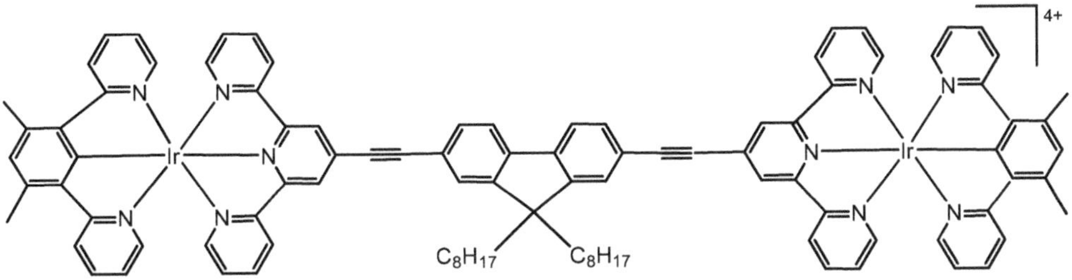

Figure 13. Complex with 1,3-dipyridyl-4,6-dimethylbenzene (N^C^N) ligand [78].

Tang et al. have synthesized and reported binuclear (Figure 14a–c) and mononuclear (Figure 14d–f) Cp*Ir(III) complexes of pyrrole thioamide ligands [79]. Hybrid ligands containing S and N donors are highly adaptable ligands able to coordinate to a very extensive range of metallic centers, because of their soft-hard hybrid donor characteristics. The interactions of pentamethylcyclopentadienyl complex $[(\eta^5\text{-Cp}^*)IrCl_2]_2$ with pyrrole-2-thioamide ligands (holding several substituents on the pyrrole ring) and triethylamine base have been studied. Two series of iridium(III) complexes incorporating π-hydrocarbon ligands along with dianionic pyrrole thioamide ligands have been obtained—the first series consisted of binuclear complexes with S-bridging pyrrole thioamide bioligands, whereas the second series comprised mononuclear complexes with ancillary PPh_3 ligands and a pyrrole thioamide dianion ligand coordinating through S and the pyrrole N atoms, forming 5-membered chelating ring. Complex compounds have been evaluated by NMR, IR and ESI mass spectrometry. The iridium(III) complexes illustrated the potentially varied coordination properties of the hybrid hard/soft donor pyrrole thioamide ligands. The newly obtained complexes were tested for cytotoxicity towards adenocarcinoma human alveolar basal epithelial A549 cell line with marginal potency. The mononuclear complex, presented in Figure 14d, showed high cytotoxic activity at low concentrations.

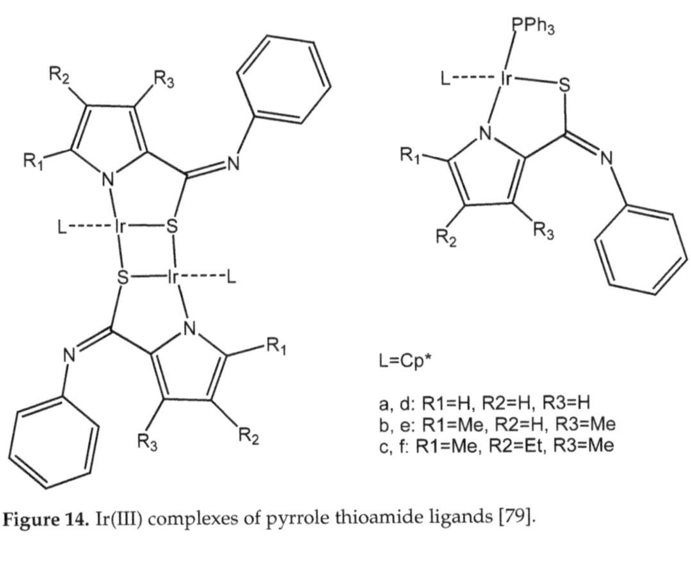

Figure 14. Ir(III) complexes of pyrrole thioamide ligands [79].

3. Heteronuclear Iridium(III) Complexes

In addition to mononuclear complexes, hetero-multinuclear complexes have attracted great consideration and have been studied for cancer treatment during the last decades. Heteronuclear complexes are those complexes which have more than one metallic center by using different metal cations. Such complexes combine the advantages of different metal ions in one molecule. In the group of iridium(III) complexes apart from multinuclear homometallic iridium(III) complexes, the heteronuclear Ir(III) complexes are of great interest. The variation of the central metal cation has a critical role in the resulting properties of these compounds. The coordination of two metal cations would suggestively extend the potential of their action mode. The inclusion of two different metal ions within a single molecule could improve their efficiency as antineoplastic agents through different mechanisms. This improvement could be connected to their interactions with various biotargets or with the enhanced physicochemical characteristics of the obtained heterometallic complexes. In addition, the presence of dissimilar modes of action led to synergistic results, modulation in oxidation-reduction properties and stability changes, which might lead to improved anticancer activities and higher selectivity compared to the respective mononuclear precursors [80]. The improved potency and capability to overcome the possible resistance, projects the obtained multinuclear compounds as prominent antineoplastic candidates.

The synthesis and characterization of a novel photoactive heterometallic platinum-iridium complex, its photochemical and photobiological properties, including phototoxicity against a variety of tumor cells (human ovarian A2780, lung A549, and prostate PC3 cell lines), have been reported by Shi et al. [81]. The photoactivated chemotherapy (PACT) diazido Pt(IV) complex has been conjugated to a photodynamic therapy (PDT) cyclometalated Ir(III) complex, giving the heterometallic platinum-iridium complex. The hetero binuclear complex trans, trans, trans-[Pt(py)$_2$(N$_3$)$_2$-(OH)(OOCCH$_2$CH$_2$CONHCH$_2$-bpyMe)Ir(ppy)$_2$]Cl, presented in Figure 15, has exhibited charge transfer between the acceptor photochemotherapeutic platinum(IV) and the donor photodynamic iridium(III) parts, displaying synergistic activity.

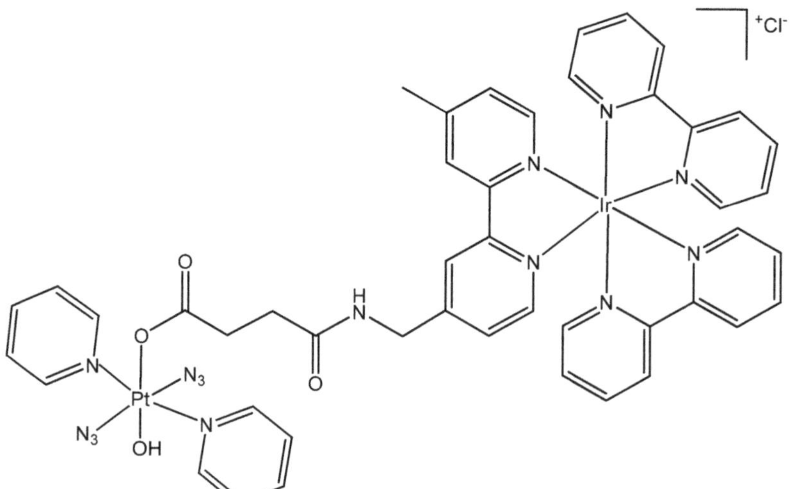

Figure 15. Heterometallic dinuclear platinum(IV)-iridium(III) complex trans, trans, trans-[Pt(py)$_2$(N$_3$)$_2$-(OH)(OOCCH$_2$CH$_2$CONHCH$_2$-bpyMe)Ir(ppy)$_2$]Cl [81].

The platinum(IV)-iridium(III) complex was stable in dark conditions, but underwent photodecomposition more quickly than the parent Pt(IV) complex trans, trans, trans-$[Pt(py)_2(N_3)_2(OH)_2]$ upon irradiation, which generated Pt(II) species, azidyl radicals, highly oxidizing excited-state Ir(III)* species and ROS ($^1O_2$) during the photodecomposition of the complex. Pt(II) species formed Pt-GMP adducts. Additionally, the Ir(III)* excited state oxidized NADH to NAD· radicals and NAD^+. The newly obtained complex was highly photocytotoxic against the tested tumor cells. In contrast, the mononuclear platinum and iridium fragments were much less active. The cellular platinum accumulation was higher for the dinuclear complex compared with the parent Pt(IV) complex possibly because of its positive charge and higher lipophilicity. Photoirradiation of the dinuclear complex in tumor cells damaged nuclei and released chromosomes. It has been found that iridium localized strongly in small cellular compartments. The luminescent platinum-iridium complex generated ROS and singlet oxygen $^1O_2$ upon irradiation. This heteronuclear bimetallic platinum-iridium complex with photodynamic therapy and photoactivated chemotherapy (PACT) motifs combined in one molecule, provided an original and promising strategy for multimodal phototherapy [81]. Various multinuclear supramolecular assemblies based on platinum(II), ruthenium(II) and iridium(III) vertices for anticancer applications have been lately reviewed by Li et al. [82].

Tripathy et al. have obtained and studied antineoplastic activity of a heterometallic dinuclear Ir–Ru complex $[\{(ppy)_2Ir\}(\mu\text{-phpy})\{Ru(p\text{-cym})Cl\}](PF_6)_2$ with pyrazino[2,3-f][1,10]phenanthroline as a bridge ligand between the metal ions with distorted octahedral and piano-stool geometry (Figure 16) [83]. The authors have utilized the advantages of η^6-arene Ru(II) and cyclometalated Ir(III) fragments along with a trinucleating polypyridyl-based ligand 2,3-di(pyridin-2-yl)pyrazino[2,3-f][1,10] phenanthroline (phpy) for the design of effective antitumor heterometallic binuclear complex. The obtained complex appeared to have more promising cytotoxic activity than the homonuclear diruthenium counterpart against various cell lines, such as breast (MCF7), ovarian (SKOV3), prostate (PC3) and endometrial (Ishikawa) cancer cells. The mechanism of cell death induced by the compound, presented in Figure 16, was further explored via cell cycle flow cytometric analysis, which exposed the stalling of the cells in the G_1 phase. Western blots have been performed to check for some fluctuations in expression levels of proteins answerable for cell apoptosis upon treatment with the complex, shown in Figure 16. Particularly, the Bcl-2 and Bax apoptotic regulators and the cleavage of poly(ADP-ribose) polymerase (PARP), included in DNA repair and apoptosis, have been examined in response to treatment with $[\{(ppy)_2Ir\}(\mu\text{-phpy})\{Ru(p\text{-cym})Cl\}](PF_6)_2$, Figure 16, [84]. No substantial growth in the levels of any of the three proteins has been recorded, demonstrating that the complex, shown in Figure 16, induced nonapoptotic cell death through the flow cytometric and Western blots analyses. Additionally, the examination of the cellular morphology revealed wide cytoplasmic vacuolization, a characteristic of autophagy [85]. The complex $[\{(ppy)_2Ir\}(\mu\text{-phpy})\{Ru(p\text{-cym})Cl\}](PF_6)_2$, Figure 16, was the first illustration of heteronuclear iridium–ruthenium complex which induced autophagic form of cell death against the resistant MCF7 cell line (IC_{50} = 0.92 μM, Table 1), however additional studies would be essential to understand the mode of action of the studied complex [83].

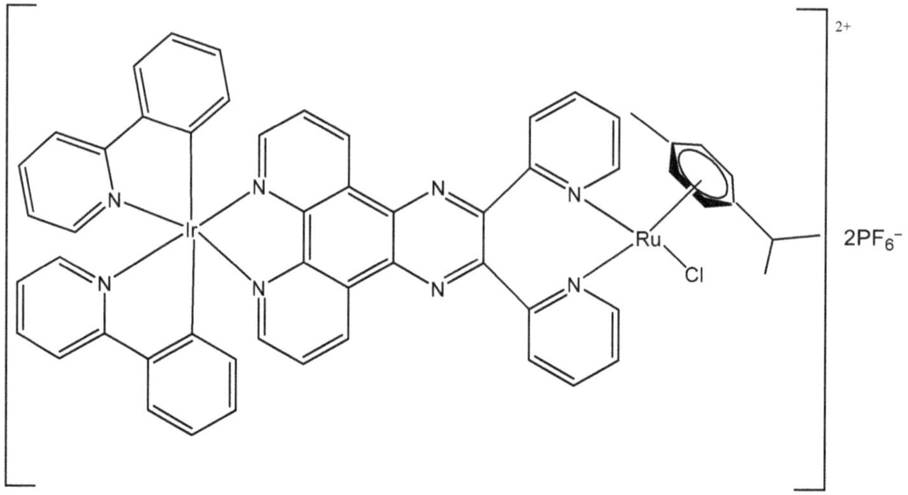

Figure 16. Structure of dinuclear Ir–Ru complex $[\{(ppy)_2Ir\}(\mu phpy)\{Ru(p\text{-}cym)Cl\}](PF_6)_2$ with pyrazino[2,3-f][1,10]phenanthroline ligand [83].

Nallas et al. have reported a bipyrimidine-bridged trimetallic complex of the formula $\{[(bpy)_2Ru(bpm)]_2IrCl_2\}^{5+}$, where bpy = 2,2′-bipyridine, and bpm = 2,2′-bipyrimidine, Figure 17, [86]. The complex coupled a catalytically active iridium(III) metal with a light-absorbing Ru(II) metal within the polymetallic basis. The detailed analysis of the spectroscopic, electrochemical, and spectrochemical properties of the polymetal complex was explained. It has been suggested that polymetal systems could be considered to function as photochemical devices, as the trimetallic complex shared the catalytical properties of Ir(III) with the photoactive properties of Ru(II) ions [87]. Within the framework of the progress of heterodimetallic complexes, similar trimetallic complex compounds of the form $\{[(bpy)_2Ru(BL)]_2MCl_2\}^{n+}$ integrating the bridging ligands BL = 2,3-bis(2-pyridyl)pyrazine (dpp), 2,3-bis(2-pyridyl)quinoxaline (dpq), and 2,3-bis(2-pyridyl)benzoquinoxaline) (dpb) and the metal Ir(III), Rh(III) or Os(II) ions have also been investigated by the same group [88,89]. These polymetallic systems exhibited unique photophysical and electrochemical properties.

Recently, luminescent 2,2′-bipyrimidine ruthenium(II)-iridium(III)–arene monometallic and homo- and hetero-dimetallic complexes have been obtained and characterized [90]. The monochlorido Ir–Ru complex $[Ir(\eta^5\text{-}Cp^*)Cl(\mu\text{-}bpm)Ru(\eta^6\text{-}pcym)Cl](PF_6)_2$, shown in Figure 18, holding tetradentate 2,2′-bipyrimidine (bpm), has been investigated, along with diruthenium and diiridium analogs, in breast carcinoma MDA-MB-468 and colon carcinoma Caco-2 cells. Although ruthenium(II)-iridium(III) complex was less active (IC_{50} = 1.9 μM) against MDA-MB-468 cell line (Table 1) than diiridium analog (IC_{50} = 1.8 μM) and diruthenium complex (IC_{50} = 0.9 μM), the heterometallic ruthenium(II)-iridium(III) compound (IC_{50} = 6.2 μM, Table 1) exceeded both homometallic analogs (IC_{50} = 32.4 and 46.0 μM for diiridium and diruthenium complexes) in Caco-2 cell line. Additionally, the Ir–Ru complex $[Ir(\eta^5\text{-}Cp^*)Cl(\mu\text{-}bpm)Ru(\eta^6\text{-}pcym)Cl](PF_6)_2$, shown in Figure 18, was more active towards the Caco-2 cell line than the respective mononuclear parts $[Ir(\eta^5\text{-}Cp^*)(bpm)Cl]PF_6$ with IC_{50} = 50.4 μM and $[Ru(\eta^6\text{-}pcym)(bpm)Cl]PF_6$ with IC_{50} = 49.6 μM.

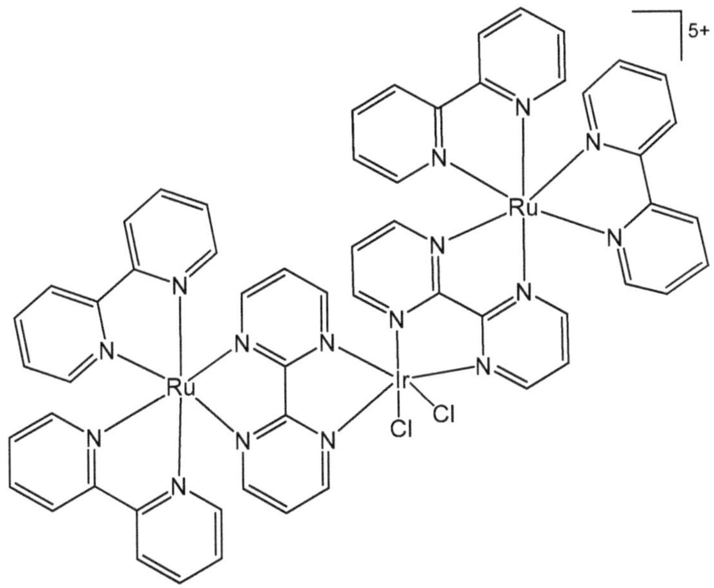

Figure 17. Structure of trimetallic complex $\{[(bpy)_2Ru(bpm)]_2IrCl_2\}^{5+}$ [86].

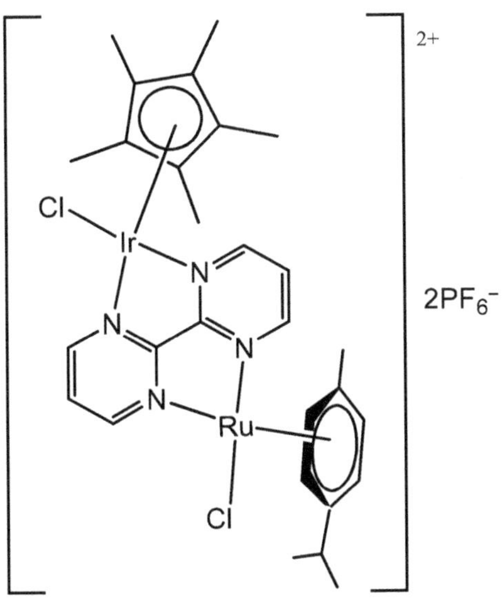

Figure 18. The heterobimetallic complex $[Ir(\eta^5\text{-}Cp^*)Cl(\mu\text{-}bpm)Ru(\eta^6\text{-}p\text{-}cym)Cl](PF_6)_2$ [90].

The same authors designed a novel type of heterobimetallic iridium-rhenium complex $[Ir(\eta^5\text{-}Cp^*)Cl(\mu\text{-}bpm)ReCl(CO)_3]Cl$ (Figure 19), by means of the conjugation of Ir(III)–Cp* and Re(I)-tricarbonyl motifs through the bridging bpm ligand, which has displayed significant anticancer potency [91]. The potency of the synthesized complex has been evaluated against triple-negative breast cancer (TNBC) MDA-MB-468 cells and normal immortalized human keratinocyte HaCaT cells. The complex displayed almost five-fold higher cytotoxic effect against the MDA-MB-468 cell line (IC_{50} = 24.1 μM, Table 1) than its dirhenium homometallic and ruthenium–rhenium heterometallic correspondents. The compound was very selective against the MDA-MB-468 cell line compared to the noncancerous HaCaT cells (IC_{50} = 234.8 μM). The iridium-rhenium complex demonstrated a sparkling binding tendency to DNA and HSA. Complex $[Ir(\eta^5\text{-}Cp^*)Cl(\mu\text{-}bpm)ReCl(CO)_3]Cl$ (Figure 19) was able to generate apoptosis of the tumor cells by inhibition of G_2/M phase in combination with a plentiful quantity of ROS production and mitochondrial dysfunction through reduction of MMP. The complex (Figure 19) revealed very low toxic effects together with strong binding efficiency with bioactive molecules.

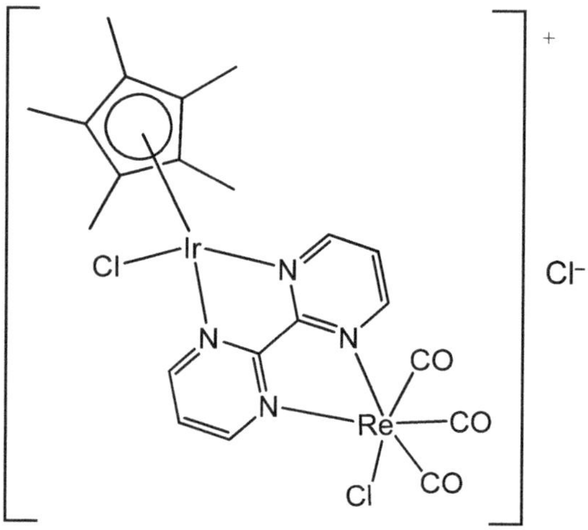

Figure 19. The structure of heterobimetallic complex $[Ir(\eta^5\text{-}Cp^*)Cl(\mu\text{-}bpm)ReCl(CO)_3]Cl$ [91].

New bimetallic Au(I)-Ir(III) complexes (Figure 20) with a bis(diphenylphosphino) methanide bridging ligand $[Ir(C\hat{}N)_2(P\hat{}P)]^+$, in coparison with the monometallic precursor $[Ir(N\hat{}C)_2(dppm)]^+$, have been reported [92]. The different Au(I) ancillary ligands, triphenylphosphine, a chloride or a thiocytosine, shown in Figure 20, did not expose any significant effects on the physical characteristics, which was principally owing to metal -ligand charge-transfers based on iridium(III). A study of cytotoxic activity, cell death mechanisms, reactive oxygen species generation, MMP integrity and cell biodistribution in tumor A549 cell line was performed for different Au(I) ancillary ligands. The gold(I) fragment, together with the ancillary ligands, appeared to be critical for the biological activity in the lung carcinoma A549 cell line versus the endothelial cell line. These bimetallic complexes have induced apoptosis with the formation of cytoplasmic vacuolization. The emission characteristics of Ir(III) fraction (green irradiation) have permitted the examination of cellular distribution, which presented the complexes more positioned in the cytoplasm, and the superimposition with red irradiation showed a mitochondrial accumulation. The existence of Au fragment has increased the ROS production and thioredoxin reductase (TrxR) inhibition compared to that demonstrated by the monoiridium complex $[Ir(N\hat{}C)_2(dppm)]^+$. Iridium peptide bioconjugates and the respective Au(I)-Ir(III) bimetallic complexes have been recently described, in which an ortho-metallated iridium(III) species of the form $[Ir(C\hat{}N)_2(N\hat{}N)]^+$ was functionalized with a peptide coupled to cytotoxic Au(I) fragment [93]. Lysosomal accumulation was detected for the studied compounds, as opposed to the expected accumulation in mitochondria triggered by the Au(I) complexes.

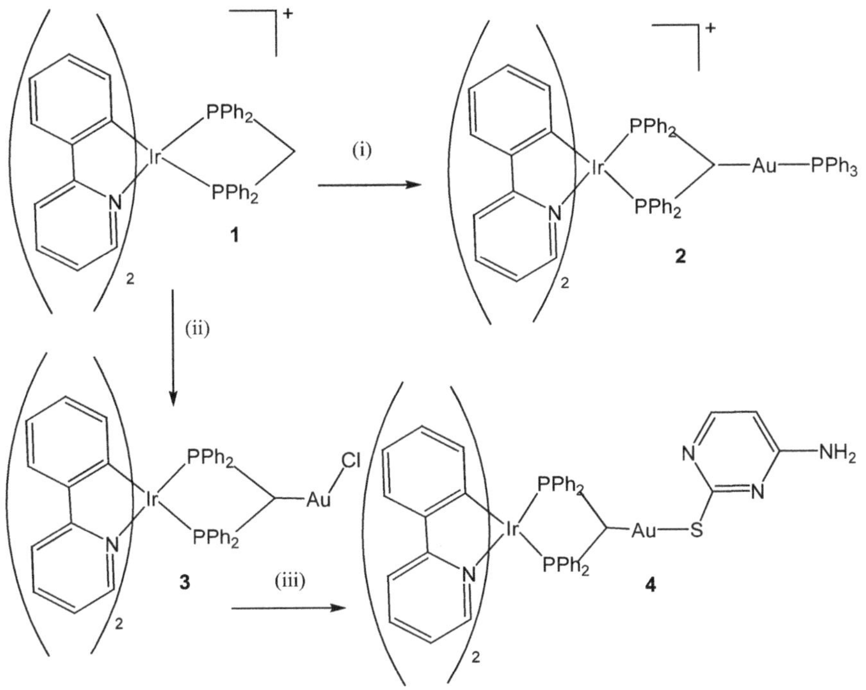

Figure 20. Synthesis and structures of bimetallic Au(I)-Ir(III) complexes (**1**)–(**4**): [[Ir(ppy)$_2$(μ-Cl)]$_2$], dppm, MeOH, reflux; (i) [Au(acac)PPh$_3$], CH_2Cl_2, RT; (ii) [AuCl(tht)], Cs_2CO_3, CH_2Cl_2, RT; (iii) Cs_2CO_3, thiocytosine, CH_2Cl_2, RT [92].

Ir(III)-Cu(II) heterobinuclear complex compounds have been synthesized by the combination of phosphanes (P-coordinated to iridium) and fluoroquinolones (O,O-coordinated to copper) as ligands [94]. Newl heterobinuclear iridium(III)–copper(II) coordination compounds with the formula [Ir(η^5-Cp*)Cl$_2$(μ-L)Cu(phen)(H$_2$O)$_n$]NO$_3$, bearing phosphines derived from fluoroquinolones (L), viz., sparfloxacin (Hsfx), ciprofloxacin (Hcfx), lomefloxacin (Hlfx) and norfloxacin (Hnfx), have been obtained and investigated as possible antitumor chemotherapeutic agents. The novel heteronuclear iridium(III)–copper(II) complexes have been obtained by stirring [Cu(phen)(NO$_3$)$_2$] with [Ir(η^5-Cp*)Cl$_2$(μ-L)] at room temperature. The piano-stool complexes, for instance, the best-acting complex, [Ir(η^5-Cp*)Cl$_2$(μ-pcfx)Cu(phen)(H$_2$O)]NO$_3$, shown in Figure 21, were expressively more active than cisplatin in different human tumor cells, but they were practically ineffective in non-carcinogenic HEK-293T cell line. The cytotoxic activity of the complexes has been evaluated in vitro against melanoma, breast, lung, and prostate cancer cells and against nontumor human embryonic kidney cells. The heteronuclear iridium(III)–copper(II) complexes displayed higher cytotoxic activity than cisplatin against the tested cell lines (A549, MCF7, DU145) except for the WM2664 cells. IC_{50} values have been measured by the MTT method in two different methods (after 24 or 24 + 48 h). Particularly the antineoplastic potency of [Ir(η^5-Cp*)Cl$_2$(μ-pcfx)Cu(phen)(H$_2$O)]NO$_3$ in DU145 cell line was remarkable not only owing to the very low IC_{50} in picomolar range ($IC_{50} = 1.3 \times 10^{-6}$ μM at 24 h exposure), see Table 1, but also due to the unusual recovery of the tested cells (IC_{50} = 125.7 μM at 24 h exposure and 48 h recovery time in a drug-free setting). Analogous results have been obtained for the same compound and its analogs in the MCF-7 cell line, while the contrary effect has been detected in the most sensitive lung carcinoma A549 cell line (IC_{50} = 35.5 nM for 24 h exposure, Table 1, and IC_{50} = 0.4 nM for 24 h exposure and 48 h recovery time in a drug-free setting). The studied complex [Ir(η^5-Cp*)Cl_2(μ-pcfx)Cu(phen)(H_2O)]NO_3 was further loaded into liposomes. Liposomes loaded with this complex efficiently accumulate inside adenocarcinoma and prostate carcinoma cell lines with colocalization in the nuclei. Significantly for upcoming research in the field of heterometallic antitumor Ir–Cp^x compounds, this complex displayed noticeably higher cytotoxicity in tumor cells than the mononuclear analog, Figure 21 [95]. It could be efficiently accumulated inside prostate carcinoma and lung adenocarcinoma cells with a selective location in the nuclei. Cytometric analyses have revealed the domination of apoptosis over the other cell death types along with a significant increase in ROS generation.

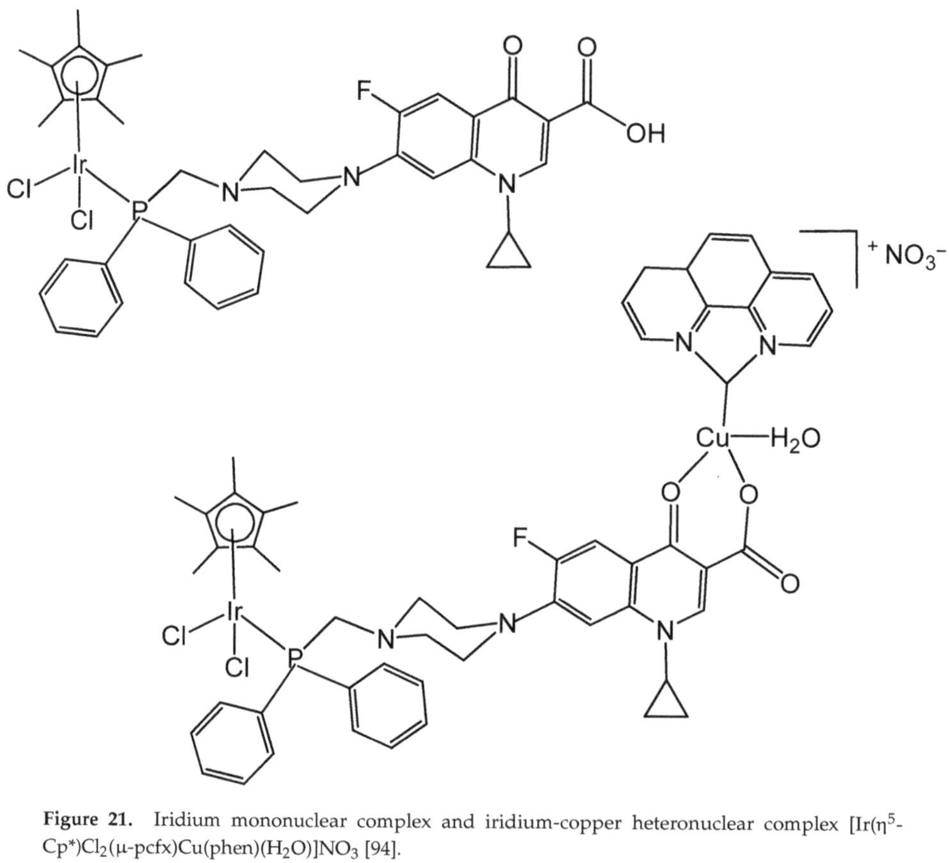

Figure 21. Iridium mononuclear complex and iridium-copper heteronuclear complex [Ir(η^5-Cp*)Cl_2(μ-pcfx)Cu(phen)(H_2O)]NO_3 [94].

The structure-activity relationships for a series of iridium(III) complexes covering a ferrocenyl moiety have been discovered by Tabrizi et al. [96]. The existence of ferrocene might improve the bioactivity of antineoplastic agents, as reported earlier [97]. Ir(III) arene complex compounds of naphthoquinone derivatives with the general formula [Ir(η^6-L1)(L2)(3,5-$(NO_2)_2$pcyd)](PF_6), shown in Figure 22, where L1 = *p*-methylphenyl)ethynylferrocene; L2 = lawsone (Figure 22a), lapachol (Figure 22b), juglone (Figure 22c), plumbagin (Figure 22d) and 3,5-$(NO_2)_2$pcyd = 3,5-dinitrophenylcyanamide, have been synthesized and studied for their appropriateness as possible antitumor drugs [96]. The DNA-binding interactions of the compounds with calf thymus DNA have been investigated by absorption, emission, and viscosity measurements. Their cytotoxic activity against the tumor cells including liver hepatocellular carcinoma HepG-2, colon adenocarcinoma (HT-29), colon carcinoma HCT-8, breast MCF-7 and ovary A2780 carcinoma cells have been studied. Remarkably, almost all the tested complexes exhibited substantial cytotoxic activity against the tumor cells and the complex with lapachol, Figure 22b, appeared as the most cytotoxic compound in comparison with the other derivatives. The structure-activity relationships showed that the lipophilicity of tested complexes was the most important factor which determined their cytotoxic activity. Consequently, the most lipophilic complex with lapachol, Figure 22b, gave rise to the best antineoplastic activity based on IC_{50} values ranging from 4 to 8 μM in cases of different tumor cells. Complexes in Figure 22 induced early- and late-stage apoptosis in the breast cancer cell line. The compounds also induced the formation of intracellular ROS in the case of the MCF-7 cell line. The origin of the ROS production was connected with thioredoxin reductase, that played a key role in preserving the cell redox status. The studied complexes were found to induce high levels of tumor cellular death in an apoptotic way which was in correlation with the effective inhibition of thioredoxin reductase at nanomolar concentrations with additional strengthening of the ROS generation contributing to cytotoxic activity.

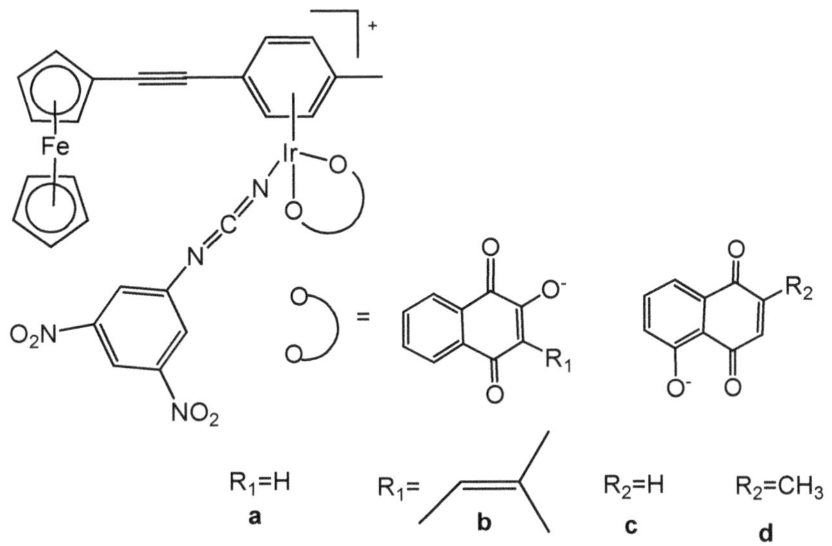

Figure 22. Ir(III) arene compounds of naphthoquinones [Ir(η^6-L1)(L2)(3,5-$(NO_2)_2$pcyd)](PF_6), where L1 = *p*-methylphenyl)ethynylferrocene and L2 = lawsone (**a**), lapachol (**b**), juglone (**c**), plumbagin (**d**) [96].

Ferrocene-attached half-sandwich Ir(III) phenylpyridine complex compounds have been synthesized and characterized [98]. The complexes have shown improved anticancer activity than the clinically used cisplatin and their rhodium corresponding complexes. Complex [Ir(η^5-Cp*)Cl(ppyfc)] (Hppyfc = ferrocene-modified 2-phenylpyridine) (Figure 23) was highly potent, but did not display any selectivity with comparable cytotoxic activity at the tumor cells and the normal BEAS-2B cells (IC_{50} = 3.4 μM). The ferrocene-iridium(III) complexes could efficiently inhibit cell migration and colony formation. Complexes could interact with protein and transport through serum protein, efficiently catalyzing the nicotinamide–adenine dinucleotide oxidation and inducing the reactive oxygen species (ROS, $^1O_2$) accumulation, confirming the antitumor mechanism of oxidation. Additionally, laser scanning confocal detection indicated that the compounds could enter cells followed by a non-energy-dependent cell uptake mechanism, efficiently accumulating in the lysosomes, leading to lysosome damage, and reduction of the mitochondrial membrane potential. Therefore, ferrocene-appended Ir(III) compounds possessed the prospect of becoming original multifunctional therapeutics, involving lysosome-targeted imaging agents and antineoplastic drugs [98].

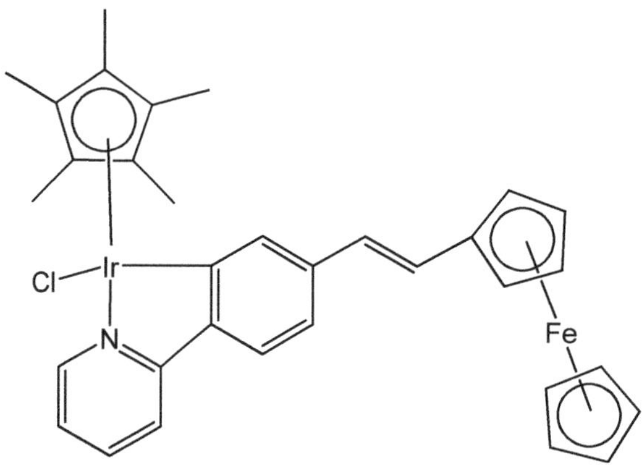

Figure 23. The structure of the complex [Ir(η^5-Cp*)Cl(ppyfc)] [98].

Palao et al. have obtained biscyclometalated Ir(III) complex compounds, Figure 24a–c, including borondipyrromethene (BODIPY)-based ancillary ligands as smarter fluorescent photosentizers [99], where BODIPY unit was grafted to different chelating cores (acetylacetonate and bipyridine) by BODIPY meso-position. The boron–dipyrromethene (BODIPY)–Ir(III) complexes in Figure 24a,b exhibited higher absorption coefficients, higher fluorescence emission and also more effective $^1O_2$ generation in visible light, than that shown by the complex in Figure 24c. In vitro photodynamic therapy activity of the complexes in Figure 24a,b in the HeLa cell line showed that these complexes are capably internalized into the tumor cells, demonstrating high photocytotoxic activity, even at low concentrations, making them hypothetically appropriate candidates in theranostic applications [99].

An important approach to improve the bioactivity of iridium organometallic compounds is the inclusion of active bioorganic ligands with recognized functions to design novel metal-based compounds with good therapeutic characteristics. In these cases, the combination of highly fluorescent BODIPY (boron dipyrromethene) ligands with an oil-soluble compound like avobenzone (1-(4-tert-butylphenyl)-3-(4-methoxyphenyl) propane-1,3-dione, AVBH), might be used as sunray blocking agents as important ingredients in lotions or creams have been obtained by Gupta et al., which gives a new platform for developing novel dicationic metal supramolecules as antineoplastic agents [100]. Recent investigations have revealed that AVBH has notable cytotoxicity against various human tumor cells [101] Moreover, the fluorescent dye BODIPY possesses exceptional photo-physicochemical properties. Therefore, BODIPY ligands and their complexes have been chosen because of their crucial application in tumor treatment on account of their emission and absorption effects which facilitated the bio-tagging trials. The structures of BODIPY (BDP = 4-dipyridine boron dipyrromethene and BDPCC = 4-ethynylpyridine boron dipyrromethene) iridium bimetallic complexes $[Ir_2(Cp^*)_2(AVB)_2BDP][2CF_3SO_3]$ and $[Ir_2(Cp^*)_2(AVB)_2BDPCC][2CF_3SO_3]$ with excellent anticancer activity and bioimaging capability are presented in Figure 25, [100]. The complexes were characterized by various analytical methods, and their structural parameters were further optimized by density functional theory (DFT). These complexes have been proven to be active against human lung A549, cervical HeLa and breast MCF-7 cancer cell lines with IC_{50} values between 1 and 5 μM. The complexes also were tested against non-malignant mouse embryo fibroblast NIH T3T cells. The BDPCC-based iridium complex was found to have stronger antineoplastic properties than the BDP-containing iridium complex with activity analogous to that of the chemotherapeutic doxorubicin. The studied iridium complexes have also been found to interact strongly with genomic DNA causing the unwinding of the double helix. The BDPCC-based complex $[Ir_2(Cp^*)_2(AVB)_2BDPCC][2CF_3SO_3]$ showed a comparatively better activity.

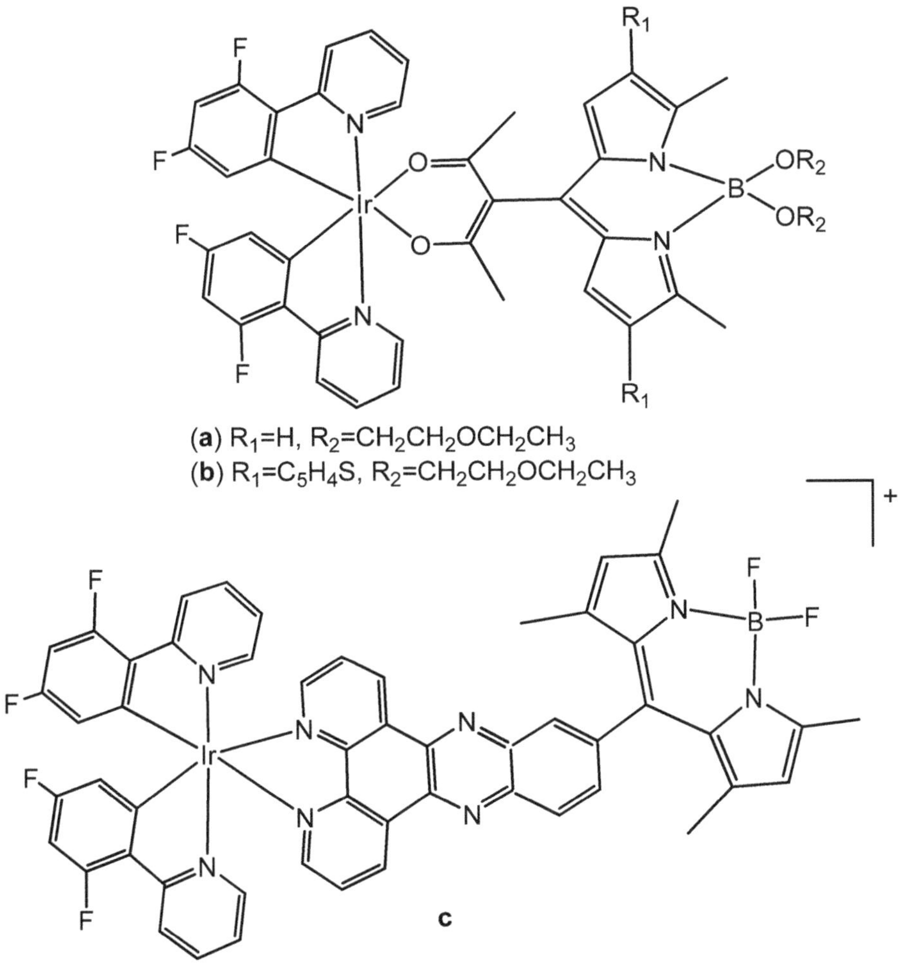

Figure 24. Structures of biscyclometalated iridium complexes [99].

Tetranuclear square BODIPY metallocycles have shown selective toxicity against MCF-7 and U87 cells [102]. The binding study has revealed the capability of the complexes to interact with proteins [103].

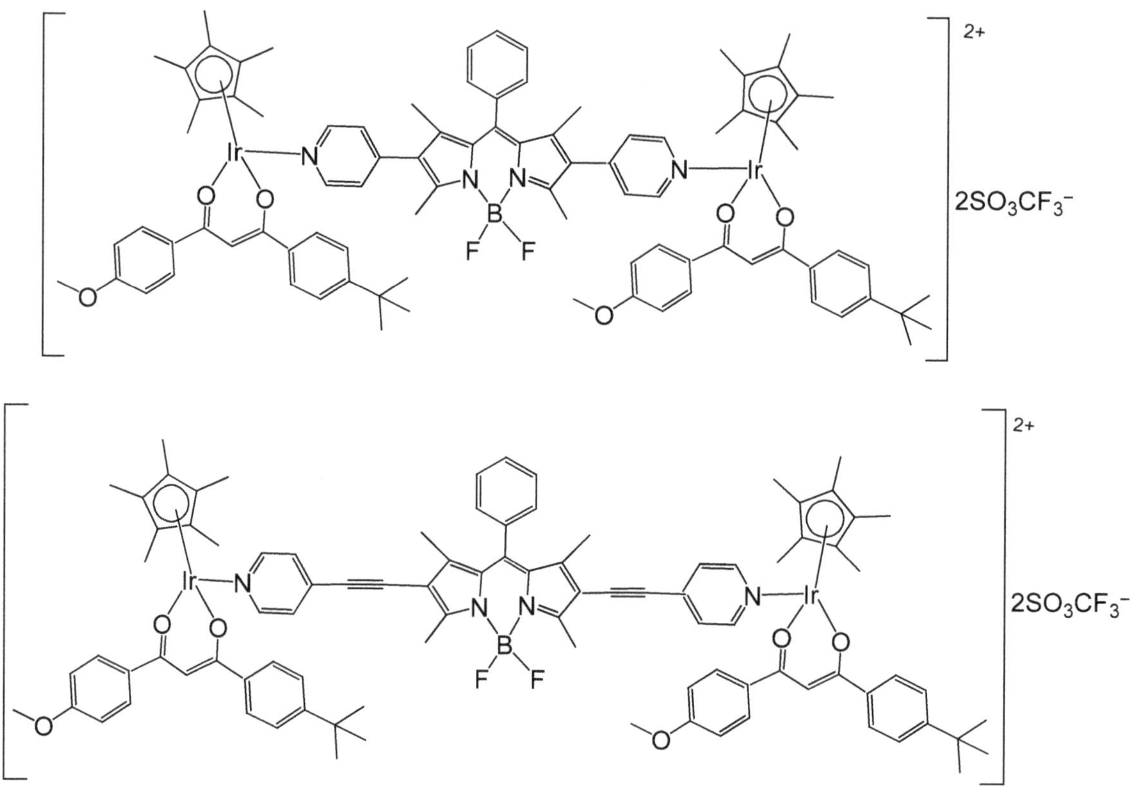

Figure 25. Structures of BODIPY (BDP and BDPCC) based iridium dimetallic complexes $[Ir_2(Cp^*)_2(AVB)_2BDP][2CF_3SO_3]$ and $[Ir_2(Cp^*)_2(AVB)_2BDPCC][2CF_3SO_3]$ [100].

4. Metallacages

The development of metal-based macrocycles using bioactive ligands represents a key direction for the design of novel antineoplastic agents. In contrast to platinum(II) and ruthenium(II) assemblies, the progress of iridium-based macrocycles incorporating cyclometalated or half-sandwich Ir(III) complexes is limited. Nevertheless, some Ir(III) assemblies have shown higher effectiveness than cisplatin against the proliferation of numerous tumor types. In recent decades, the investigations involving metalla-macrocycles, tested as antitumor candidates, have increased [82], including the reported hexanuclear complexes [104], and the 5,8-dihydroxy-1,4-naphthoquinonato tetranuclear metallacycles with bipyridyl linkers [105].

Three new pentamethylcyclopentadienyl Ir(III) metalla-rectangles, Figure 26, have been obtained by a self-assembly approach using the embelin-derived metallaclips $(\eta^5\text{-}C_5Me_5)_2Ir_2(\mu_4\text{-}C_6HRO_4\text{-}\kappa O)Cl_2$ ($R = (CH_2)_{10}CH_3$)). Two additional ditopic bridging linear ligands, containing pyrazine (Figure 26a), 4,4′-bipyridine (Figure 26b), or 1,2-bis(4-pyridyl)ethylene (Figure 26c), have been used to complete the metal-based rectangular structures. The compounds have been studied by IR, ^1H- and ^{13}C-NMR, and ESI-MS spectral and elemental analyses. Tetranuclear iridium complexes, Figure 26, were assessed for antiproliferative action against lung, prostate, and cervical cancerous and noncancerous (HEK-293) cells [106]. The iridium(III) fragments have been linked through the hydrophobic benzoquinone embelin, purposefully chosen because of its cellular permeability [107]. Higher and biologically potential selectivity against tumor over healthy cells was found for the most cytotoxic tetranuclear iridium complex $[Ir_4(\mu\text{-emb})_2(\mu dpee)_2(\eta^5\text{-}Cp^*)_4](CF_3SO_3)_4$, depicted in Figure 26c, which displayed markedly lower IC_{50} values in tumor cells (0.6 μM in HeLa) compared with HEK-293 cell line (70.8 μM), Table 1, [106]. The capability of the tetracationic metallamacrocycles to accumulate in mitochondria has been established by flow cytometry and confocal microscopy investigations. Cell cycle examination displayed accumulation in the sub-G_1 phase, signifying that treatment with such complexes led to DNA fragmentation. The complexes induced apoptosis in the early and late stages. The studied metalla-rectangles appeared to interact with DNA and mitochondria membranes, most probable due to the positive charge of the metallamacrocycles and the existence of lipophilic side chains, therefore giving these compounds the required characteristics for possible in vivo studies.

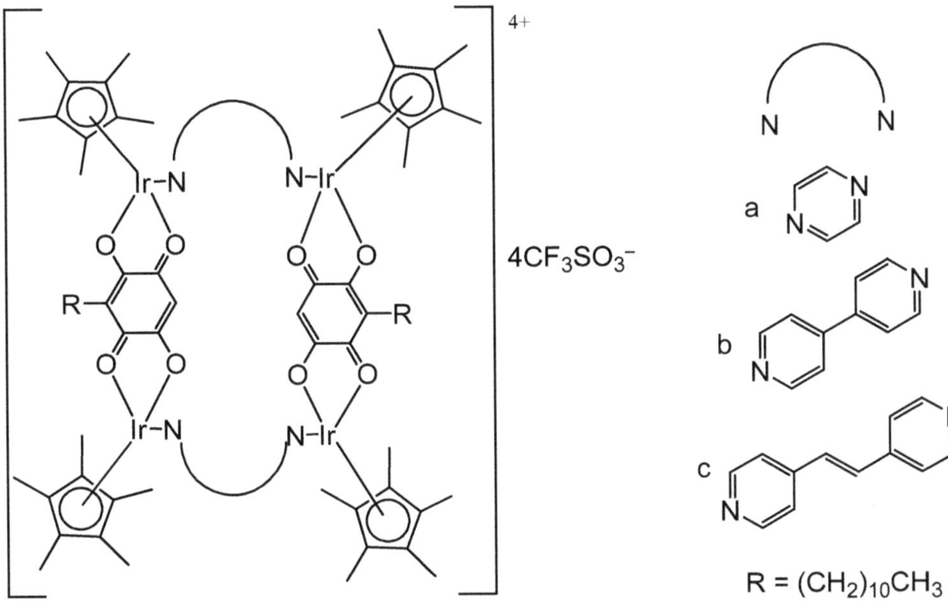

Figure 26. Pentamethylcyclopentadienyl Ir(III) metalla-rectangles $(\eta^5\text{-}C_5Me_5)_2M_2(\mu_4\text{-}C_6HRO_4\text{-}\kappa O)Cl_2$ with R = $(CH_2)_{10}CH_3$, comprising pyrazine (**a**), 4,4′-bipyridine (**b**), or 1,2-bis(4-pyridyl)ethylene (**c**) [106].

Ir(III)-Cp* metalla-prismatic structures (Figure 27) $[(Cp^*Ir)_6(tpt)_2(dhnq)_3]^{6+}$, where tpt = 2,4,6-tri-(pyridin-4-yl)-1,3,5-triazine and dhnq = 5,8-dihydroxy1,4-naphthoquinonato ligands, have been obtained and identified by different analytical methods. The cytotoxic and photodynamic activities have been assessed on human HT-29 colon cancer cells, displaying IC_{50} around 1 μM, which decreased to nanomolar concentrations at irradiation, demonstrating a perfect synergistic effect between the cytotoxic and phototoxic activity [108]. Photoactive supramolecular metallamacrocycles integrating Ir(III) complexes have been reported by Martir et al. [109].

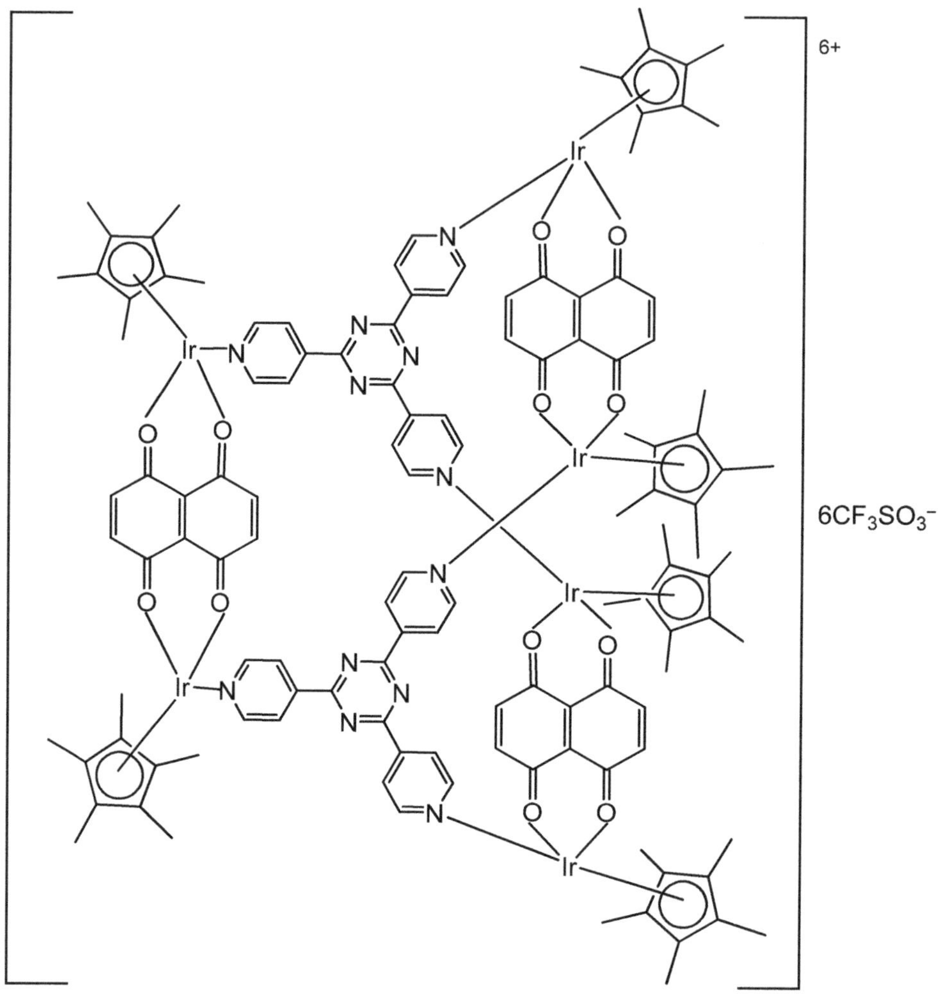

Figure 27. Ir(III) metalla-prism $[(Cp^*Ir)_6(tpt)_2(dhnq)_3]^{6+}$ [108].

The octanuclear iridium metallocube (Figure 28) has been obtained by incorporating Cp*-Ir building fragments bridged by an alkyl-functionalized ligand, and connected by tetrapyridyl porphyrin fragments [110]. The cubic supramolecular macrocycle has shown strong antitumor activity toward MCF-7, B16, and A549 cancer cells with IC_{50} under 0.1 μM and good selectivity over the normal NIH-3-T3 cell line. The proposed mechanism of action was apoptotic induction with DNA interaction.

Ir(III) metalla-rectangle IrBAnPy, shown in Figure 29, has been obtained by the coordination of an anthracene-functionalized dipyridyl ligand $BAnPy_2$, resulting in the formation of a robust rectangular [111]. Its antineoplastic properties have been examined, revealing promising cytotoxic activity against A549 and H460 cancer cells in comparison with the non-malignant MRC5 cells. The detected IC_{50} were 16.24 μM for H460 and 2.38 μM for A549, respectively. In contrast, the IC_{50} for non-malignant MRC5 cell line was 29.44 μM. Micronucleus and cell death assay have been performed to determine the damage occurring to the cancer cellular nucleus.

Table 1. IC_{50} values of polynuclear iridium complexes.

Compound	IC_{50}	Cell Lines	Ref.
$[Ir_2(m\text{-pzpy})(\eta^5\text{-Cp*})_2Cl_3]PF_6$	>100 μM	HT-29	[65]
$[\{(\eta^5\text{-Cp*})Ir(dppz)\}_2(\mu\text{-pyz})]^{4+}$	3.5 μM 1.8 μM	MCF-7 HT-29	[66]
$[\{(\eta^5\text{-C}_5Me_5)Ir(dppz)\}_2(\mu\text{-4,4'-bpy})]^{4+}$	3.1 μM 3.8 μM	MCF-7 HT-29	
$[\{(\eta^5\text{-Cp*})Ir(dppz)\}_2(\mu\text{-dpee})]^{4+}$	0.61 μM	MCF-7	[67]
$[\{(\eta^5\text{-Cp*})Ir(dppz)\}_2(\mu\text{-dpey})]^{4+}$	0.49 μM	MCF-7	
$[\{(\eta^5\text{-Cp*})Ir(dppz)\}_2(\mu\text{-dpeb})]^{4+}$	2.2 μM	MCF-7	
$[Ir_2(\mu\text{-L*})(\eta^5\text{-Cp*})_2Cl_2](PF_6)_2$	3.1 μM 6.0 μM	A2780 MCF-7	[69]
$[Ir_2(\mu\text{-L**})(\eta^5\text{-Cp*})_2Cl_2](PF_6)_2$	12.6 μM 16.5 μM	HepG2 A549	[70]
$[\{(ppy)_2Ir\}(\mu\text{-phpy})\{Ru(p\text{-cym})Cl\}](PF_6)_2$	0.92 μM	MCF7	[83]
$[Ir(\eta^5\text{-Cp*})Cl(\mu\text{-bpm})Ru(\eta^6\text{-p-cym})Cl](PF_6)_2$	1.9 μM 6.2 μM	MDA-MB-468 Caco-2	[90]
$[Ir(\eta^5\text{-Cp*})Cl(\mu\text{-bpm})ReCl(CO)_3]Cl$	24.1 μM	MDA-MB-468	[91]
$[Ir(\eta^5\text{-Cp*})Cl_2(\mu\text{-pcfx})Cu(phen)(H_2O)]NO_3$	1.3 pM 35.5 nM	DU145 A549	[94]
$[Ir_4(\mu\text{-emb})_2(\mu\text{-dpee})_2(\eta^5\text{-Cp*})4](CF_3SO_3)_4$	0.6 μM	HeLa	[106]
$[(Cp^*Ir)_6(tpt)_2(dhnq)_3]^{6+}$	1 μM	HT-29	[108]

Abbreviations: pzpy = N′-[(1Z)-1-(pyridin-2-yl)ethylidene]pyrazine-2-carbohydrazonamide); dppz = dipyrido[2,3-a:2′,3′-c]phenazine; pyz = pyrazine; 4,4′-bpy = 4,4′-bipyridine; dpee = 4-[(*E*)-2-(4-pyridinyl)ethenyl]pyridine; dpey = 4-(2-pyridin-4-ylethynyl)pyridine; dpeb = 1,4-di(2-pyridin-4-ylethynyl)benzene; L* = N,N′-(biphenyl-4,4′-diyldimethylidyne)bis-2-(pyridin-2-yl)ethanamine; L** = (1E)-(6-methylpyridin-2-yl)-N-(4-(4-(((E)-(6-methylpyridin2-yl)methylidene)amino)benzyl)-phenyl)methanimine; bpm = 2,2′-bipyrimidine; pcfx = ciprofloxacin; fcdpm = 5-ferrocenyldipyrromethene; ppy = pyridine; p-cym = p-cymene; phen = phenanthroline; emb = embelin; phpy = 2,3-di(pyridin-2-yl)pyrazino[2,3-f][1,10] phenanthroline; tpt = 2,4,6-tri-(pyridin-4-yl)-1,3,5-triazine; dhnq = 5,8-dihydroxy1,4-naphthoquinone.

Metal-organic cages (MOCs) have shown well-defined structural shapes and cavities, and possess promising applications in many fields [112,113].

Ir_8Pd_4-heteronuclear metal-organic cage Ir_8Pd_4-MOC (MOC-51) has been constructed from bipodal metal-based ligand $[Ir(ppy)_2(qpy)(BF_4)]$ (qpy = 4,4′:2′,2′′:4′′,4′′′-quaterpyridine; ppy = 2-phenylpridine) with a palladium(II) salt [114]. The cubic barrel-shaped metal-organic cage showed large $^1O_2$ quantum yields under visible irradiation, consequently displaying great potential in cell imaging and PDT. In comparison with the Ir(III) metallo-based ligand, the Ir_8Pd_4-MOC has shown less dark toxicity toward HeLa cells and higher mitochondria-targeting efficiency.

Liu et al. have recently described $[Pd_4Ir_8]^{16+}$ supramolecular cubic cage MOC-53 containing multiple iridium(III) metallo-based ligands, localized in mitochondria with high $^1O_2$ production activating efficiently apoptosis [115]. The bimetallic supramolecular cage has shown superior antineoplastic activity, bright phosphorescence and excellent photostability. The reported supramolecular cubic cage has shown higher phototoxicity toward the HeLa cell line.

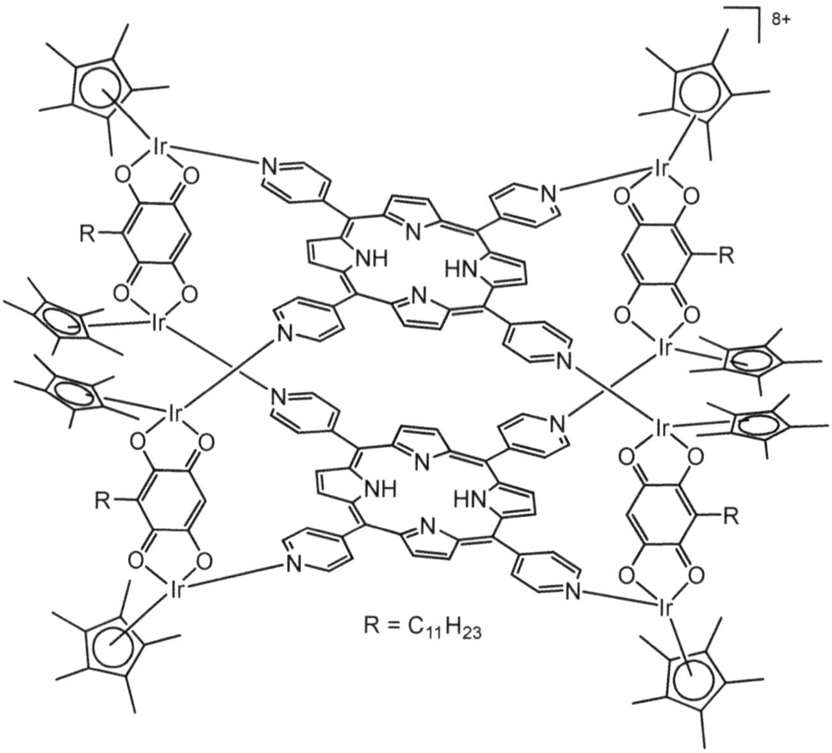

Figure 28. Octanuclear iridium metallocube with tetrapyridyl porphyrin fragments [110].

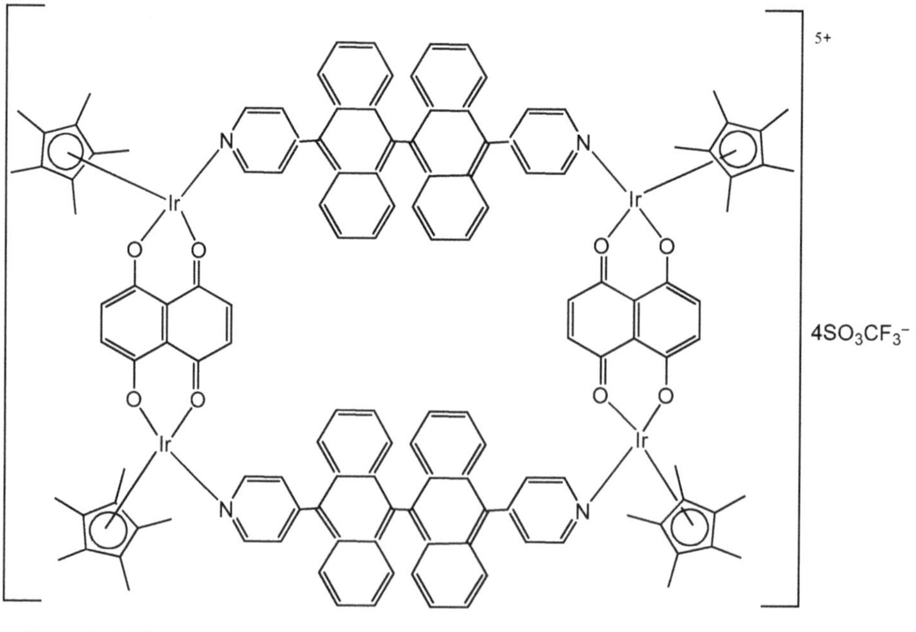

Figure 29. Iridium metalla-rectangle IrBAnPy [111].

5. Conclusions and Perspective

This review surveyed iridium(III) multinuclear or heteronuclear complexes and their use in anticancer drug discovery. Although promising mononuclear iridium(III) complexes have been extensively studied and provided a positive hope for developing effective antineoplastic agents, the studies on the antitumor application of homo- and hetero-multinuclear Ir(III) complexes are insufficient and this area needs further examination. This research field is currently extremely active and it has been expanded over the last years. Numerous research teams have developed various multinuclear iridium complexes and have investigated their properties in different conditions (diverse cell lines, concentrations and incubation time, etc.), thus a complete validation of the structure-activity relationship is difficult to achieve. It is greatly appreciated that the common range of IC_{50} values of the newly reported iridium(III) multinuclear complexes is less than 100 μM. In most of the cases, these complexes have demonstrated selectivity and good tolerance by healthy cells. Their anticancer mechanism of action has been found to be different from that of platinum drugs, offering cytotoxicity against cisplatin-resistant cancer cells. Multinuclear iridium(III) complexes were either derived from their mononuclear analogs or were prepared with similar ligands of diverse denticities, which permitted the coordination of a different number of metallic centers. These strategies have enabled a comparison of the effect of metal cations on the resulting antineoplastic activity and, in most cases, multinuclear complexes displayed better activity than the mononuclear ones. Obviously, the studies, reported in the literature, have confirmed that the design of multinuclear Ir(III) complexes became a viable strategy and a suitable research direction. Nevertheless, as demonstrated in this review, it is difficult to compare complexes with different nuclearity and a systematic study associated with the possible biological applications of these complexes is still lacking. It is expected that this review will offer vital guidance for the further development of these promising anticancer candidates.

Funding: This research received no external funding.

Institutional Review Board Statement: Not applicable.

Informed Consent Statement: Not applicable.

Data Availability Statement: No new data were created or analyzed in this study.

Acknowledgments: The administrative support received by the European Union-Next Generation EU, through the National Recovery and Resilience Plan of the Republic of Bulgaria, project No. BG-RRP-2.004-0004-C01 is greatly acknowledged.

Conflicts of Interest: The author declares no conflicts of interest.

References

1. Sharma, A.; Sudhindra, P.; Roy, N.; Paira, P. Advances in novel iridium (III) based complexes for anticancer applications: A review. *Inorg. Chim. Acta* **2020**, *513*, 119925. [CrossRef]
2. Kostova, I. Cytotoxic Organometallic Iridium (III) Complexes. *Molecules* **2025**, *30*, 801. [CrossRef] [PubMed]
3. Pracharova, J.; Vigueras, G.; Novohradsky, V.; Cutillas, N.; Janiak, C.; Kostrhunova, H. Exploring the effect of polypyridyl ligands on the anticancer activity of phosphorescent iridium(III) complexes: From proteosynthesis inhibitors to photodynamic therapy agents. *Chemistry* **2018**, *24*, 4607–4619. [CrossRef] [PubMed]
4. Lucas, S.; Lord, R.; Wilson, R.; Phillips, R.; Sridharan, V.; McGowan, P. Synthesis of iridium and ruthenium complexes with (N, N), (N, O) and (O, O) coordinating bidentate ligands as potential anti-cancer agents. *Dalton Trans.* **2012**, *41*, 13800–13802. [CrossRef]
5. Cao, R.; Jia, J.; Ma, X.; Zhou, M.; Fei, H. Membrane localized iridium(iii) complex induces endoplasmic reticulum stress and mitochondriamediated apoptosis in human cancer cells. *J. Med. Chem.* **2013**, *56*, 3636–3644. [CrossRef]

6. Cao, J.J.; Tan, C.P.; Chen, M.H.; Wu, N.; Yao, D.Y.; Liu, X.G. Targeting cancer cell metabolism with mitochondria-immobilized phosphorescent cyclometalated iridium(iii) complexes. *Chem. Sci.* **2017**, *8*, 631–640. [CrossRef] [PubMed]
7. Qin, Q.-P.; Meng, T.; Tan, M.-X.; Liu, Y.-C.; Luo, X.-J.; Zou, B.-Q. Synthesis and in vitro biological evaluation of three 4′-(4-methoxyphenyl)-2, 2′:6′, 2″-terpyridine iridium(iii) complexes as new telomerase inhibitors. *Eur. J. Med. Chem.* **2018**, *143*, 1387–1395. [CrossRef]
8. Hao, L.; Li, Z.-W.; Zhang, D.-Y.; He, L.; Liu, W.; Yang, J. Monitoring mitochondrial viscosity with anticancer phosphorescent ir(III) complexes via twophoton lifetime imaging. *Chem. Sci.* **2019**, *10*, 1285–1293. [CrossRef]
9. Redrado, M.; Minana, M.; Coogan, M.P.; Concepcion Gimeno, M.; Fernandez-Moreira, V. Tunable emissive ir(iii) benzimidazole-quinoline hybrids as promising theranostic lead compounds. *Chemmedchem* **2022**, *2022*, e202200244. [CrossRef]
10. Li, Y.; Liu, B.; Lu, X.-R.; Li, M.-F.; Ji, L.-N.; Mao, Z.-W. Cyclometalated iridium(III) n-heterocyclic carbene complexes as potential mitochondrial anticancer and photodynamic agents. *Dalton Trans.* **2017**, *46*, 11363–11371. [CrossRef]
11. Monti, F.; Kessler, F.; Delgado, M.; Frey, J.; Bazzanini, F.; Accorsi, G. Charged bis-cyclometalated iridium(III) complexes with carbene-based ancillary ligands. *Inorg. Chem.* **2013**, *52*, 10292–10305. [CrossRef] [PubMed]
12. Ye, R.-R.; Tan, C.-P.; Ji, L.-N.; Mao, Z.-W. Coumarin-appended phosphorescent cyclometalated iridium(iii) complexes as mitochondria-targeted theranostic anticancer agents. *Dalton Trans.* **2016**, *45*, 13042–13051. [CrossRef]
13. Hearn, J.M.; Romero-Canelón, I.; Qamar, B.; Liu, Z.; Hands-Portman, I.; Sadler, P.J. Organometallic iridium (III) anticancer complexes with new mechanisms of action: NCI-60 screening, mitochondrial targeting, and apoptosis. *ACS Chem. Biol.* **2013**, *8*, 1335–1343. [CrossRef]
14. Liu, Z.; Romero-Canelón, I.; Qamar, B.; Hearn, J.M.; Habtemariam, A.; Barry, N.P.; Sadler, P.J. The potent oxidant anticancer activity of organoiridium catalysts. *Angew. Chem.* **2014**, *126*, 4022–4027. [CrossRef]
15. Liu, Z.; Sadler, P.J. Formation of glutathione sulfenate and sulfinate complexes by an organoiridium (III) anticancer complex. *Inorg. Chem. Front.* **2014**, *1*, 668–672. [CrossRef]
16. Zhang, W.Y.; Banerjee, S.; Hughes, G.M.; Bridgewater, H.E.; Song, J.I.; Breeze, B.G.; Sadler, P.J. Ligand-centred redox activation of inert organoiridium anticancer catalysts. *Chem. Sci.* **2020**, *11*, 5466–5480. [CrossRef]
17. Starha, P.; Habtemariam, A.; Romero-Canelon, I.; Clarkson, G.J.; Sadler, P.J. Hydrosulfide adducts of organo-iridium anticancer complexes. *Inorg. Chem.* **2016**, *55*, 2324–2331. [CrossRef]
18. Prathima, T.S.; Choudhury, B.; Ahmad, M.G.; Chanda, K.; Balamurali, M.M. Recent developments on other platinum metal complexes as target-specific anticancer therapeutics. *Coord. Chem. Rev.* **2023**, *490*, 215231. [CrossRef]
19. Wu, Y.; Liu, J.; Shao, M.; Zhang, P.; Song, S.; Yang, G. Cyclometalated iridium(iii) dithioformic acid complexes as mitochondriatargeted imaging and anticancer agents. *J. Inorg. Biochem.* **2022**, *233*, 111855. [CrossRef]
20. Xiong, K.; Zhou, Y.; Lin, X.; Kou, J.; Lin, M.; Guan, R. Cyclometalated iridium(III) complexes as mitochondria-targeting photosensitizers against cisplatin-resistant cells. *Photochem. Photobiol.* **2022**, *98*, 85–91. [CrossRef]
21. Wang, F.-X.; Chen, M.-H.; Hu, X.-Y.; Ye, R.-R.; Tan, C.-P.; Ji, L.-N. Ester-modified cyclometalated iridium(iii) complexes as mitochondria-targeting anticancer agents. *Sci. Rep.* **2016**, *6*, 38954. [CrossRef]
22. Wang, L.; Guan, R.; Xie, L.; Liao, X.; Xiong, K.; Rees, T.W. An er-targeting iridium(III) complex that induces immunogenic cell death in non-small-cell lung cancer. *Angew. Chem. Int. Ed. Engl.* **2021**, *60*, 4657–4665. [CrossRef]
23. Zhang, P.; Huang, H.; Banerjee, S.; Clarkson, G.J.; Ge, C.; Imberti, C. Nucleus-targeted organoiridium–albumin conjugate for photodynamic cancer therapy. *Angew. Chem. Int. Ed. Engl.* **2019**, *58*, 2350–2354. [CrossRef] [PubMed]
24. Huang, C.; Liang, C.; Sadhukhan, T.; Banerjee, S.; Fan, Z.; Li, T. Invitro and in-vivo photocatalytic cancer therapy with biocompatible iridium(iii) photocatalysts. *Angew. Chem. Int. Ed. Engl.* **2021**, *60*, 9474–9479. [CrossRef] [PubMed]
25. Liu, L.; Chen, J.; Wang, M.M.; Huang, Y.; Qian, Y.; Xue, X. The cyclometalated iridium (III) complex based on 9-anthracenecarboxylic acid as a lysosomal-targeted anticancer agent. *J. Inorg. Biochem.* **2022**, *235*, 111913. [CrossRef]
26. Acuña, M.I.; Rubio, A.R.; Martínez-Alonso, M.; Busto, N.; Rodríguez, A.M.; Davila-Ferreira, N.; Domínguez, F. Targets, mechanisms and cytotoxicity of half-sandwich Ir(III) complexes are modulated by structural modifications on the benzazole ancillary ligand. *Cancers* **2023**, *15*, 107. [CrossRef] [PubMed]
27. Fan, Z.; Rong, Y.; Sadhukhan, T.; Liang, S.; Li, W.; Yuan, Z. Singlecell quantification of a highly biocompatible dinuclear iridium(III) complex for photocatalytic cancer therapy. *Angew. Chem. Int. Ed. Engl.* **2022**, *61*, e202202098. [CrossRef]
28. Ouyang, C.; Chen, L.; Rees, T.W.; Chen, Y.; Liu, J.; Ji, L. A mitochondriatargeting hetero-binuclear Ir(III)–Pt (II) complex induces necrosis in cisplatin-resistant tumor cells. *Chem. Commun.* **2018**, *54*, 6268–6271. [CrossRef]
29. Zhang, C.; Guan, R.; Liao, X.; Ouyang, C.; Rees, T.W.; Liu, J. A mitochondria-targeting dinuclear ir–ru complex as a synergistic photoactivated chemotherapy and photodynamic therapy agent against cisplatin-resistant tumour cells. *Chem. Commun.* **2019**, *55*, 12547–12550. [CrossRef]

30. Thomas, S.; Balónová, B.; Cinatl, J.; Wass, M.; Serpell, C.; Id, O. Thiourea and guanidine compounds and their iridium complexes in drug-resistant cancer cell lines: Structure-activity relationships and direct luminescent imaging. *Chemmedchem* **2022**, *15*, 349–353. [CrossRef]
31. Xiong, K.; Chen, Y.; Ouyang, C.; Guan, R.-L.; Ji, L.-N.; Chao, H. Cyclometalated iridium(iii) complexes as mitochondria-targeted anticancer agents. *Biochimie* **2016**, *125*, 186–194. [CrossRef] [PubMed]
32. Li, Y.; Tan, C.-P.; Zhang, W.; He, L.; Ji, L.-N.; Mao, Z.-W. Phosphorescent iridium(iii)-bis-n-heterocyclic carbene complexes as mitochondria-targeted theranostic and photodynamic anticancer agents. *Biomaterials* **2015**, *39*, 95–104. [CrossRef]
33. Ouyang, M.; Zeng, L.; Huang, H.; Jin, C.; Liu, J.; Chen, Y. Fluorinated cyclometalated iridium(iii) complexes as mitochondria-targeted theranostic anticancer agents. *Dalton Trans.* **2017**, *46*, 6734–6744. [CrossRef]
34. Song, X.-D.; Kong, X.; He, S.-F.; Chen, J.-X.; Sun, J.; Chen, B.-B. Cyclometalated iridium(iii)-guanidinium complexes as mitochondria-targeted anticancer agents. *Eur. J. Med. Chem.* **2017**, *138*, 246–254. [CrossRef] [PubMed]
35. Tang, B.; Wan, D.; Wang, Y.-J.; Yi, Q.-Y.; Guo, B.-H.; Liu, Y.-J. An iridium (III) complex as potent anticancer agent induces apoptosis and autophagy in b16 cells through inhibition of the akt/mtor pathway. *Eur. J. Med. Chem.* **2018**, *145*, 302–314. [CrossRef]
36. Yi, Q.-Y.; Wan, D.; Tang, B.; Wang, Y.-J.; Zhang, W.-Y.; Du, F. Synthesis, characterization and anticancer activity in vitro and in vivo evaluation of an iridium (III) polypyridyl complex. *Eur. J. Med. Chem.* **2018**, *145*, 338–349. [CrossRef] [PubMed]
37. He, L.; Wang, K.-N.; Zheng, Y.; Cao, J.-J.; Zhang, M.-F.; Tan, C.-P. Cyclometalated iridium(iii) complexes induce mitochondria-derived paraptotic cell death and inhibit tumor growth in vivo. *Dalton Trans.* **2018**, *47*, 6942–6953. [CrossRef]
38. Millett, A.J.; Habtemariam, A.; Romero-Canelon, I.; Clarkson, G.J.; Sadler, P.J. Contrasting anticancer activity of half-sandwich iridium(III) complexes bearing functionally diverse 2-phenylpyridine ligands. *Organometallics* **2015**, *34*, 2683–2694. [CrossRef]
39. Křikavová, R.; Romanovová, M.; Jendželovská, Z.; Majerník, M.; Masaryk, L.; Zoufalý, P.; Nemec, I. Impact of the central atom and halido ligand on the structure, antiproliferative activity and selectivity of half-sandwich Ru(II) and Ir(III) complexes with a 1,3,4-thiadiazole-based ligand. *Dalton Trans.* **2023**, *52*, 12717–12732. [CrossRef]
40. Novohradsky, V.; Zerzankova, L.; Stepankova, J.; Kisova, A.; Kostrhunova, H.; Liu, Z.; Brabec, V. A dual-targeting, apoptosis-inducing organometallic half-sandwich iridium anticancer complex. *Metallomics* **2014**, *6*, 1491–1501. [CrossRef]
41. Guo, L.; Li, P.; Li, J.; Gong, Y.; Li, X.; Wen, T.; Liu, Z. Potent half-sandwich 16-/18-electron iridium (III) and ruthenium(II) anticancer complexes with readily available amine–imine ligands. *Inorg. Chem.* **2023**, *62*, 21379–21395. [CrossRef]
42. Mou, Z.D.; Deng, N.; Zhang, F.; Zhang, J.; Cen, J.; Zhang, X. Halfsandwich Schiff-base Ir(III) complexes as anticancer agents. *Eur. J. Med. Chem.* **2017**, *138*, 72–82. [CrossRef]
43. Li, J.; Guo, L.; Tian, Z.; Tian, M.; Zhang, S.; Xu, K. Novel halfsandwich iridium(III) imino-pyridyl complexes showing remarkable in vitro anticancer activity. *Dalton Trans.* **2017**, *46*, 15520–15534. [CrossRef]
44. Yang, T.; Zhu, M.; Jiang, M.; Yang, F.; Zhang, Z. Current status of iridium-based complexes against lung cancer. *Front. Pharmacol.* **2022**, *13*, 1025544. [CrossRef] [PubMed]
45. Štarha, P. Anticancer iridium (III) cyclopentadienyl complexes. *Inorg. Chem. Front.* **2025**, *12*, 897–954. [CrossRef]
46. Ma, L.; Li, L.; Zhu, G. Platinum-containing heterometallic complexes in cancer therapy: Advances and perspectives. *Inorg. Chem. Front.* **2022**, *9*, 2424–2453. [CrossRef]
47. Notaro, A.; Gasser, G. Monomeric and dimeric coordinatively saturated and substitutionally inert Ru (II) polypyridyl complexes as anticancer drug candidates. *Chem. Soc. Rev.* **2017**, *46*, 7317–7337. [CrossRef] [PubMed]
48. Higgins, S.L.; White, T.A.; Winkel, B.S.; Brewer, K.J. Redox, spectroscopic, and photophysical properties of Ru−Pt mixed-metal complexes incorporating 4, 7-diphenyl-1, 10-phenanthroline as efficient DNA binding and photocleaving agents. *Inorg. Chem.* **2011**, *50*, 463–470. [CrossRef]
49. López-Hernández, J.E.; Nayeem, N.; Cerón-Carrasco, J.P.; Ahad, A.; Hafeez, A.; León, I.E.; Contel, M. Platinum(IV)–Gold(I) Agents with Promising Anticancer Activity: Selected Studies in 2D and 3D Triple-Negative Breast Cancer Models. *Chem. Eur. J.* **2023**, *29*, e202302045. [CrossRef]
50. Wenzel, M.; Bigaeva, E.; Richard, P.; Le Gendre, P.; Picquet, M.; Casini, A.; Bodio, E. New heteronuclear gold(I)-platinum(II) complexes with cytotoxic properties: Are two metals better than one? *J. Inorg. Biochem.* **2014**, *141*, 10–16. [CrossRef]
51. Pete, S.; Roy, N.; Kar, B.; Paira, P. Construction of homo and heteronuclear Ru (II), Ir (III) and Re (I) complexes for target specific cancer therapy. *Coord. Chem. Rev.* **2022**, *460*, 214462. [CrossRef]
52. Mishra, S.; Tripathy, S.K.; Paul, D.; Laha, P.; Santra, M.K.; Patra, S. Asymmetrically Coordinated Heterodimetallic Ir–Ru System: Synthesis, Computational, and Anticancer Aspects. *Inorg. Chem.* **2023**, *62*, 7003–7013. [CrossRef] [PubMed]
53. Serratrice, M.; Maiore, L.; Zucca, A.; Stoccoro, S.; Landini, I.; Mini, E.; Massai, L.; Ferraro, G.; Merlino, A.; Messori, L. Cytotoxic properties of a new organometallic platinum(ii) complex and its gold(i) heterobimetallic derivatives. *Dalton Trans.* **2016**, *45*, 579–590. [CrossRef] [PubMed]

54. Bellam, R.; Jaganyi, D.; Robinson, R.S. Heterodinuclear Ru–Pt Complexes Bridged with 2,3-Bis(pyridyl)pyrazinyl Ligands: Studies on Kinetics, Deoxyribonucleic Acid/Bovine Serum Albumin Binding and Cleavage, In Vitro Cytotoxicity, and In Vivo Toxicity on Zebrafish Embryo Activities. *ACS Omega* **2022**, *7*, 26226–26245. [CrossRef] [PubMed]
55. Bellam, R.; Jaganyi, D.; Mambanda, A.; Robinson, R. Role of a 2,3-bis(pyridyl)pyrazinyl chelate bridging ligand in the reactivity of Ru(ii)–Pt(ii) dinuclear complexes on the substitution of chlorides by thiourea nucleophiles—A kinetic study. *New J. Chem.* **2018**, *42*, 12557–12569. [CrossRef]
56. Askari, B.; Rudbari, H.A.; Micale, N.; Schirmeister, T.; Maugeri, A.; Navarra, M. Anticancer study of heterobimetallic platinum(II)-ruthenium(II) and platinum(II)-rhodium(III) complexes with bridging dithiooxamide ligand. *J. Organomet. Chem.* **2019**, *900*, 120918. [CrossRef]
57. Ćoćić, D.; Jovanović-Stević, S.; Jelić, R.; Matić, S.; Popović, S.; Djurdjević, P.; Baskić, D.; Petrović, B. Homo- and hetero-dinuclear Pt(ii)/Pd(ii) complexes: Studies of hydrolysis, nucleophilic substitution reactions, DNA/BSA interactions, DFT calculations, molecular docking and cytotoxic activity. *Dalton Trans.* **2020**, *49*, 14411–14431. [CrossRef]
58. Zhang, C.; Guan, R.; Liao, X.; Ouyang, C.; Liu, J.; Ji, L.; Chao, H. Mitochondrial DNA targeting and impairment by a dinuclear Ir–Pt complex that overcomes cisplatin resistance. *Inorg. Chem. Front.* **2020**, *7*, 1864–1871. [CrossRef]
59. Atrián-Blasco, E.; Sáez, J.; Rodriguez-Yoldi, M.J.; Cerrada, E. Heteronuclear complexes with promising anticancer activity against colon cancer. *Biomedicines* **2024**, *12*, 1763. [CrossRef]
60. Štarha, P. Multinuclear biologically active Ru, Rh, Os and Ir arene complexes. *Coord. Chem. Rev.* **2021**, *431*, 213690. [CrossRef]
61. Maeda, H. The enhanced permeability and retention (EPR) effect in tumor vasculature: The key role of tumor-selective macromolecular drug targeting. *Adv. Enz. Regul.* **2001**, *41*, 189–207. [CrossRef] [PubMed]
62. Redrado, M.; Fernández-Moreira, V.; Gimeno, M.C. Theranostics Through the Synergistic Cooperation of Heterometallic Complexes. *ChemMedChem* **2021**, *16*, 932–941. [CrossRef] [PubMed]
63. Muhamedi, A.; Batsanov, A.S. Phosphorescent mono- and diiridium(III) complexes cyclometalated by fluorenyl- or phenylpyridino ligands with bulky substituents, as prospective OLED dopants. *Acta Crystallogr. Sect. E Crystallogr. Commun.* **2020**, *76*, 392–399. [CrossRef]
64. De Palo, A.; Draca, D.; Murrali, M.G.; Zacchini, S.; Pampaloni, G.; Mijatovic, S.; Marchetti, F. A comparative analysis of the in vitro anticancer activity of iridium (III){η5-C5Me4R} complexes with variable R groups. *Intern. J. Mol. Sci.* **2021**, *22*, 7422. [CrossRef]
65. Lapasam, A.; Pinder, E.; Phillips, R.M.; Kaminsky, W.; Kollipara, M.R. Synthesis, structure and bonding modes of pyrazine based ligands of Cp* Rh and Cp* Ir complexes: The study of in-vitro cytotoxicity against human cell lines. *J. Organomet. Chem.* **2019**, *899*, 120887. [CrossRef]
66. Nazif, M.A.; Bangert, J.A.; Ott, I.; Gust, R.; Stoll, R.; Sheldrick, W.S. Dinuclear organoiridium (III) mono-and bis-intercalators with rigid bridging ligands: Synthesis, cytotoxicity and DNA binding. *J. Inorg. Biochem.* **2009**, *103*, 1405–1414. [CrossRef]
67. Nazif, M.A.; Rubbiani, R.; Alborzinia, H.; Kitanovic, I.; Wölfl, S.; Ott, I.; Sheldrick, W.S. Cytotoxicity and cellular impact of dinuclear organoiridium DNA intercalators and nucleases with long rigid bridging ligands. *Dalton Trans.* **2012**, *41*, 5587–5598. [CrossRef]
68. Parveen, S.; Hanif, M.; Leung, E.; Tong, K.; Yang, A.; Astin, J.; De Zoysa, G.H.; Steel, T.R.; Goodman, D.; Movassaghi, S.; et al. Anticancer organorhodium and -iridium complexes with low toxicity in vivo but high potency in vitro: DNA damage, reactive oxygen species formation, and haemolytic activity. *Chem. Commun.* **2019**, *55*, 12016–12019. [CrossRef]
69. Štarha, P.; Hošek, J.; Trávníček, Z.; Dvořák, Z. Cytotoxic dimeric half-sandwich Ru(II), Os(II) and Ir(III) complexes containing the 4,4′-biphenyl-based bridging ligands. *Appl. Organomet. Chem.* **2020**, *34*, e5785. [CrossRef]
70. Masaryk, L.; Koczurkiewicz-Adamczyk, P.; Milde, D.; Nemec, I.; Sloczynska, K.; Pękala, E.; Štarha, P. Dinuclear half-sandwich Ir(III) complexes containing 4,4′-methylenedianiline-based ligands: Synthesis, characterization, cytotoxicity. *J. Organomet. Chem.* **2021**, *938*, 121748. [CrossRef]
71. Steel, T.R.; Tong, K.K.; Söhnel, T.; Jamieson, S.M.; Wright, L.J.; Crowley, J.D.; Hartinger, C.G. Homodinuclear organometallics of ditopic N,N-chelates: Synthesis, reactivity and in vitro anticancer activity. *Inorg. Chim. Acta* **2021**, *518*, 120220. [CrossRef]
72. Kavukcu, S.B.; Ensarioğlu, H.K.; Karabıyık, H.; Vatansever, H.S.; Türkmen, H. Cell death mechanism of organometallic ruthenium(II) and iridium(III) arene complexes on HepG2 and Vero cells. *ACS Omega* **2023**, *8*, 37549–37563. [CrossRef]
73. Shao, M.; Yao, M.; Liu, X.; Gao, C.; Liu, W.; Guo, J.; Liu, Z. In vitro and in vivo of triphenylamineappended fluorescent half-sandwich iridium(III) thiosemicarbazones antitumor complexes. *Inorg. Chem.* **2021**, *60*, 17063–17073. [CrossRef]
74. Sava, G.; Zorzet, S.; Perissin, L.; Mestroni, G.; Zassinovich, G.; Bontempi, A. Coordination metal complexes of Rh (I), Ir (I) and Ru (II): Recent advances on antimetastatic activity on solid mouse tumors. *Inorg. Chim. Acta* **1987**, *137*, 69–71. [CrossRef]
75. Ishii, Y.; Sakaguchi, S. A novel catalysis of [{IrCl (Cod)} 2] Complex in Organic syntheses. *Bull. Chem. Soc. Jpn.* **2004**, *77*, 909–920. [CrossRef]

76. Wang, J.; Lu, Y.; McCarthy, W.; Conway-Kenny, R.; Twamley, B.; Zhao, J.; Draper, S.M. Novel ruthenium and iridium complexes of N-substituted carbazole as triplet photosensitisers. *Chem. Commun.* **2018**, *54*, 1073–1076. [CrossRef]
77. Pandrala, M.; Sundaraneedi, M.K.; Ammit, A.J.; Woodward, C.E.; Wallace, L.; Keene, F.R.; Collins, J.G. Differential anticancer activities of the geometric isomers of dinuclear iridium (III) complexes. *Eur. J. Inorg. Chem.* **2015**, *34*, 5694–5701. [CrossRef]
78. Liu, B.; Monro, S.; Lystrom, L.; Cameron, C.G.; Colón, K.; Yin, H.; Kilina, S.; McFarland, S.A.; Sun, W. Photophysical and Photobiological Properties of Dinuclear Iridium(III) Bis-Tridentate Complexes. *Inorg. Chem.* **2018**, *57*, 9859–9872. [CrossRef] [PubMed]
79. Tang, H.; Saunders, G.C.; Raymond, O.; Ma, X.; Henderson, W. Pyrrole thioamide complexes of organo-rhodium(III), -iridium(III) and -ruthenium(II). *Inorg. Chim. Acta* **2020**, *506*, 119545. [CrossRef]
80. Giorgi, E.; Binacchi, F.; Marotta, C.; Cirri, D.; Gabbiani, C.; Pratesi, A. Highlights of New Strategies to Increase the Efficacy of Transition Metal Complexes for Cancer Treatments. *Molecules* **2023**, *28*, 273. [CrossRef]
81. Shi, H.; Carter, O.W.; Ponte, F.; Imberti, C.; Gomez-Gonzalez, M.A.; Cacho-Nerin, F.; Sadler, P.J. A Photodynamic and Photochemotherapeutic Platinum-Iridium Charge-Transfer Conjugate for Anticancer Therapy. *Ang. Chem. Intern. Ed.* **2024**, *63*, e202400476. [CrossRef] [PubMed]
82. Li, X.; Zhao, X.; Wang, W.; Shi, Z.; Zhang, Y.; Tian, Q.; Duan, C. Biomedical applications of multinuclear Pt (II)/Ru (II)/Ir (III) metallo-supramolecular assemblies for intensive cancer therapy. *Coord. Chem. Rev.* **2023**, *495*, 215366. [CrossRef]
83. Tripathy, S.K.; De, U.; Dehury, N.; Pal, S.; Kim, H.S.; Patra, S. Dinuclear $[\{(p\text{-cym})\ RuCl\}_2(\mu\text{-phpy})](PF_6)_2$ and heterodinuclear $[(ppy)_2Ir(\mu\text{-phpy})Ru(p\text{-cym})Cl](PF_6)_2$ complexes: Synthesis, structure and anticancer activity. *Dalt. Trans.* **2014**, *43*, 14546–14549. [CrossRef]
84. Wang, Y.Q.; Wang, P.Y.; Wang, Y.T.; Yang, G.F.; Zhang, A.; Miao, Z.H. An update on poly (ADP-ribose) polymerase-1 (PARP-1) inhibitors: Opportunities and challenges in cancer therapy. *J. Med. Chem.* **2016**, *59*, 9575–9598. [CrossRef]
85. Kaur, J.; Debnath, J. Autophagy at the crossroads of catabolism and anabolism. *Nat. Rev. Mol. Cell Biol.* **2015**, *16*, 461–472. [CrossRef] [PubMed]
86. Nallas, G.N.A.; Jones, S.W.; Brewer, K.J. Bipyrimidine-bridged mixed-metal trimetallic complexes of ruthenium (II) with rhodium (III) or iridium (III), $\{[(bpy)_2Ru\ (bpm)]_2MCl_2\}^{5+}$. *Inorg. Chem.* **1996**, *35*, 6974–6980. [CrossRef]
87. Kar, B.; Roy, N.; Pete, S.; Moharana, P.; Paira, P. Ruthenium and iridium based mononuclear and multinuclear complexes: A Breakthrough of Next-Generation anticancer metallopharmaceuticals. *Inorg. Chim. Acta* **2020**, *512*, 119858. [CrossRef]
88. Bridgewater, J.S.; Vogler, L.M.; Molnar, S.M.; Brewer, K.J. Tuning the spectroscopic and electrochemical properties of polypyridyl bridged mixed-metal trimetallic ruthenium (II), iridium (III) complexes: A spectroelectrochemical study. *Inorg. Chim. Acta* **1993**, *208*, 179–188. [CrossRef]
89. Molnar, S.M.; Jensen, G.E.; Vogler, L.M.; Jones, S.W.; Laverman, L.; Bridgewater, J.S.; Brewer, K.J. Photochemical properties of mixed-metal supramolecular complexes. *J. Photochem. Photobiol. A Chem.* **1994**, *80*, 315–322. [CrossRef]
90. Roy, N.; Sen, U.; Moharana, P.; Babu, L.T.; Kar, B.; Vardhan, S.; Paira, P. 2,2′- Bipyrimidine-based luminescent Ru(II)/Ir(III)–arene monometallic and homo- and hetero-bimetallic complexes for therapy against MDA-MB-468 and Caco-2 cells. *Dalton Trans.* **2021**, *50*, 11725–11729. [CrossRef]
91. Roy, N.; Shanavas, S.; Kar, B.; Thilak Babu, L.; Das, U.; Vardhan, S.; Paira, P. G2/M-Phase-inhibitory mitochondrial-depolarizing Re(I)/Ru(II)/Ir(III)-2,2′-bipyrimidine-based heterobimetallic luminescent complexes: An assessment of in vitro antiproliferative activity and bioimaging for targeted therapy toward human TNBC cells. *ACS Omega* **2023**, *8*, 12283–12297. [CrossRef]
92. Redrado, M.; Benedi, A.; Marzo, I.; García-Otín, A.L.; Fernández-Moreira, V.; Concepción Gimeno, M. Multifunctional Heterometallic IrIII−AuI Probes as Promising Anticancer and Antiangiogenic Agents. *Chem. Eur. J.* **2021**, *27*, 9885–9897. [CrossRef] [PubMed]
93. Luengo, A.; Marzo, I.; Reback, M.; Daubit, I.M.; Fernández-Moreira, V.; Metzler-Nolte, N.; Gimeno, M.C. Luminescent bimetallic IrIII/AuI peptide bioconjugates as potential theranostic agents. *Chem. Eur. J.* **2020**, *26*, 12158–12167. [CrossRef]
94. Komarnicka, U.K.; Koziel, S.; Pucelik, B.; Barzowska, A.; Siczek, M.; Malik, M.; Wojtala, D.; Niorettini, A.; Kyziol, A.; Sebastian, V.; et al. Liposomal binuclear Ir(III)-Cu(II) coordination compounds with phosphino-fluoroquinolone conjugates for human prostate carcinoma treatment. *Inorg. Chem.* **2022**, *61*, 19261–19273. [CrossRef] [PubMed]
95. Kozieł, S.; Komarnicka, U.K.; Ziółkowska, A.; SkórskaStania, A.; Pucelik, B.; Płotek, M.; Sebastian, V.; Bieńko, A.; Stochel, G.; Kyzioł, A. Anticancer potency of novel organometallic Ir(III) complexes with phosphine derivatives of fluoroquinolones encapsulated in polymeric micelles. *Inorg. Chem. Front.* **2020**, *7*, 3386–3401. [CrossRef]
96. Tabrizi, L.; Chiniforoshan, H. Designing new iridium (III) arene complexes of naphthoquinone derivatives as anticancer agents: A structure–activity relationship study. *Dalton Trans.* **2017**, *46*, 2339–2349. [CrossRef]
97. Ornelas, C. Application of ferrocene and its derivatives in cancer research. *New J. Chem.* **2011**, *35*, 1973–1985. [CrossRef]

98. Ge, X.; Chen, S.; Liu, X.; Wang, Q.; Gao, L.; Zhao, C.; Liu, Z. Ferrocene-appended iridium (III) complexes: Configuration regulation, anticancer application, and mechanism research. *Inorg. Chem.* **2019**, *58*, 14175–14184. [CrossRef]
99. Palao, E.; Sola-Llano, R.; Tabero, A.; Manzano, H.; Agarrabeitia, A.R.; Villanueva, A.; Ortiz, M.J. AcetylacetonateBODIPY-biscyclometalated iridium(III) complexes: Effective strategy towards smarter fluorescent photosensitizer agents. *Chemistry* **2017**, *23*, 10139–10147. [CrossRef]
100. Gupta, G.; Cherukommu, S.; Srinivas, G.; Lee, S.W.; Mun, S.H.; Jung, J.; Lee, C.Y. BODIPY-based Ru (II) and Ir (III) organometallic complexes of avobenzone, a sunscreen material: Potent anticancer agents. *J. Inorg. Biochem.* **2018**, *189*, 17–29. [CrossRef]
101. Pettinari, R.; Marchetti, F.; Petrini, A.; Pettinari, C.; Lupidi, G.; Smolenski, P.; Dyson, P.J. From sunscreen to anticancer agent: Ruthenium (II) arene avobenzone complexes display potent anticancer activity. *Organometallics* **2016**, *35*, 3734–3742. [CrossRef]
102. Gupta, G.; Das, A.; Ghate, N.B.; Kim, T.; Ryu, J.Y.; Lee, J.; Lee, C.Y. Novel BODIPY-based Ru (II) and Ir (III) metalla-rectangles: Cellular localization of compounds and their antiproliferative activities. *Chem. Commun.* **2016**, *52*, 4274–4277. [CrossRef]
103. Gupta, G.; Das, A.; Park, K.C.; Tron, A.; Kim, H.; Mun, J. Self-assembled novel BODIPY-based palladium supramolecules and their cellular localization. *Inorg. Chem.* **2017**, *56*, 4615–4621. [CrossRef]
104. Gupta, G.; Mahesh Kumar, J.; Garci, A.; Rangaraj, N.; Nagesh, N.; Therrien, B. Anticancer Activity of Half-Sandwich RhIII and IrIII Metalla-Prisms Containing Lipophilic Side Chains. *ChemPlusChem* **2014**, *79*, 610–618. [CrossRef]
105. Gupta, G.; Murray, B.S.; Dyson, P.J.; Therrien, B. Synthesis, molecular structure and cytotoxicity of molecular materials based on water soluble half-sandwich Rh (III) and Ir (III) tetranuclear metalla-cycles. *Materials* **2013**, *6*, 5352–5366. [CrossRef] [PubMed]
106. Gupta, G.; Kumar, J.M.; Garci, A.; Nagesh, N.; Therrien, B. Exploiting natural products to build metallaassemblies: The anticancer activity of embelin-derived Rh (III) and Ir(III) metalla-rectangles. *Molecules* **2014**, *19*, 6031–6046. [CrossRef]
107. Nikolovska-Coleska, Z.; Xu, L.; Hu, Z.; Tomita, Y.; Li, P.; Roller, P.P.; Wang, S. Discovery of embelin as a cell-permeable, small-molecular weight inhibitor of XIAP through structure-based computational screening of a traditional herbal medicine three-dimensional structure database. *J. Med. Chem.* **2004**, *47*, 2430–2440. [CrossRef] [PubMed]
108. Gupta, G.; Denoyelle-Di-Muro, E.; Mbakidi, J.P.; Leroy-Lhez, S.; Sol, V.; Therrien, B. Delivery of porphin to cancer cells by organometallic Rh (III) and Ir (III) metalla-cages. *J. Organomet. Chem.* **2015**, *787*, 44–50. [CrossRef]
109. Martir, D.R.; Zysman-Colman, E. Photoactive supramolecular cages incorporating Ru (II) and Ir (III) metal complexes. *Chem. Commun.* **2019**, *55*, 139–158. [CrossRef]
110. Gupta, G.; Oggu, G.S.; Nagesh, N.; Bokara, K.K.; Therrien, B. Anticancer activity of large metalla-assemblies built from half-sandwich complexes. *CrystEngComm.* **2016**, *18*, 4952–4957. [CrossRef]
111. Gupta, G.; Patel, P.; Lee, J.; Kaushik, N.; Choi, E.H.; Kaushik, N.K.; Lee, C.Y. Bi-anthracene based ruthenium and iridium metalla-rectangles: Coordination driven self-assembly and anti-cancer activities. *Inorg. Chem. Commun.* **2025**, *174*, 113916. [CrossRef]
112. Li, X.; Wu, J.; He, C.; Zhang, R.; Duan, C. Multicomponent self-assembly of a pentanuclear Ir–Zn heterometal–organic polyhedron for carbon dioxide fixation and sulfite sequestration. *Chem. Commun.* **2016**, *52*, 5104–5107. [CrossRef] [PubMed]
113. Oldknow, S.; Martir, D.R.; Pritchard, V.E.; Blitz, M.A.; Fishwick, C.W.; Zysman-Colman, E.; Hardie, M.J. Structure-switching M3L2 Ir (III) coordination cages with photo-isomerising azo-aromatic linkers. *Chem. Sci.* **2018**, *9*, 8150–8159. [CrossRef] [PubMed]
114. Li, C.; Wang, Y.; Lu, Y.; Guo, J.; Zhu, C.; He, H.; Su, C. An iridium (III)-palladium (II) metal-organic cage for efficient mitochondria-targeted photodynamic therapy. *Chin. Chem. Lett.* **2020**, *31*, 1183–1187. [CrossRef]
115. Liu, Y.Y.; Yu, H.J.; Wang, Y.P.; Li, C.J.; Wang, X.F.; Ye, C.G.; Su, C.Y. A photoactive Ir–Pd bimetallic cage with high singlet oxygen yield for efficient one/two-photon activated photodynamic therapy. *Mater. Chem. Front.* **2022**, *6*, 948–955. [CrossRef]

MDPI AG
Grosspeteranlage 5
4052 Basel
Switzerland
Tel.: +41 61 683 77 34

Inorganics Editorial Office
E-mail: inorganics@mdpi.com
www.mdpi.com/journal/inorganics

www.ingramcontent.com/pod-product-compliance
Lightning Source LLC
Chambersburg PA
CBHW061616210926
51850CB00009B/2338